Mélissa Morel

Iron Metallurgy in Northeastern Madagascar

Mélissa Morel

Iron Metallurgy in Northeastern Madagascar

Study of Rasikajy Metallurgical Production between the 11th and 15th Centuries

LIBRUM Publishers & Editors LLC

The pre-press work for this publication was supported by the Swiss National Science Foundation. We also thank the Schweizerisch-Liechtensteinische Stiftung für archäologische Forschungen im Ausland (SLSA) for their financial support.

Layout and typesetting: Mélissa Morel
Cover: Katja von Ruville, Frankfurt a. M.
Proofreading: Julie Cordell, London / Frankfurt a. M.

ISBN: 978-3-906897-86-8
DOI: 10.19218/3906897868

Open-access at

LIBRUMOPEN
www.librumopen.com

Cover image: View from the southeast of the working area and the two MBR140 slag heaps at Amboronala.

Department of Geosciences
University of Fribourg (Switzerland)

Iron Metallurgy in Northeastern Madagascar

Study of *Rasikajy* Metallurgical Production between the 11th and 15th Centuries

THESIS

presented to the Faculty of Science and Medicine
of the University of Fribourg (Switzerland)
in consideration for the award of the academic grade of
Doctor of Philosophy in Earth Sciences

by

MÉLISSA MOREL

from
Cannes (France)

Thesis No: 5197
UniPrint
Fribourg
2022

Abstract

In medieval times, a population the so-called *Rasikajy* was settled on the northeastern coast of Madagascar. This population was in contact with the Indian Ocean trading network, as attested by the presence of imported goods from China, India or Persia in some tombs in the necropolis of Vohémar, excavated in the first half of the 20th century. This population remains under-studied and little is known about their material culture or their way of life.

The presence of iron slag in northeastern Madagascar has long been documented in archaeological literature, but no in-depth study had been carried out. The aims of this doctoral project were therefore to describe and better understand these metallurgical remains and the associated technical tradition. Three excavation campaigns and four additional survey campaigns were carried out between 2017 and 2021, completely renewing our knowledge on this area of Madagascar. Approximately 150 slag heaps spread over twenty locations have been described, representing about 450 tons of slag. However, these sites are concentrated in the southern half of the study area, thus defining a spatially delimited metallurgical district. The metallurgical production could be dated by radiocarbon dating to between the 11th and 14th century CE. Before this period, the *Rasikajy* already knew and used iron tools, but they imported this metal via the Indian Ocean trade.

The combined approach of fieldwork and laboratory work has enabled the reconstruction of metallurgical practices, despite the fragile and poorly preserved remains. The *Rasikajy* technical tradition used a furnace in the form of a simple elliptical bowl dug directly into the sandy substratum, without any clay lining. No clay superstructure could be identified either. However, small walls were built of loose sand, sometimes reinforced with a few stone blocks. A single tuyere, made of stone or clay, was set into this sandy wall and connected to bellows. Lateritic ores, in the form of ferruginous concretions with remarkably high iron content, were reduced in these small structures. When an excess of slag was produced for the capacity of the bassin, the slag was drained off through a small channel dug in the sand.

This technical tradition is observed throughout the metallurgical district. However, each site is slightly different, which shows a local adaptation of the technique, probably depending on the availability of raw materials, e.g. to make the tuyere.

An detailed chemical (XRF) and mineralogical (XRD, optical microscopy and SEM-EDS) study revealed a high variability in slag composition. This variability is partly due to a high contamination of sand, which is the construction material of the furnace. Mass balance calculations also show that iron production was irregular from one smelting operation to another. In some cases smelting operations even failed, producing no iron at all. The technique appears to be poorly controlled, although the structure of the furnace certainly does not allow for complete control of each smelting operation. However, the variability is such that we believe that the metallurgists were non-specialised workers, producing iron only occasionally when metal ran out. This is clearly not a mass production but rather a small-scale production with occasional smelting campaigns to meet local iron needs.

Finally, a reflection on the origin of this unusual technical tradition has been carried out. As the *Rasikajy* were in contact with other populations in the Indian Ocean, a transfer of technology could have taken place. In this case, parallels of this technique would have been found elsewhere in the Indian Ocean. No similar technique has been documented so far in archaeological, historical or ethnographic sources. Hence, it is possible that the *Rasikajy* reinvented a basic smelting technique from smithing techniques.

Résumé

A l'époque médiévale, une population appelée les *Rasikajy* était installée sur la côte nord-est de Madagascar. Cette population était en contact le réseau commercial de l'Océan Indien comme l'atteste la présence de biens d'importation en provenance de Chine, d'Inde ou de Perse dans certaines sépultures de la nécropole de Vohémar, fouillée dans la première moitié du XXe siècle. Cette population reste encore sous étudiée et on connait mal leur culture matérielle ou leur mode de vie.

La présence des scories de fer dans le nord est de Madagascar est attestée depuis longtemps dans la littérature archéologique, mais aucune étude approfondie n'avait été menée à bien. Les objectifs de ce projet de thèse étaient donc de décrire et de mieux comprendre ces vestiges métallurgiques et la technique associée. Trois campagnes de fouilles et quatre campagnes de prospection ont pu être menées à bien entre 2017 et 2021 renouvelant complètement les connaissances sur cette région de l'île. Environ 150 amas de scorie répartis sur une vingtaine de sites ont été localisés, ce qui représente environ 450 tonnes de scorie. Ces sites sont cependant concentrés dans la moitié sud de la zone d'étude, définissant ainsi un district métallurgique délimité dans l'espace. La production métallurgique a pu être datée grâce à des datations radiocarbone entre le XIe et le XIVe siècle de notre ère. Avant cette période, les *Rasikajy* connaissaient et utilisaient déjà des outils en fer, mais ils importaient ce métal via le commerce de l'Océan Indien.

L'approche combinée des travaux de terrain et des travaux de laboratoire ont permis de reconstruire les pratiques métallurgiques, malgré des vestiges fragiles et mal conservés. La tradition technique Rasikajy utilisait un fourneau en forme d'une simple cuvette elliptique creusée directement dans le substrat sableux, sans aucun revêtement en argile. Aucune superstructure en argile n'a pu être identifiée non plus. Par contre, des petits murs étaient construits en sable meuble, parfois renforcés de quelques blocs de pierres. Une tuyère unique, fabriquée en pierre ou en argile selon les cas, était calée dans cette paroi sableuse et reliée à des soufflets. Des minerais latéritiques, sous forme de concrétions ferrugineuses avec teneurs en fer remarquablement hautes, étaient réduits dans ces petites structures. Lorsqu'un excès de scorie était produit pour la contenance de la cuve, la scorie était écoulée par un petit canal creusé dans le sable.

Cette tradition technique est observée dans tout le district métallurgique. Chaque site est cependant légèrement différent, ce qui montre une adaptation locale de la technique, probablement en fonction de la disponibilité des matières premières, pour fabriquer les tuyères par exemple.

Une étude chimique (XRF) et minéralogique (XRD, microscopie optique et SEM-EDS) approfondie a permis de mettre en évidence une forte variabilité de la composition des scories. Cette variabilité est en partie dû à une forte contamination en sable, qui est le matériau de construction du fourneau. Les calculs de balance de masse montrent également que la production en fer était irrégulière d'une opération de réduction à l'autre. Dans certains cas les opérations de réduction ont même échoué, ne produisant aucun fer. La technique semble globalement être mal maitrisée, même si la structure du fourneau ne permet certainement pas de maitriser complètement chaque opération de réduction. Cependant, la variabilité est telle que nous pensons que les métallurgistes étaient des travailleurs non spécialisés ne produisant que ponctuellement du fer pour renouveler les stocks lorsque le métal venait à manquer. Il ne s'agit clairement pas d'une production de masse mais plutôt d'une production à petite échelle avec des campagnes de réduction ponctuelles pour subvenir aux besoins locaux en fer.

Enfin, une réflexion sur l'origine de cette tradition technique inhabituelle a été menée. Les *Rasikajy* étant en contacts avec d'autres populations de l'Océan Indien, un transfert de technologie aurait pu avoir lieu. On retrouverait donc des parallèles de cette technique ailleurs dans l'Océan Indien. Aucune technique similaire n'a été documentée jusqu'à présent dans les sources archéologiques, historiques ou ethnographiques. Il est possible que les *Rasikajy* aient réinventé une technique de réduction basique à partir de techniques de forge.

Abstrakt

Im Mittelalter siedelte eine Bevölkerung namens Rasikajy an der Nordostküste Madagaskars. Diese Bevölkerung stand in Kontakt mit dem Handelsnetz des Indischen Ozeans, was durch die Entdeckung von Importgütern aus China, Indien und Persien in einigen Gräbern der Nekropole von Vohémar, die in der ersten Hälfte des 20. Jahrhunderts ausgegraben wurde, belegt wird. Diese Bevölkerung ist noch nicht ausreichend erforscht und man weiß wenig über ihre materielle Kultur und ihren Lebensstil.

Das Vorkommen von Eisenschlacke im Nordosten Madagaskars ist seit langem in der archäologischen Literatur belegt, aber es wurden noch keine umfassenden Studien durchgeführt. Die Ziele dieses Dissertationsprojekts bestanden daher darin, diese metallurgischen Überreste und die damit verbundene Technik zu beschreiben und besser zu verstehen. Zwischen 2017 und 2021 konnten drei Grabungskampagnen und vier Prospektionskampagnen durchgeführt werden, die das Wissen über diese Region der Insel vollständig erneuerten. Es wurden um die 150 Schlackenhaufen an rund 20 Standorten lokalisiert, was etwa 450 Tonnen Schlacke entspricht. Diese Fundstellen konzentrieren sich jedoch auf die südliche Hälfte des Untersuchungsgebiets und definieren somit einen räumlich abgegrenzten metallurgischen Distrikt. Die metallurgische Produktion konnte durch Kohlenstoffdatierungen auf das 11. bis 14. Jahrhundert n. Chr. datiert werden. Vor dieser Zeit kannten und verwendeten die Rasikajy bereits Eisenwerkzeuge, importierten das Metall jedoch über den Handel im Indischen Ozean.

Der kombinierte Ansatz aus Feld- und Laborarbeit ermöglichte es, die metallurgischen Praktiken trotz der fragilen und schlecht erhaltenen Überreste zu rekonstruieren. Die technische Tradition der Rasikajy verwendete einen Ofen in Form einer einfachen elliptischen Grube, die direkt in das sandige Substrat gegraben wurde, ohne jegliche Lehmauskleidung. Auch ein Oberbau aus Lehm konnte nicht identifiziert werden. Stattdessen wurden kleine Mauern aus lockerem Sand errichtet, die manchmal mit einigen Steinblöcken verstärkt waren. Eine einzelne Belüftung, die je nach Fall aus Stein oder Lehm hergestellt wurde, war in dieser Sandwand verkeilt und mit Blasebälgen verbunden. Lateritische Erze in Form von eisenhaltigen Konkretionen mit bemerkenswert hohem Eisengehalt wurden in diesen kleinen Strukturen reduziert. Wenn zu viel Schlacke für das Fassungsvermögen des Behälters produziert wurde, wurde die Schlacke durch einen kleinen, in den Sand gegrabenen Kanal abgeleitet.

Diese technische Tradition ist im gesamten Bezirk zu beobachten. Jeder Standort ist jedoch etwas anders, was auf eine lokale Anpassung der Technik hindeutet, die wahrscheinlich von der Verfügbarkeit der Rohstoffe abhängt, um z. B. die Belüftungsröhren herzustellen.

Eine gründliche chemische (XRF) und mineralogische (XRD, Lichtmikroskopie und REM-EDS) Untersuchung ergab eine starke Variabilität der Schlackenzusammensetzung. Diese Variabilität ist zum Teil auf eine starke Verunreinigung mit Sand zurückzuführen, der das Baumaterial des Ofens ist. Die Berechnungen der Massenbilanz zeigen auch, dass die Eisenproduktion von einem Reduktionsvorgang zum nächsten unregelmäßig war. In einigen Fällen scheiterten die Reduktionsvorgänge sogar und es wurde kein Eisen produziert. Die Technik scheint insgesamt mittelmässig angewandt worden zu sein, auch wenn die Struktur des Ofens es sicherlich nicht erlaubt, jeden Reduktionsvorgang vollständig zu beherrschen. Die Variabilität ist jedoch so groß, dass wir davon ausgehen, dass es sich bei den Metallurgen um nicht-spezialisierte Arbeiter handelte, die nur gelegentlich Eisen herstellten, um die Bestände zu erneuern, wenn das Metall knapp wurde. Es handelt sich eindeutig nicht um eine metallurgische Massenproduktion, sondern eher um eine Produktion in kleinem Maßstab mit punktuellen Reduktionskampagnen um den lokalen Eisenbedarf zu decken.

Acknowledgements

During my years at university and especially during these five years of doctoral research, I had the opportinuty to an exchange with many people who made me learn and progress.

First of all, I would like to thank my supervisor, Prof. Vincent Serneels. A big thank you for the patience and passion you have shown throughout these years during our long discussions, for making me go a little further and for answering my endless questions. I am grateful to have been able to learn so much.

I would also like to thank the archaeometry research group of the University of Fribourg. Thanks to Ildiko Katona Serneels for her advice on the entire analytical part, especially the SEM, and for her support in all circumstances. A huge thank you to Christoph Nitsche, my thesis partner. Thank you for our scientific exchanges, for your help and for putting up with my complaints, over a beer or at the office. I am happy to have shared this thesis adventure with you, it would not have been the same without you. Thank you also to Jean Rodier who was also an excellent companion during my thesis and who challenged me scientifically on many occasions. Finally, I would like to thank the master students and in particular to Raphael Kunz.

My gratitude goes to the Malagasy teams without whom nothing of this thesis would have been possible. Thanks to Chantal Radimilahy and Jean-Aimé Rakotoarisoa for their help in the field and their availability throughout my thesis. Thanks also to Bako Rasoarifetra. A huge thank you to the teams of students who worked with me during these three field campaigns: Fanny, Judith, Narindra, Herlain, Edinot and many others. A special thanks goes to two students who accompanied me remotely during the pandemic and who did a remarkable survey work: Walker and Hervé. Thank you to the field teams, including the drivers, Victor and Abdoul, and everyone on the camps. I would also like to thank the Swiss/European team who accompanied us in the field in Madagascar. I am thinking in particular of Guido Schreurs, Michel Mauvilly and Ladji Dianifaba.

A big thank you to the technical teams of the Institute of Earth Sciences of the University of Fribourg who helped and trained me during my laboratory work. Thanks to Patrick Dietsche, Afifé El Korh, and Christoph Neururer. A special thought for Alexandre Salzmann who was always available to help me throughout my thesis.

Thanks to all the other collaborators of the Geosciences Department of the University of Fribourg for the years spent together. Thanks to all those who had the patience to explain to me the basics of geology over a cup of tea.

Of course, a thesis would not be a thesis without the other doctoral or post-doctoral buddies, some of whom have already been thanked above. Thanks to Olivia, Sandra, Valentin, Valentina, Bruno, Kévin, Fabio, Yann and Stephanie for the more or less serious discussions, for the *pétanque* or Mario kart and for the Thursday evenings. You really made the difference.

Thanks to the friends I met here and there, Morgane, Nicolas, Héloïg, Armel, Agathe, Clem, who in addition to our lively scientific debates have been an incredible support these last years. A special mention to my roommates and friends who lived with and supported me during the writing of my thesis, Vincent, Sabrina and Emilie. You have been incredible. Thank you to Elisa, who agreed to read and correct most of this thesis. Thank you for your support and your kindness.

A very special thank you to Xavier who, in addition to accompanying me during the last years of my thesis, never hesitated to challenge my ideas. Thank you for handling my template and data processing problems, you really made everything easier.

Thank you to my family and in particular to my parents for always encouraging me to pursue this improbable path that is archaeometry. Thank you for giving me this interest in history and for teaching me to be curious and to want to dig into things. Thanks also to my sister for always being there and especially for taking my mind off things. Thank you, all three of you, for always believing in me and for your love.

Contents

List of Figures

Introduction 1

Madagascar is the largest island in the Indian Ocean, stretching over a lenght of 1500 km and a width of 500 km. It is located a few hundred kilometres from the southeastern African coast, 420 km at the closest. Despite the efforts of Malagasy and international teams, the history of Madagascar remains poorly understood, although the potential for research is huge.

The northeastern coast of Madagascar had already been noticed in the literature for its archaeological potential. This area is mainly known through the reports of European researchers from the first half of the 20th century (Vernier and Millot, 1971), and, later, from the detailed work of Pierre Vérin (1986) and of Dewar and Wright (1993). About ten sites along the coast were located before our project started.

The northeastern part of Madagascar was involved in the Indian Ocean trading network from the second half of the first millennium CE. There are a few indications about Madagascar in the early Arabic written sources, even if these remain difficult to interpret (Vérin, 1986:41-45). Moreover, archaeological records repeatedly show the presence of imported goods, such as Indian beads, Chinese bronze mirrors, Chinese or Arabic ceramics, etc (Serneels et al., 2018; Radimilahy, 1998; Vérin, 1986; Battistini and Verin, 1966; Gaudebout and Vernier, 1941).

The existence of a population settled on the northeastern coast *ca.* between 500 and 1500 CE was highlighted during the excavation of the Vohémar graveyard (Gaudebout and Vernier, 1941) as well as during the study of the walled town of Mahilaka (Radimilahy, 1998). We use conventionally the name "*Rasikajy*" to designate this population. The blooming of the *Rasikajy* population took place between the 9th and the 16th centuries CE. During this period, there were numerous small settlements along the coast, they imported goods through the Indian Ocean network and they developped their own production and know-how (chlorite schist, ceramics and iron).

There are cultural similarities between the *Rasikajy* and the Swahili world. The "Swahili culture", although it is a controversial term, commonly refers to the populations living along the eastern African coast, from the 8th-9th centuries onwards (Wyne-Jones and LaViolette, 2018). These populations were also active participants in the Indian Ocean trade. The Swahili people share a mixed culture between the local African culture and the culture of the Muslim traders from the Arab or Persian world. The *Rasikajy* have however also their own cultural specificities and do not share all the characteristics of the Swahili populations.

This thesis contributes to a better understanding on *Rasikajy* society, their culture and their way of life. More specifically, the aim is to characterize the material remains of the iron smelting technique of the *Rasikajy*. Indeed, slag heaps had already been located by previous researchers in northeastern Madagascar (Radimilahy, 1998; Clist, 1995; Vérin, 1986). Because the techniques implemented by a population are intrinsically linked to its cultural identity (Killick and Fenn, 2012; Roux, 2009; Killick, 2004; Gosselain, 2000), the study on the metallurgical remains and on the iron-working technique gave us the opportunity to learn more about the *Rasikajy*, especially as there are no written records available on this population. We also studied the socio-economic organisation of iron production, in order to better understand the organisation of *Rasikajy* society.

This thesis started in 2017 and was carried out in the framework of the project "Pots, Stones and Iron by the *Rasikajy*" funded by the Swiss National Science Foundation (SNF) and by the Swiss-Liechtenstein Foundation for Archaeological Research Abroad (SLSA). This project was directed by Vincent Serneels (University of Fribourg, Switzerland), in collaboration with the University of Antananarivo (Madagascar). The project aims to investigate the ancient exploitation of mineral resources in northeastern Madagascar through field and laboratory work. We studied how external cultural influences impact on the development of different techniques within the same population, the *Rasikajy* (ceramic, chlorite schist and iron). Two doctoral theses were conducted in parallel and in collaboration as part of this project. One was carried out by Christoph Nitsche on the exploitation of softstones, and this one, which focuses on iron metallurgy.

Our research took place over five years, between 2017 and 2021. Three field campaigns were carried out by an international team (2017-2018-2019) where three iron smelting workshops were excavated and 13 test pits in slag heaps were conducted to collect charcoal for dating. The coastline was also intensively surveyed, especially the area between Bemanevika and Antalaha (Figure 1.1). During the following years (2020-2021), the global health situation prevented us from going back on the field and collecting more samples. Four surveys campaigns could nevertheless be carried out by students from the University of Antsiranana and the University of Antalaha, either as part of their bachelor's or master's degree.

Approximately 500 kilometres of coastline were surveyed during the course of this work, which significantly improved the available

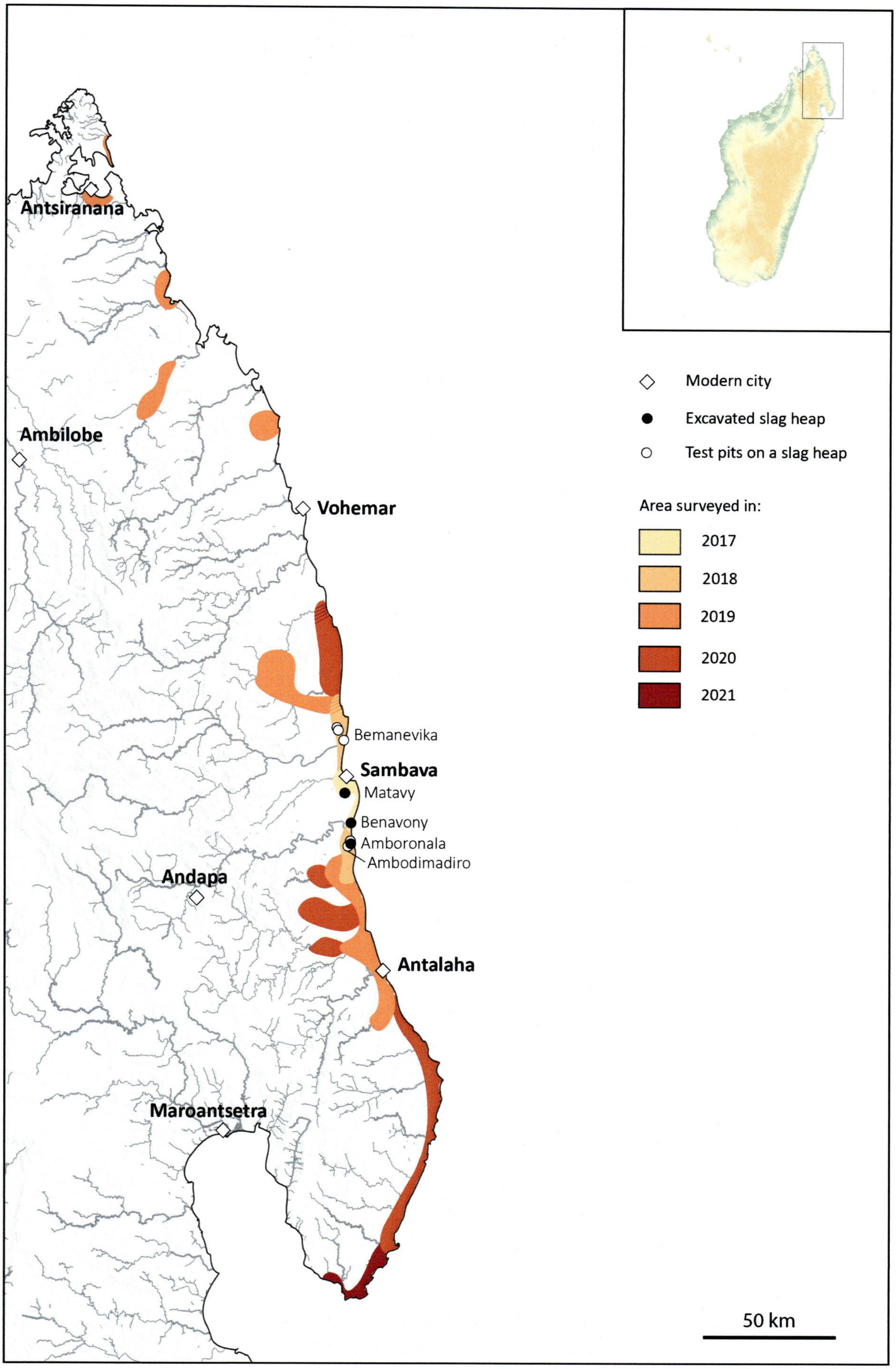

Figure 1.1: Topographic map of northeastern Madagascar with the areas surveyed and excavated between 2017 and 2021 as part of the study of iron metallurgy (Sources: USGS GMTED2010 Elevation Data, OpenStreetMap).

knowledge on the *Rasikajy* and redefined what we knew about this population. Among the many remains identified, a total of 150 slag heaps were located at about 20 different locations. Surveys could however not be conducted along the whole coastline with the same intensity and accuracy, due to time and accessibility. However, the area between Bemanevika and Antalaha has been intensively covered. For this reason, we think our knowledge of this area is representative of the archaeological remains.

Field observations combined with in-depth studies of the metallurgical waste have led to the definition of a single technical tradition of iron ore smelting implemented by the *Rasikajy* from the 11th to the 14th century (Chapter 4 and Chapter 5). These were simple shallow pits dug directly into the sandy substratum without any clay lining, with a wall also built out of sand. A single tuyere, often made of stone, was installed horizontally in the sandy wall. This was connected to bellows. This single technical tradition, with some local variations, seems to have been used throughout the study area and throughout the period.

Laboratory work (224 samples analysed, including 136 slag samples) has shown a huge variability of the slag chemistry (Chapter 6) and mineralogy (Chapter 7), even among samples coming from the same slag heap. Slag pieces are generally poorly melted and metal losses are high. This smelting technique was poorly controlled and poorly mastered. The people involved were not specialized craftsmen and practised iron smelting only sporadically. This type of small-scale, non-specialized production has rarely been described in an archaeological context.

Finally, we sought to place this unusual technique in a wider context, firstly within a Malagasy history of techniques (Chapter 8). For this purpose, an in-depth study of historical, archaeological and ethnographic sources related to iron in Madagascar was carried out.

To conclude, the *Rasikajy* technique was compared with other known technical traditions around the Indian Ocean to try to identify the origin of this technique. Indeed, the history of the *Rasikajy* has to be placed in the wider context of the Indian Ocean, especially as the *Rasikajy* were part of these networks. The Indian Ocean trading network at that time extended from Southeast Asia to the southeastern coast of Africa, including important actors such as China, India and the Arab world (Beaujard, 2005).

There was a wide variety of goods exchanged within the Indian Ocean: textiles, spices, metal, ceramics, slaves, etc. However, trade does not only involve the circulation of goods, but also of people, ideas, values and technologies. More than a commercial route, this great network facilitates cultural exchanges as well as transfers of knowledge. Two populations that meet can pass on some technical knowledge from one to the other, leading to the evolution of techniques. This is called technology transfer. Production techniques are therefore impacted by cultural contacts and population movements.

Iron metallurgy was a widespread skill around the Indian Ocean during the second half of the first millennium CE. The *Rasikajy* production technique could therefore benefit from the know-how of many populations outside Madagascar. The study of some of the archaeological, historical and ethnographic sources has therefore led us to a more general reflection on the origin of the *Rasikajy* iron smelting technique.

Finally, although it was never the purpose of this work, we hope here to contribute to the tough debate on the origin of the Malagasy population. The populating of Madagascar is one of the most debated and yet least understood issues in Malagasy archaeology (Beaujard, 2005). At the present state of knowledge, it is not possible to know where and when the first human group settled in Madagascar and where they came from.

However, it is known thanks to genetic and linguistic studies that Madagascar was populated by successive waves of migration, of Bantu people from East Africa, and of Austronesian people from insular Southeast Asia (Pierron et al., 2017; Tofanelli et al., 2009; Hurles et al., 2005; Beaujard, 2003; Adelaar, 1995). The island of Madagascar has therefore been populated by people with diverse cultural origins and with different technological backgrounds. The study of the history of techniques in Madagascar could therefore provide some new data and useful insights into this question.

General Approach to Iron Archaeometallurgy | 2

The theoretical approach to the production line, or *chaîne opératoire*, of iron metallurgy has already been described and studied by many previous authors. This thesis benefits from all the methodological achievements in archaeometallurgical studies over the last thirty years, and in particular from the development of new laboratory methods (Mangin et al., 2004; Serneels, 1993).

Iron is a metal employed in several everyday life situations, such as agriculture, handicrafts and daily tools, as well as in the manufacture of weapons and as a building material. Its use can therefore be essential for many aspects. Iron can be produced locally, imported and exported. Hence, studies on iron metallurgy allow insights into economic history by means of localization, dating and quantification of the production. Provenance studies, which will not be discussed in this thesis, also allow the reconstruction of some of the trading networks (Charlton, 2015; Leroy, 2010). A history of techniques can also be established through the investigation of different iron making techniques and their comparison to ethnographic and archaeological records. It is then possible to investigate the development of iron metallurgy in some regions and the emergence of technological innovations, the diffusion of these techniques and to address the issues of technology transfer from one population to another (Juleff, 2009; Killick, 2009).

2.1 The production line of iron

The practice of iron metallurgy can be presented as a *chaîne opératoire* (Figure 2.1). The general concept of *chaîne opératoire* can be applied to all production systems (ceramics, glass, stone tools, fabric, etc.). It describes the successive stages of transformation of the materials for manufacturing a functional object, as well as the raw materials needed and eventually the manpower involved (Pétrequin and Pétrequin, 2000; Perlès, 1987; Lemonnier, 1976; Leroi-Gourhan, 1964). This concept places the production of an object in a cultural framework and takes into account the social and cognitive specificities of a human group. An object is indeed manufactured with the aim of meeting a specific pre-existing need and within a precise socio-cultural context.

The *chaîne opératoire* of iron metallurgy in particular is complex and segmented. From the extraction of the ore and the manufacture of charcoal to a finished iron object, many stages are necessary, involving many people. Each step generates typical wastes and remains that can be identified in the archaeological record (Figure 2.1).

Moreover, each stage requires specific know-how. For example, in most cases, the smelting stage is not carried out by the same person or in the same place as the smithing stage. The people involved in iron production may or may not be specialised workers. Iron making requires a labor force that varies according to the production scale. Small-scale production can be achieved with a small and versatile workforce, while mass production requires a large workforce with specialized workers and work supervision.

There is considerable diversity in iron production techniques both geographically and chronologically (Robion-Brunner, 2020; Pleiner, 2000; Rostoker and Bronson, 1990; Cline, 1937). Each step of the *chaîne opératoire* can be slightly adapted, which leads to a great variability in these techniques. A specific and particular succession of technical gestures, leading to the production of iron, depends on many different criteria, including the raw materials available in the immediate environment, the amount of metal one wants to produce or the degree of specialisation of a population. But the way in which one produces iron also depends on how one has learned to make iron, and therefore on a cultural identity (Rehren et al., 2007; Killick, 2004; Pfaffenberger, 1992; Gosselain, 2000). In most cases, craftsmen reproduce the gestures they have been taught, even if these are not optimal and could be improved.

Different populations can therefore set up different *traditions techniques*, to follow the French terminology (Gallay et al., 2012), or form a community of practice, to follow the English terminology (Lave and Wenger, 1991). Multiplying the study of specific cases allows the definition of the variability of iron working and thus the study of cultural practices.

2.1.1 Iron ore: chemical variability, extraction and preparation

Iron ores are rocks rich in iron oxides or hydroxides (magnetite Fe_3O_4, hematite Fe_2O_3, goethite $FeO.OH$) and sometimes in iron carbonate (siderite $FeCO_3$). A rock is considered to be an iron ore when it is technically and economically attractive to process it to recover the iron. An ore that was exploited in the past will thus not necessarily be one nowadays, due to the evolution of metallurgical techniques.

There is a very large geological variability of iron-rich rocks on Earth whose use as iron ore can be evidenced (Zitzmann, 1977): laterites (Serneels et al., 2015; Celis, 1991), ferruginous sands (Wagner,

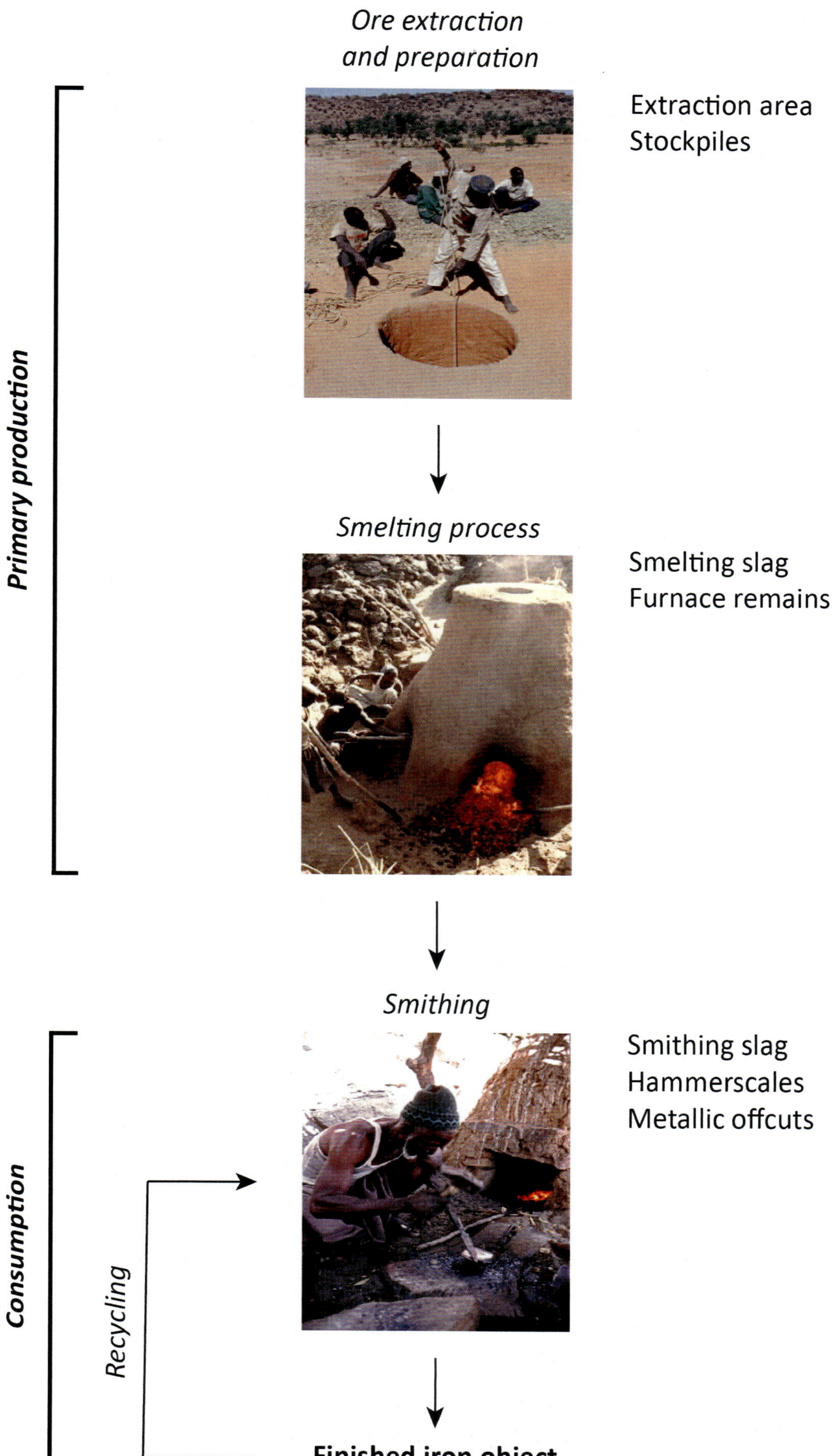

Figure 2.1: Simplified diagram of the main steps of the *chaîne opératoire* for iron metallurgy, using the direct smelting process (Images from Huysecom, 1996, showing metallurgists from the Dogon Country (Mali) at work).

2008; Brown, 1995; Buchanan, 1807), ferruginous oolitic limestone (Leroy et al., 2015; Serneels, 1993), etc. There is therefore, a very high chemical variability of available iron ores depending on the type of rocks. Chemical variability can also be significant at the scale of a single deposit. The formation processes of these rocks are indeed heterogeneous, for example in relation to the circulation of fluids. The totality of a deposit is therefore not necessarily interesting to exploit.

Not all rocks rich in iron oxides are necessarily easy to use as iron ore. A rock with a content of less than 50 wt.% Fe_2O_3 is too poor to be smelted in most bloomery furnaces. It can be seen, however, that in the absence of good ore, if the need for iron was great or urgent, the populations would use local resources of any quality. Some minor elements can also be undesirable in an ore, such as phosphorus which makes iron brittle when it is in high concentrations in the metal (>1%) or sulfur which is also considered as having a negative effect in cast iron (Serneels, 2005).

The first step of the iron metallurgy *chaîne opératoire* is the selection and collection of iron ore. Different collection methods were used depending on the techniques and the type of outcrop, requiring different level of expertise and a more or less complex organisation of labor. The ores were extracted in underground mines, which required organised removal of the rubble and air circulation. They could therefore be very organised structures. Ore may also be extracted from simple pits a few metres deep or even collected directly on the ground (Brown, 1995; Celis, 1991).

The ore is then prepared and beneficiated to produce a concentrated ore for optimum smelting. The treatment can be mechanical and the ore grains are then sorted, separated from the gangue and washed. They can also be crushed into smaller grains to facilitate the sorting and to optimise the contact surface with the reducing gas in the furnace.

Heat treatment can also be undertaken. The ore is roasted in a furnace at a few hundred degrees under oxidising conditions. It removes volatile compounds, such as sulphur from sulphides, carbon dioxide from carbonates and water from hydroxides, in the form of gas. The iron then combines with the oxygen from the air to form oxides. On the one hand, this treatment concentrates the ore in iron, and eliminates the sulphur. But the main effect of this heating is the creation of micro-fractures in the ore grains (Mangin et al., 2004; Killick and Gordon, 1988). This facilitates the penetration of reducing gases during the metallurgical process and increases the contact surface.

These treatments are important and even essential in some cases as they enable deposits with low iron content to become usable ores. Indeed, a final concentrate with less than 60 wt.% Fe_2O_3 is very difficult to use.

There is very little evidence of this stage in our study area. No ore extraction area has been found. This study will therefore not focus on this stage of the *chaîne opératoire*.

2.1.2 The smelting process

The role of a bloomery furnace is twofold: to reduce the iron oxides contained in the ore to metallic iron and to separate the unreduced components from the metal by draining the liquid slag.

Figure 2.2: Modern charcoal kilns from Benavony, northeastern Madagascar.

The fuel used in early metallurgical practices was mainly charcoal, which has a higher calorific value than dry wood. Mineral coal does not seem to have been used, or only anecdotally, before the Industrial Revolution. Charcoal is made by slow calcination of dry wood in a smothering process. There are many technical processes for making charcoal. Concerning Malagasy metallurgy, no studies have been carried out on charcoal. Today, charcoal kilns are built in sand pits where wood is piled up (Figure 2.2). Once calcination has begun, the wood is covered with sand until the process is complete.

A bloomery furnace is a structure that has, among others, the function of reducing the iron oxides contained in the ores into metallic iron. Ore and charcoal are loaded into the furnace, often in alternating layers. The entire stock ore can be loaded at once in some furnaces, while for other techniques the furnace is loaded several times as the operation progresses.

Whether the ventilation is a natural draught or with bellows, air is supplied through the lower part of the furnace. It is in this part of the furnace that the charcoal burns and the temperature is highest, reaching temperatures of between 1100 and 1400°C. The combustion of the carbon in the charcoal releases carbon monoxide:

$$C + O_2 \longrightarrow CO_2 \tag{2.1}$$
$$CO_2 + C \longrightarrow 2CO \tag{2.2}$$

The carbon monoxide then reacts with the iron oxides to form metallic iron:

$$3Fe_2O_3 + CO \longrightarrow 2Fe_3O_4 + CO_2 \tag{2.3}$$
$$Fe_3O_4 + CO \longrightarrow 3FeO + CO_2 \tag{2.4}$$
$$FeO + CO \longrightarrow Fe + CO_2 \tag{2.5}$$

The melting temperature of metallic iron (1536°C) is never reached in bloomery furnaces. This oxidation-reduction reaction therefore takes place at the solid state. It takes place in the areas of the ore particle that are in contact with the reducing gas, i.e. at the surface of the grain, but also in the fractures.

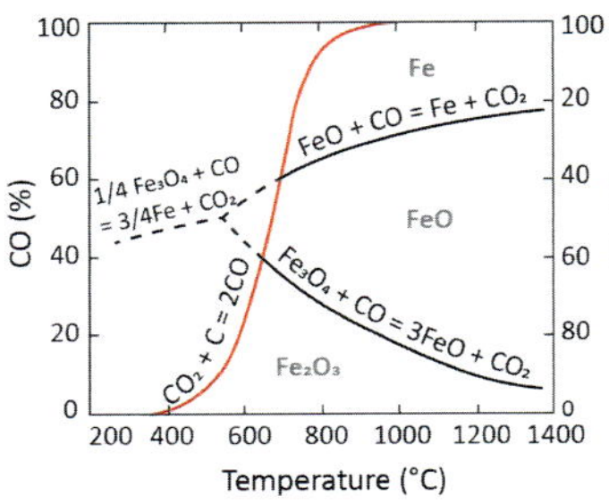

Figure 2.3: Chaudron diagram illustrating the reaction between iron oxides and a CO/CO_2 gas mixture depending on temperature.

The reduction reaction starts around 650°C and becomes dominant at 1000°C (Figure 2.3). The higher the temperature and the more abundant the carbon monoxide, the more easily this reaction takes place (Figure 2.5). Because this reaction is not instantaneous, the longer the contact time between the reducing gas and the iron oxides, the more time the reaction has to take place and the more metal is produced (Figure 2.4). The zones where the iron becomes metallic grow progressively. These zones weld together to form a metallic bloom.

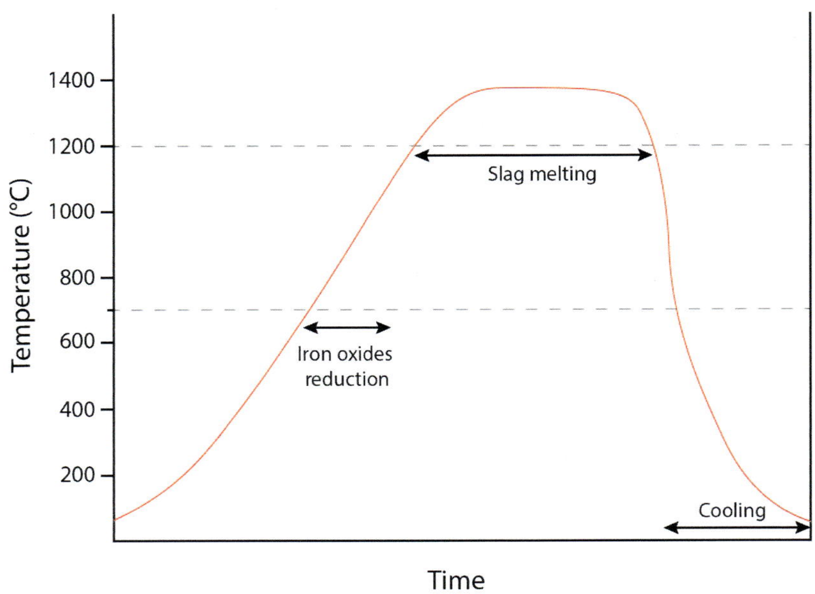

Figure 2.4: Diagram of the temperature cycle in a bloomery furnace during iron smelting. Temperatures in a furnace are not homogeneous, the highest temperatures are reached near the air intakes. Depending on the techniques and the mastery of the smelters, the temperature cycles can vary. One can imagine a long plateau around 1000°C to optimise smelting and then a rise in temperature to tap the slag. This diagram is therefore only one theoretical representation of temperature variations among many possibilities.

However, the ore is not composed exclusively of iron oxides, but also of other oxides such as SiO_2, Al_2O_3, TiO_2, CaO, MnO, etc. These oxides are reduced under higher temperatures and redox conditions than iron (Figure 2.5) and are therefore not, or only very partially, reduced in a bloomery furnace. This gangue is progressively depleted of iron as the iron oxides are reduced. After a while, the temperature is high enough for the gangue to melt and flow as slag at the base or outside the furnace. The second essential role of the smelting process is therefore to melt this slag and separate it from the bloom. If the slag is poorly separated from the metal, then it is difficult to recover much iron. Once the slag melts and flows, the reduction operation stops.

The melting temperature of the slag varies according to its chemical composition. There are two optimum chemical compositions where the melting temperature is lowest (Figure 2.8). However, if too many iron oxides are reduced, the melting temperature quickly

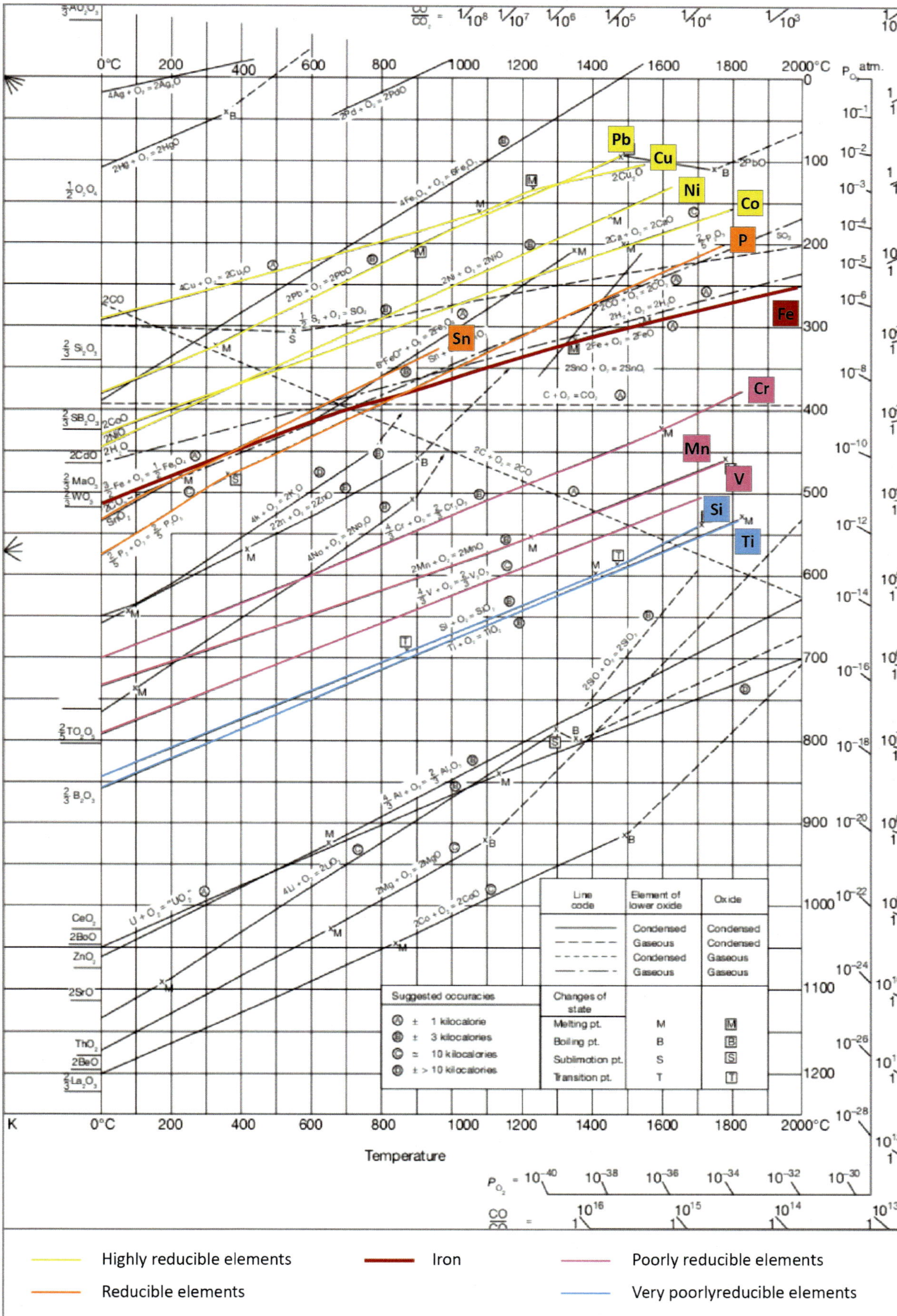

Figure 2.5: Ellingham diagram (Ellingham, 1944) adapted to the reflection on iron bloomery process. The minor elements in the ore behave differently from each other. The yellow or orange elements are usually reduced with the iron and are therefore found in the resulting metal. The pink and blue elements are more difficult to smelt and require particularly strong smelting conditions for them to be present in the metal. These elements are therefore found in the slag.

becomes very high and impossible to reach in a bloomery furnace. The complexity of the process and the know-how of the smelters is therefore to manage to stay long enough below the melting temperature of the slag to reduce as much iron as possible, while taking care not to reduce too much to still be able to melt the slag.

Many parameters must be considered to control the smelting process, including the type of ore (iron content, grain size, porosity), the type of fuel (and size of the particles), the materials available for building the furnace, the morphology of the furnace, the slag flowing system, the air system, etc. This partly explains the great diversity of the *traditions techniques* developed over time.

2.1.3 Post-smelting smithing

Once the bloom has been removed from the furnace, the metal must be separated from any impurities such as charcoal or slag (purification) and the bloom must often be compacted. Sometimes, the bloom is worked directly from the furnace while still hot, whereas some other times it is reheated. It is also possible to carry out a cold hammering before heating up the metal to remove the main impurities. This step makes the metal denser and more malleable. The result is a semi-finished product, which can be immediately transformed into an object or sold as raw material.

Finally comes the smithing of the object. The purpose of smithing is not only to shape an object (shaping) but also to modify and control the physical properties of the metal (carbon content and mineralogical structure). To that end, the blacksmith has a whole range of actions and techniques at his disposal. The implementation of these actions depends on various criteria such as the type of object, the raw material available, the blacksmith's know-how, etc.

Smithing is done in a hearth under more or less oxidising conditions. The surface of the metal tends to oxidise under these conditions which can represent a significant loss of metal. To limit this effect, the blacksmith can protect the surface of his object with a product such as sand. This product will melt under the effect of heat and create a protective layer.

The object can also undergo so-called finishing operations, either to improve the object technically (sharpening, polishing) or for aesthetic reasons.

The object will need to be repaired during its lifetime, which is also the role of the blacksmith. Not all people who use iron know how to produce it or how to make complex objects. On the other hand, they all share simple smithing techniques to repair their tools or weapons.

There is not much evidence of the post-smelting stage in northeastern Madagascar. This stage of the *chaîne opératoire* has therefore been little studied in this thesis.

2.1.4 How to differentiate smelting from smithing slag?

On several occasions during the course of this PhD research, both in the fieldwork and in the study of archaeological and historical sources, it has been necessary to distinguish between primary production and consumption sites, and therefore between smelting and smithing slag. There is no absolute and unambiguous criterion for distinguishing the two types of slag. A combination of observations is necessary (Figure 2.6). Moreover, it is always difficult to make a diagnosis for a single piece, it is much easier when there is an assemblage of slag and an archaeological context. It is also always difficult to be confident with only naked-eye observations. It is easier with the support of laboratory techniques, such as microscopic observations or elementary chemical analyses.

The typical assembly for diagnosing smelting slag is the following:

- A smelting site is most of the time isolated far from the settlement, or at least in the periphery of the settlement.
- Smelting slag assemblages represent large quantities of slag, sometimes several tons or more (Robion-Brunner, 2008). Most of the time, smelting slag accumulates in heap but it can also be the spreading of slag over a large area (Serneels et al., 2014).
- A smelting workshop may present remains in varying conditions of preservation. However, the presence of furnace wall fragments or ore stockpiles are generally good indicators of a primary production workshop. Pieces of partially smelted ore may sometimes be caught in a smelting slag.
- The morphology of smelting slag pieces can vary greatly depending on the furnace and its implementation. Nevertheless, all smelting slags have melted and flowed to be separated from the iron bloom. The two main morphological categories are slag that has flowed and accumulated inside the furnace and slag that has flowed outside the furnace.
- The chemical compositions of smelting slag are ore related. The unreduced chemical elements of the ore are found in the slag (Charlton et al., 2010; Dillmann and L'Héritier, 2007; Coustures et al., 2003; Serneels, 1993). These are essential

chemical markers in the understanding of the ore/slag relationships. Moreover, smelting slags are rarely rich in free iron oxides; metal is much more frequent in smithing slag.

Equally, a number of criteria can provide enough evidence to identify smithing slag:

- In most cases, smithing slag pieces are found inside settlements. However, there are situations where smithing workshops are rather directly next to a smelting area and on the periphery of settlements or inside a specialised site.
- Smithing slag is found in smaller quantities than smelting slag. Concentrations of smithing slag generally reach a few dozen kilograms to a few hundred kilograms of slag. They rarely form heaps but are found scattered or filling pits. In some cases however, smithing slag can be recovered in very substantial quantities, for example on some important Roman sites (Schucany, 2006; Serneels and Perret, 2003),
- The most characteristic assemblage to identify a smithing workshop is the association of slag with hammerscales and metallic debris, especially offcuts.
- Smithing slag pieces generally show a plano-convex bottom slag morphology. They can also look like small rounded bits (Soulignac, 2017; Serneels, 1993).
- The chemistry of a smithing slag is characterized by high iron content. Iron can be both in the form of oxide or of metal. It may also be contaminated by a smithing flux, generally siliceous (Michler, 2016). Finally, it may be stratified, each layer representing a different stage of the smithing process (Soulignac, 2017; Eschenlohr and Serneels, 1991).

	Workshop location	Amount of slag	Associated materials	Slag morphology	Slag chemistry
Smelting	Outside from settlement	Large quantities	Ore, furnace	Internal and external flowing slag	Related to ore
Smithing	Inside settlement	Smaller quantities	Hammerscale, offcuts	Plano-convex bottom slag, stratified	High iron content

Figure 2.6: Table summarising the characteristics used to differentiate smithing slag from smelting slag.

2.2 The efficiency of a iron production technique

This section presents a general reflection on the efficiency of a production technique. This reflection can be applied to all production techniques and in particular to those involving a furnace or a hearth (ceramics, glass, metals). There are many ways to characterise a production technique: according to the morphology of the production structures, the quantity of materials produced, the relationship of this technical practice with the society in which it is integrated, etc.

However, if we focus on iron metallurgy, although there are sophisticated approaches to describe and define an iron production technique, there is no satisfactory method for comparing several techniques. The morphologies of the furnaces and their functioning are too diverse to establish a straightforward classification (Robion-Brunner, 2018; Martinelli, 1993). A classification only based on morphological criteria provides valuable informations on the history of the techniques and technology transfers (Juleff, 2009) and one based on the amount of iron produced only compares the scale of production and its economic implications.

However, these classifications do not allow us to understand why some techniques are maintained over very long periods or, on the contrary, why one technique replaces another. To better understand these questions, it is necessary to take into account the role of a technique in a given society and the social impacts of metallurgical production. We propose here to question the concept of efficiency of a smelting technique.

Efficiency is a not well-defined concept in the literature, even though there have already been unsatisfactory attempts to use it as a criterion for comparing technologies (Charlton et al., 2010). It would actually be more relevant to define three different categories of efficiency: technical, economic and social efficiency.

It should be pointed out that, in the case of ancient societies, and particularly African ones, the goal of metal production was not necessarily to produce more and more by constantly improving yields. If a technique fulfils its social role, it can be maintained over time without need to improve it. A comparison of these different technical, economic and human factors could provide a new understanding of the diversity of iron smelting techniques.

2.2.1 Technical efficiency

The technical efficiency would be the balance between the mass of iron present in the ore and the mass of iron which was really produced, i.e. the amount of iron oxides that could be reduced from a given ore.

Technical efficiency is often estimated qualitatively by measuring the iron oxides content in the slag and therefore the amount of unreduced iron, or the amount of metallic iron lost in slag, using bulk chemical analysis (Charlton et al., 2010) and LOI (Loss Of

Ignition - see Section 6.1.2). However, this approach does not take into account the iron content of the chosen ore.

Another approach consists in examining the mineralogy of the slag and considering that a slag rich in free iron oxides (wustite - FeO) is the result of a less efficient smelting process than a slag poor in wustite. These iron oxides could indeed be directly available to be transformed into metal with a better smelting process. This approach however does not take into account the presence of chemical elements (Al_2O_3, TiO_2) which can lead to the formation of other iron-rich minerals such as spinels (Figure 2.7). Slag with little or no wustite may therefore also contain high Fe_2O_3 content and not necessarily be the result of technically efficient smelting processes.

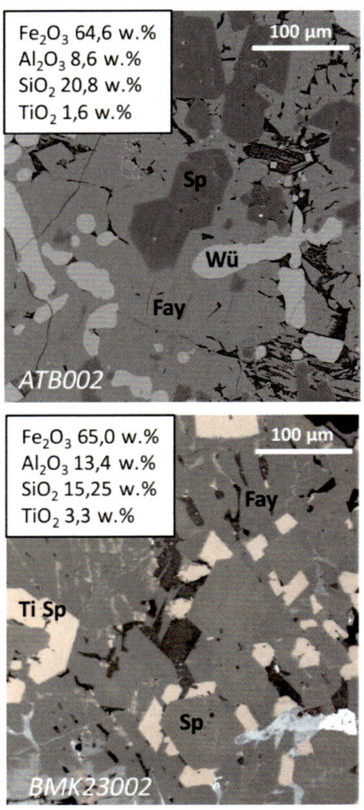

Figure 2.7: Comparison of two smelting slag samples from north-eastern Madagascar with different bulk chemical compositions but similar Fe_2O_3 content (Optical metallographic microscope images). **(Above)** Antaimby-Sambava (ATB): Texture with aluminous spinels (Sp), wustite dendrites (Wü) and chunky fayalite crystals (Fay). **(Bottom)** Bemanevika (BMK): Texture with aluminous spinels (Sp), titaniferous spinels (Ti Sp) and chunky fayalite crystals (Fay).

In order to estimate the technical efficiency and especially to be able to compare two technical traditions, we propose comparing the yield, which is the ratio between the metallic iron produced and the iron content of the ore:

$$\text{Yield (\%)} = \frac{Produced\ metallic\ iron}{Fe\ in\ ore} \tag{2.6}$$

The yield takes into account the ore loaded into the furnace. Iron production can be estimated through mass balance calculations and a detailed study of the relationship between ore, furnace wall contamination and slag (Chapter 6).

There are many criteria that can influence the yield, and thus the technical efficiency of a furnace: ore quality, ore particle size, open or closed structure of the furnace, size of the furnace, number of air inlets, slag fluidity, etc. For example, in a furnace with a high chimney, the reducing gas is in contact with the ore particles for a longer period of time, and therefore the smelting process has more time to take place. But on the other hand, with a high chimney follows a stronger draught of air and therefore a faster combustion of the charcoal. The ore then falls down the chimney of the furnace more quickly, leaving less time for the smelting reaction to take place. A technically efficient smelting tradition therefore manages to balance all these factors by adapting to a given ore.

The ore is indeed probably a key aspect in understanding how an iron smelting technique works. Not all iron smelting techniques are able to produce iron from poor quality ore, or on the contrary, some extract only a very small amount of the iron available in an excellent ore.

More importantly, the smelters should not carry out the smelting process the same way if they use an ore with a low or high iron content. To understand this, it is necessary to look at the phase

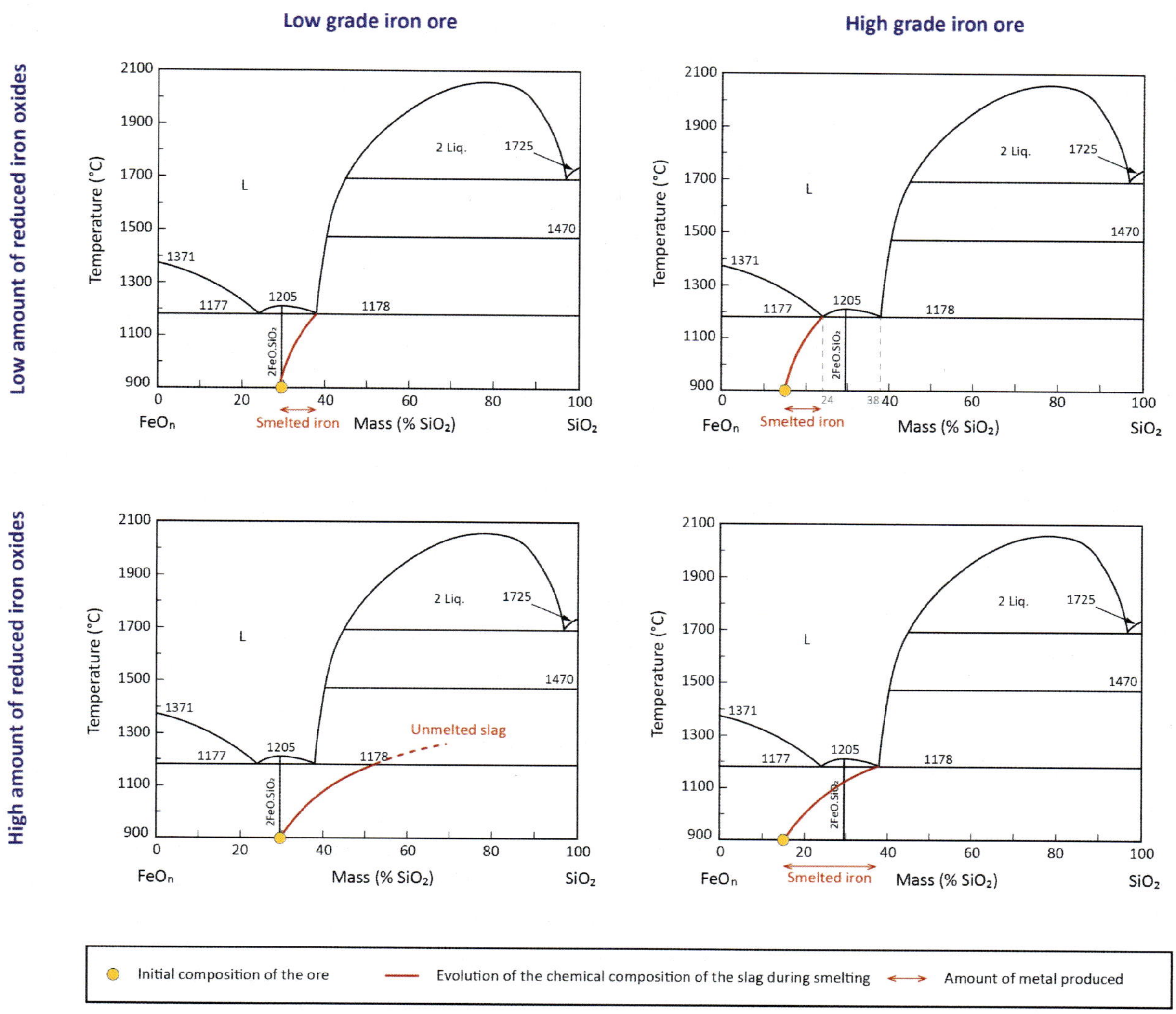

Figure 2.8: $SiO_2 - FeO$ phase diagram: Modelling of the evolution of the chemical composition of the slag (Kowalski et al., 2000) according to the initial iron content of the ore (orange point), time and the rate of temperature rise in the bloomery furnace.

diagrams of the chemical systems of the slag, and in particular the bivariate $SiO_2 - FeO$ phase diagram, if we consider a simplified chemical system and neglect the other major elements (Figure 2.8). Similar observations can however be made on more complex diagrams, such as the $SiO_2 - FeO - Al_2O_3$ phase diagram. The ore before the smelt has a given chemical composition (Figure 2.8 - orange point). From this composition iron oxides are reduced during the smelting process and transformed into metal. The composition of the remaining ore therefore changes as the iron content decreases. At the same time, the temperature in the furnace increases until the melting point of the iron-depleted ore is reached, which results in slag. The slope of the curve representing the evolution of the chemical composition of the ore during smelting (Figure 2.8 - red line) depends on the speed of the temperature rise, the duration of the process and the quality of the reducing atmosphere in the furnace. As seen previously, the more time allowed for smelting to take place before melting the slag, the more iron is smelted.

However, it can be noted that for iron content below 62 wt.%, the melting point of the slag becomes very high and impossible to reach in a bloomery furnace (Figure 2.8). If too much iron is reduced, it is no longer possible to melt the slag and separate it from the metal.

The optimal treatment of an ore is therefore not the same for iron-rich and iron-poor ores:

- A good ore (85 wt.% Fe_2O_3) has the potential to provide a significant amount of metal. For this to happen, the slag must not melt quickly. Ideally, the smelting process should last long enough, while remaining below the melting temperatures of the slag, to hope to recover as much metal as possible.
- On the contrary, in the case of a poor quality ore (70 wt.% Fe_2O_3), it is in the smelter's interest to rise rapidly in temperature and reach the melting temperature. Otherwise, the temperature required to melt the slag becomes too high.

Thus, a smelter who is used working with bad ores, and therefore to increase the furnace temperature quickly, will not necessarily produce more iron with a better ore. The operating mode is linked to the type of ore used and has a direct influence on the quantity of iron produced, and therefore on yield and technical efficiency.

It is therefore essential that, in order to address the question of technical efficiency, archaeometallurgical studies also address the understanding of the ores, and not only of the smelting structures.

2.2.2 Economic efficiency

The economic efficiency corresponds to the balance between the value of the production and its costs. Production can be calculated from mass balance calculations (Chapter 6). The production is the mass of iron produced per unit mass of ore loaded into the furnace, or per unit mass of slag (the amount of slag and ore being related):

$$\text{Production (\%)} = \frac{Mass\ of\ produced\ metallic\ iron}{Mass\ of\ ore\ in\ the\ furnace} \tag{2.7}$$

Depending on the smelting traditions, the production can range from 1-2 kilograms of iron at a time to approx. 100 kilograms. This quantity of metal must be profitable in comparison with the work invested to produce it (extraction of the ore, construction of the furnace, charcoal production...).

The economic efficiency depends greatly on the natural and human resources available to a given society. Production costs vary according to the conditions of production. For example, a population living close to a forest can afford an iron smelting technique that consumes a lot of charcoal since wood is abundant. Similarly, some societies practise slavery, thus minimising labour costs.

The value of metal in a given human group is difficult to estimate archaeologically. It depends on the quality of the metal, the rarity of the product and, as described, the production costs.

A smelting technique can have sub-optimal or poor technical efficiency but be economically very efficient. It would then justify maintaining this technique over time. However, there are cases, such as war situations where metal supply is essential, even at the expense of technical and economic efficiencies. The key issue is to produce metal.

2.2.3 Social efficiency

The social efficiency corresponds to the ability of a technology to provide the quantity and quality of iron necessary for the proper functioning of a human group. A technology that produces little iron in a society with low iron consumption or that can be supplied with metal through trade is a technically socially efficient.

Moreover, social efficiency can also be related to the ability of a technique to fulfil either its ritual role, to maintain the labour force at work, to justify the existence of social castes or to maintain a group identity. All these social considerations can justify the existence of a technology that would not be efficient or profitable from another point of view. Social efficiency, and especially the ritual dimension, is very difficult to assess archaeologically but is essential if we intend to understand the place of production in the organisation of a society.

The Northeastern Coast of Madagascar

3

3.1 Landscapes in northeastern Madagascar

The studied area is 450 kilometres long, from Cape of Ambre (*Tanjon'i Bobaomby*), at the extreme north of the island, to Cape Masoala (*Tanjon'i Masoala*), at the southern edge of the peninsula that closes the Antongil Bay (Figure 3.10). The topography, climate and geology change between the North and the South and very different types of landscape follow one another (Figure 3.2 and Figure 3.1).

3.1.1 General geology of northeastern Madagascar

Madagascar became an island some 90 million years, but its geological record covers more than 3 billion years of Earth's history. The basement is formed of old rocks of Precambrian age (Tucker et al., 2014). They outcrop in the central highlands and the escarpment down to the eastern coast. The western side of the island is covered by the whole range of Phanerozoic sediments from Cambrian to Quaternary.

Madagascar was part of the supercontinent Gondwana until its breakup around 160 million years ago (de Wit, 2003). Together with India and Antarctica, it drifted southward before separating from India during the Cretaceous period. These rifting episodes were accompanied by extensive volcanism. Parallel to the coast, the rocks from this episode, such as basalts and rhyolites, form a discontinuous strip with a width up to 15 km.

To the south of the Andrafiamena depression, the hinterland is made of the Precambrian basement: gneisses, schists and metamorphosed plutonic rocks. Finally, Quaternary reworked sands are deposited along the shore line, rarely exceeding 10 km width. The many small rivers flowing down the hills create lakes and marshes in this low and flat topography.

In the hinterland, the rocks of the basement and the Cretaceous volcanics have been deeply affected by a long-lasting process of alteration under tropical conditions leading to the development of a lateritic formation. The landscape is covered by a thick layer of red clays. Locally, this residual layer is enriched in iron oxides and contains iron-rich concretions of various size and shape.

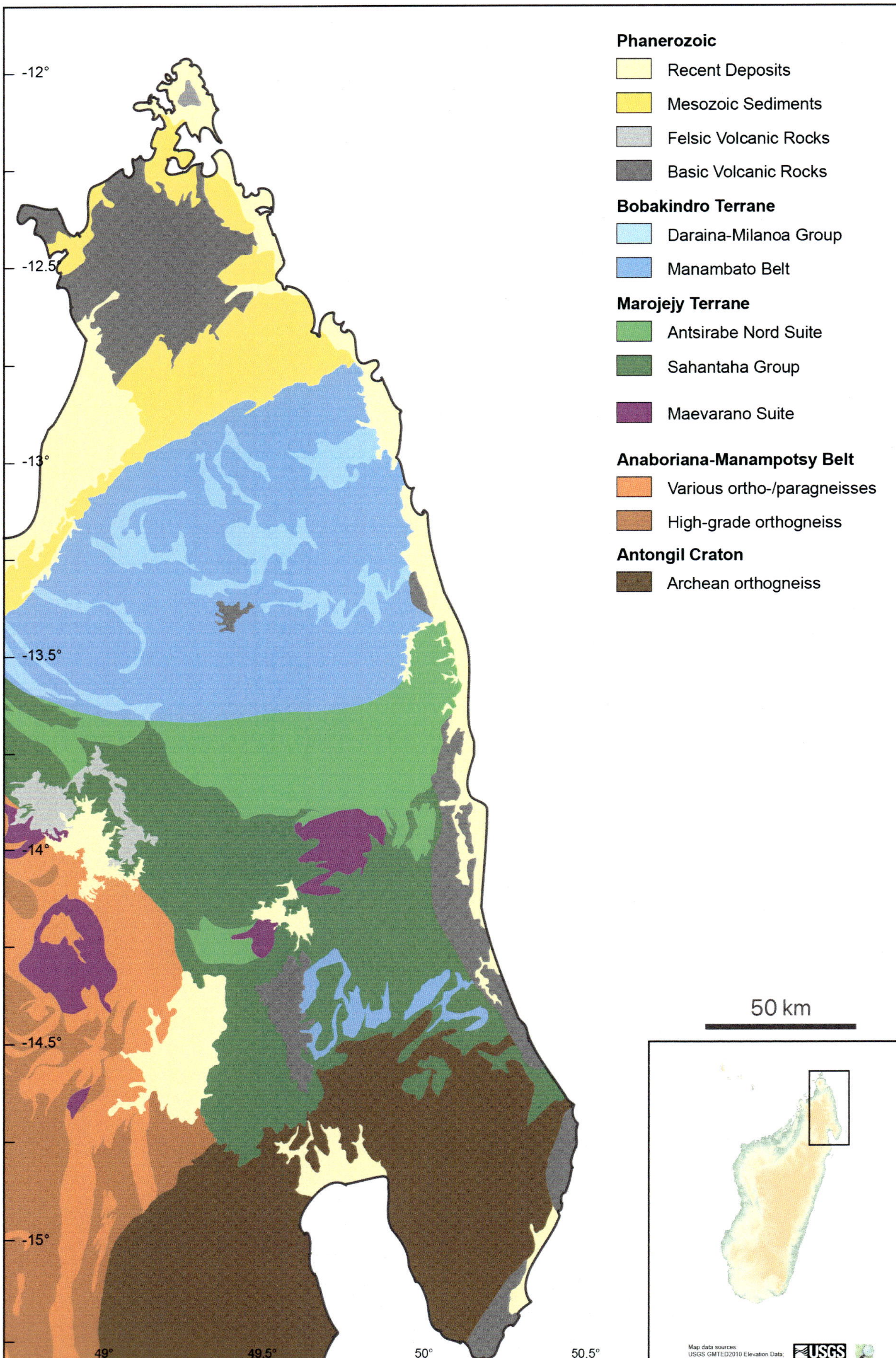

Figure 3.1: Geological map of northeastern Madagascar (Adapted from Nitsche et al., 2022; Armistead et al., 2019).

Along the eastern shore, the continental plateau formed by the coastal erosion during the Quaternary era, is covered by loose sands frequently reworked as a function of the variation of the sea level and oceanic current.

3.1.2 Climates and landscapes

In the northernmost part of the study area, the sedimentary cover outcrops from Cap of Ambre to the Andrafiamena depression, where the rivers Andrevo and Lokia flow. This 100 km long coastline is indeed covered by an important sedimentary sequence, dating from the Tertiary for the northernmost sediments, to the Perno-Triassic for the southernmost old sediments of this area. Inland, the Montagne d'Ambre is a very recent volcanic complex of Cenozoic age which culminates at 1500 metres above sea level. The morphology of the shore is characterized by a rather rugged coastline with numerous coastal islands and bays, such as the Antsiranana bay where the main modern city of the region, Antsiranana, is located, as well as numerous coral reefs. The climate is very dry in this area which limits the possibilities for agriculture for the local population, the Antakarana, who are pastoralists (Figure 3.2a). No iron smelting slag has been recovered in this area.

A second climate zone can be defined along the next 100 km, from the Andrafiamena depression to the town of Analovana, 40 km south of Vohémar. To the north of this area begins the Bemarivo domain which is characterised by the outcrop of the rocks from the basement, which are Cryogenian igneous rocks, without any sedimentary cover. Cretaceous volcanic rocks begin to outcrop south of Vohémar. The morphology of the shore evolves from a rugged to a more regular coastline with coral reefs, for example in the bay of Vohémar. Large rivers run through the hinterland, such as the Manambato or the Manambery. The landscape is shaped by a dry climate with dry forests or grasslands, where the herds can graze. In this portion of the study area again, no slag heap could be located during our surveys.

An abrupt climatic boundary then makes the transition from the dry climate to a sub-humid climate. The area from Analovana to Antalaha is still geologically located in the Bemarivo domain (Tucker et al., 2014). The landscape is characterized by well developed Cretaceous volcanic edifices, 10 to 20 km wide, on the eastern edge of the Proterozoic old basement, forming small hills (Figure 3.2b). The coastline is constituted by a discontinuous coastal strip, from 1 to 10 km wide, made up of Quaternary sand dunes. There is an alternation between sandy strips and marshy areas, where nowadays the Betsimisaraka people set up their rice fields.

Figure 3.2: Main lanscapes from northeastern Madagascar. **(a)** Dry landscape typical from the northern part of the study area (Irodo). **(b)** Landscape with sand dunes in the foreground, swampy areas below and the first volcanic hills in the background (Amboronala). **(c)** Typical red soil with accumulation of lateritic concretions (Ambodipont Limite). **(d)** View on the Ankavanana river and on the actual village of Masindrano and the rice fields.

Moreover, cyclones hit the shores every year, reshaping the sandy coastline. The climate is much more humid than in the northern parts of the study area and is favourable to tropical vegetation. Numerous rivers flow from west to east, from the crests of the central mountains (2500 metres above sea level) to the ocean. These rivers are natural ways to penetrate the hinterland. They can indeed be sailed upstream using pirogues (Figure 3.2d). Anchoring of boats is also possible in the river mouths, sheltered from the sea swell. Most of the slag heaps we studied and all the sites we excavated are located in this area.

To the south, the remaining 50 km of the coastline corresponds to the Masoala peninsula. The Cretaceous volcanic edifices are in very close proximity to the shoreline leaving only a thin sandy coastal strip, bordered by numerous coral reefs. The basement rocks are Neo Archean granitoids and belong to the Masoala suite (Antongil craton - Tucker et al., 2014). Only small rivers flow into the ocean and the tropical climate is very humid. Little evidence of iron smelting was located during surveys.

3.2 The *Rasikajy*

The North-East of Madagascar is probably one of the most studied areas of Malagasy archaeology. The earliest evidence of human occupation in this region is attested in *Lakaton'i Anja* cave where lithic artefacts dating from the first millennium CE were uncovered (Dewar et al., 2013).

Most of the studies however focus on a population called *Rasikajy* living in this area from the end of the first millennium to the first half of the second millennium. The oldest *Rasikajy* occupation was indeed attested on settlements and dates back to the 9th century CE (Battistini and Verin, 1966). Occupation is then well described between the 11th and 16th centuries and this society seems to flourish particularly in the 15th and 16th centuries (Vérin, 1986).

3.2.1 Who are the *Rasikajy*?

The term *Rasikajy* was used by local people to designate a previous population. Many legends or oral traditions are associated with this designation, but the historical part cannot be identified. It was later taken up by European authors of the 19th and 20th centuries. Gaudebout and Vernier (1941) for example refer to them as overseas people who carved objects from soapstone. We however still do not know whether the people living along the northeastern coast of Madagascar at that time considered themselves as a single cultural or ethnic entity. We know very little about them, their origins and their way of living (Schreurs and Rakotoarisoa, 2011; Vérin, 1986 ; Vernier and Millot, 1971). It is also very difficult to estimate how many of them were settled.

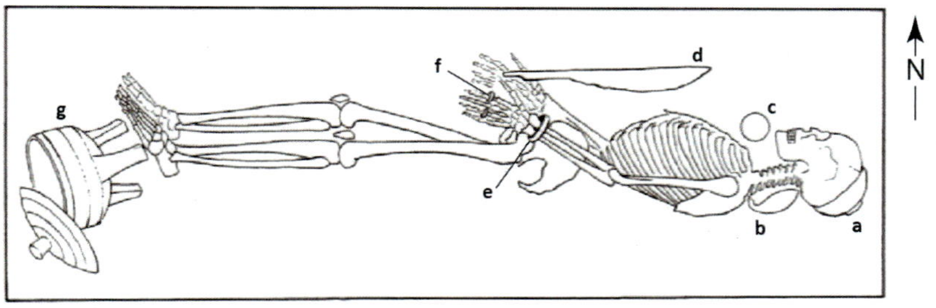

Figure 3.3: Drawing of a tomb excavated in the necropolis of Vohémar (from Vérin, 1986), inspired from sketch of P. Gaudebout. **a.** Large Chinese porcelain bowl. **b.** Mother of pearl spoon. **c.** Small porcelain bowl. **d.** Abattis sword. **e.** Bracelet. **f.** Rings. **g.** Chlorite schist cooking pot and its cover.

Most of the material culture of the *Rasikajy* is known through the excavation of the necropolis of Vohémar which delivered the largest collection of *Rasikajy* artefacts (Vernier and Millot, 1971, Figure 3.3). By comparing this documentation with the archaeological records (Vérin, 1986) and artefacts from our fieldwork, it would seem that the material culture is relatively homogeneous throughout the study area. For this reason, in order to simplify the discussion, we will refer to the population living in the study area between the 9th and 16th centuries CE as the *Rasikajy*[1].

1: We are aware that it is too simplistic to define the identity of a social group on the basis of a few elements of material culture alone, but in the current state of research, one can hardly do much better.

The *Rasikajy* were probably Islamized as attested by the presence of mosques along the coast (Vérin, 1986). More than 600 graves were excavated at Vohémar, revealing bodies oriented following Islamic traditions. The bodies are indeed oriented east-west, lying on their side with the face looking in the direction of Mecca, i.e. towards the north. Objects were found in the graves, which is contrary to Muslim burial traditions (Figure 3.3). It was therefore probably a kind of syncretism which was practised.

3.2.2 Material culture of the *Rasikajy*

Figure 3.4: Fragments of local and imported ceramics observed during the visit of the Antaimby district inside the town of Sambava.

The *Rasikajy* are mainly known for their softstone artefacts made of a stone called "chlorite schist" in the archaeological documentation. The most common artefacts are tripod cooking pots with lids which were lathe turned. There are also rare bowls and incense burners. It is also quite common to find pierced circular fragments of these stone pots during archaeological excavations (Figure 3.5a). It is assumed that these discs, which have a diameter of 2-5 cm and a central hole, were reused as spindle whorl discs (Radimilahy, 2011).

Most of the recovered artefacts are local ceramics and sherds are found on the surface of many sites, or in modern coalfields (Figure 3.4). Unfortunately, during the excavation of the necropolis of Vohémar, no local ceramic fragments were collected. Moreover, to date, no typology of these ceramic assemblages has been established for the Northeast of Madagascar. The excavations we carried out, in particular at Benavony, have provided a large collection of local ceramic fragments (Serneels et al., 2018). The range of objects is limited as it seems that there are either large bowls for food presentation, with a diameter ranging between 20 to 40 centimetres, or cooking pots/storage jars, with or without a carina, which are 10 to 20 centimetres wide and 15 to 30 centimetres high. The sherds may present decorations or engobe (Figure 3.5c). Ceramics are currently under study and in particular the raw materials are being investigated in the framework of a Master's dissertation at the University of Fribourg, by Josip Lujic[2]. This study has shown that the *Rasikajy* ceramics were manufactured following a low and heterogeneous firing process. The pastes are made of sandy clay to which sand has rarely been added. Each site has its own local production and uses the surrounding raw materials. There is limited exchange between sites. There are also some special and unusual artefacts like some talc tempered ceramics, which were found in the lower level of Benavony (Serneels et al., 2018). These ceramics are imported, in particular because the talc powder incorporated in the clay paste does not match the local chlorite schist. These ceramics give the visual effect of having been made

2: This Master thesis entitled "Etude pétrographique de céramiques médievales du nord-est de Madagascar", supervised by Vincent Serneels, was defended at the University of Fribourg in 2022.

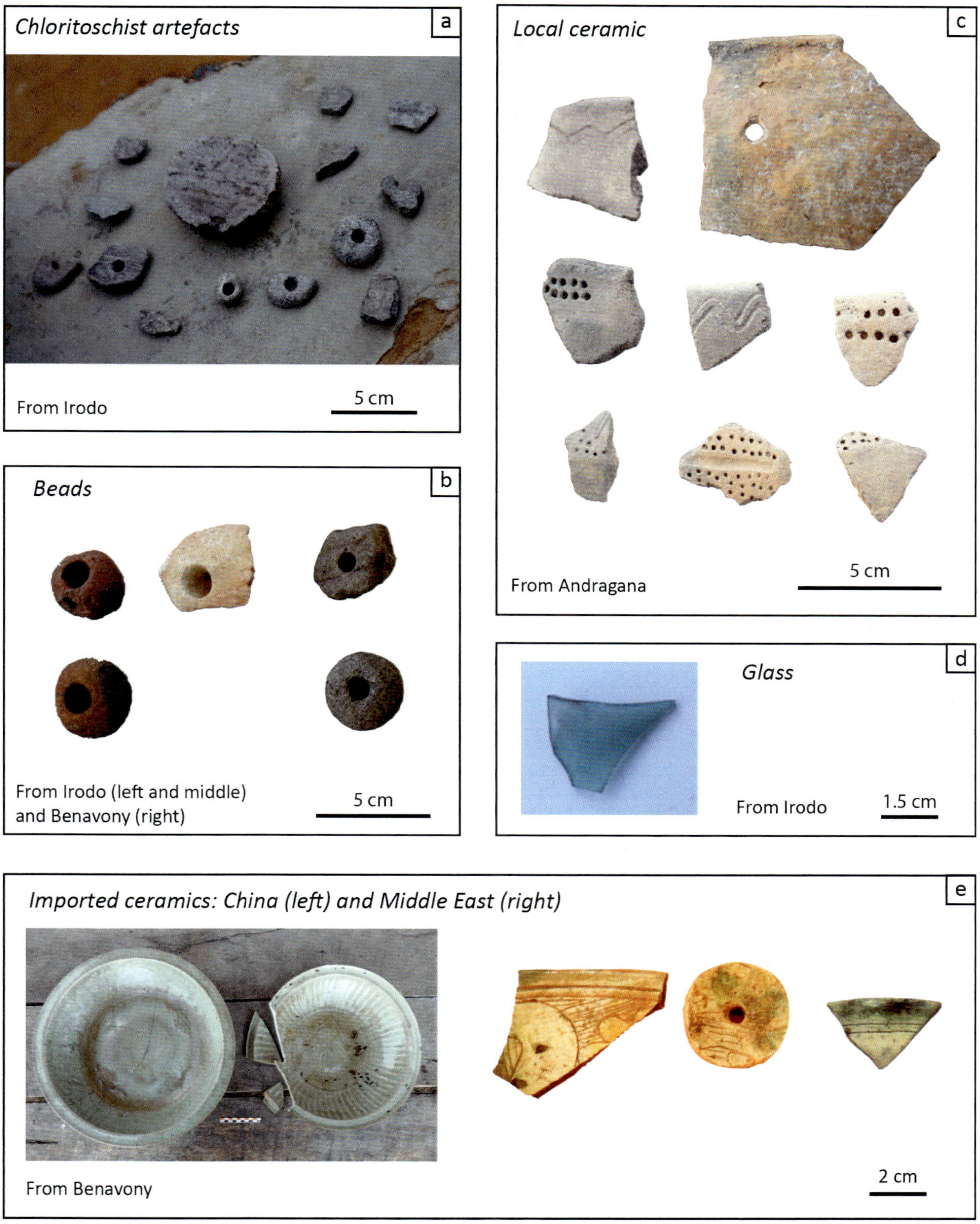

Figure 3.5: Overview of the material culture of the *Rasikajy*: artefacts typically collected during the surveys and the archaeological excavations.

of stone and remind one of chlorite schist pot fragments. Moreover, some fragments of local pottery are pierced with a hole, perhaps to be repaired or perhaps in some cases to make spindle whorls (Figure 3.5c).

Some iron objects are also found during archaeological excavations, often in a very poor state of preservation. Relatively well-preserved objects were found during the Vohémar excavation (Gaudebout and Vernier, 1941), but they have all disappeared. They are mainly *sabres d'abatti* (Figure 3.3d), which look like modern machetes, or small knives.

It is also very common to find beads on settlement sites, which can be made of a wide variety of materials: bone, ivory, shell, but also stone such as quartz, carnelian, agate or calcedony (Figure 3.5b - Rasoarifetra, 2011). Few shell spoons were also uncovered during the excavation of some graves.

Finally, the *Rasikajy* material culture is characterised by the occurrence of various imported goods, testifying to the participation of the *Rasikajy* in the medieval Indian Ocean trade. The excavations of the Vohémar necropolis, for example, have unearthed objects made of copper alloys, such as rings, bracelets, blush needles or Chinese mirors, as well as fragments of glass artefacts that could be small flasks or perfume containers (Figure 3.5d). Many fragments of imported ceramics are also mentioned in the archaeological reports and were found during our excavations and surveys (Figure 3.5e). These are mainly Chinese wares (Zhao, 2011), such as celadons or porcelain, or Islamic ceramics such as *sgraffiato* (Vérin, 1986).

3.3 Natural resources available in northeastern Madagascar

The northeast of Madagascar has many natural resources that may have been exploited by past populations. Some of these resources were used locally for the immediate needs of the population or possibly traded along the northeastern Malagasy coast. As seen above, the *Rasikajy* were connected to the Indian Ocean trading network. They imported goods, but they certainly exported others as well. The exploitation of these resources has been largely understudied, and there is little literature available.

3.3.1 Mineral resources

The island of Madagascar is known today for its rich and diverse mineral resources. In particular, the northeast of Madagascar has deposits of various metals and stones (Figure 3.7).

The exploitation of stone resources

As mentioned above, clays, sometimes of high quality, are available in the hinterland and have been exploited for the manufacture of local ceramics. Clays are difficult to find in the coastal sedimentary zone. It is indeed a product of weathering of Cretaceous volcanic rocks and of the Precambrian basement.

The *Rasikajy* are known for the exploitation of softstone called "chlorite schist" in the archaeological literature. This kind of metamorphic rock is easy to carve with iron tools. Several quarries are located and described in the hinterland of Vohémar (Figure 3.7 and Figure 3.10 - Mouren and Rouaix, 1913; Monnier, 1910). This production and its trade are the subject of a PhD thesis, as part of the project "Pots, Stones and Iron by the *Rasikajy*", also at the University of Fribourg, by Christoph Nitsche (Nitsche et al., 2022 - Figure 3.6).

Figure 3.6: Example of a chlorite schist tripod showed by local population in Amboronala.

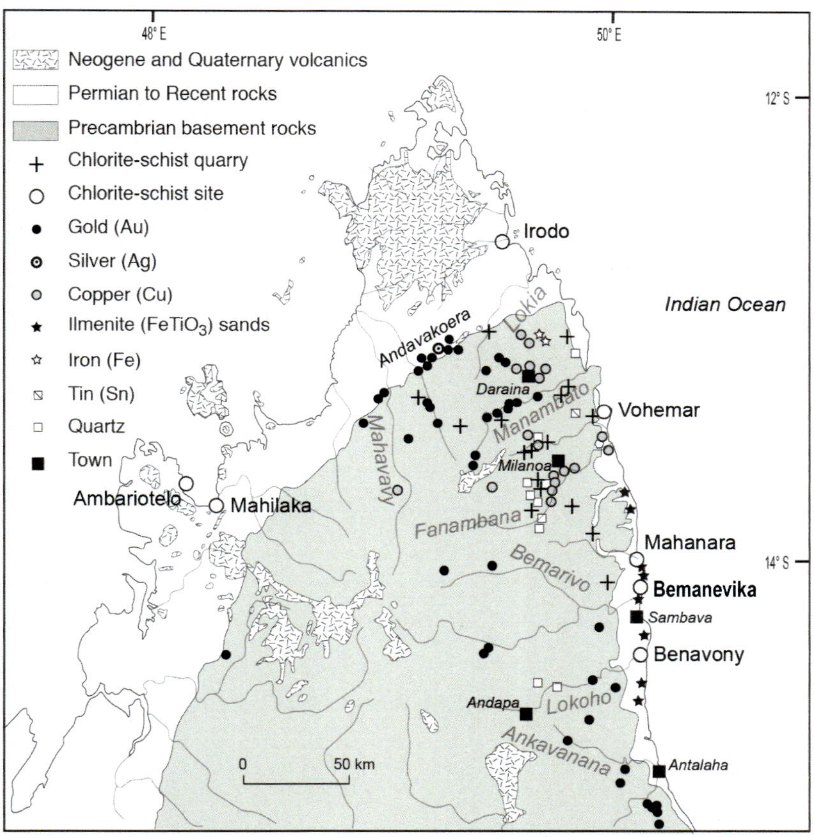

Figure 3.7: Map of some of the natural mineral resources available in northern Madagascar (from Schreurs and Rakotoarisoa, 2011).

The rock crystal exploitation is a heavily debated topic of Malagasy archaeology. Northern Madagascar is known for the presence of large quartz veins which could have been a interesting and exploitable natural resource (Figure 3.7). Historical sources describing the use of rock crystal in the Arab-Muslim world, as well as archaeological findings from sites on the eastern coast of Africa and the Comoros, indicate that Madagascar is likely to have played a major role in the rock crystal trade (Pradines, 2019; Horton et al., 2017). Several European written sources also describe the abundance and the availability of quartz (Lacroix, 1922; Grandidier and Grandidier, 1908). It seems that quartz was even massively extracted in the 19th and early 20th century. Quartz fragments have been found on several sites from this period such as Benavony (Serneels et al., 2018) or Dembeni in the Comoros (Pradines et al., 2016). Today, however, no archaeological quartz extraction areas have been uncovered yet, and no convincing provenance studies have been conducted.

Finally, northern Madagascar is also known for its precious stones, such as sapphires and emeralds. These stones may have been collected and exported as early as pre-colonial times, but no archaeological or historical work has revealed the existence of such exploitation.

The exploitation of metal resources

Gold deposits are available in northern Madagascar in most parts of the old basement but they are very disseminated (Figure 3.7 - Lacroix, 1922). Gold is also concentrated by superficial running waters in fluviatil sediments. In the northern part of our study area, most of the gold deposits are located in the Andrafiamena depression. A mine operated between 1905 and 1910 at Andavakoera and today, some mines are still active near Betsiaka. To the south, the gold deposits are available in Ambato and Andrarona, close to Antalaha. It is however not clear if these deposits were exploited before modern times. Europeans were without any doubt looking for gold when they started to explore Madagascar, but written accounts of the 16th to the 18th century do not report a significant gold production anywhere on the island. No direct archaeological evidence supports an early activity. But on the other hand, if gold had been collected by simple stream sediment washing, this would have left no trace.

If gold was sought and collected, there are only few gold objects reported in the Malagasy archaeological records. The scarcity of Malagasy gold artefacts contrasts against their abundance testified in southern Africa (Miller et al., 2000; Swan, 1994). Most of the gold objects described for northeastern Madagascar are probably imports, such as the find of a handful of Islamic coins and jewellery

(Vérin, 1986; Gaudebout and Vernier, 1941). There is no doubt that gold resources were available for early settlers and that they would have mastered the technical knowledge to exploit them. Probably, some exploitation took place, but material evidence is still lacking.

Concerning silver, no deposit is reported in the geological and historical literature. Silver artefacts are very rare in the archaeological records.

Copper is associated with Cretaceous volcanic rocks near Vohémar and Antalaha (Lacroix, 1922). Metallic copper can also be found in these rocks but it is too disseminated to be mined. It could however be collected as was gold in river sediments but this is not reported as a traditional activity. Copper artefacts are rare in the *Rasikajy* archaeological records even if some copper rings or chains are described. At least some of the copper objects are imported artefacts, such as Chinese mirrors.

High grade lateritic iron ores are available along the eastern coast of Madagascar, as a weathering product of Cretaceous basaltic rocks (Chapter 6). Black sands rich in magnetite and ilmenite form large black bands on the beaches and could be a potential source of iron ore (Figure 3.7 - Chapter 6). Iron was in any case exploited by the *Rasikajy* and this production will be detailed later in this thesis.

3.3.2 Other resources

Northeastern Madagascar, and in particular the area in the south of Vohémar, is covered by relatively dense tropical forests. Wood is therefore an abundant resource. On the contrary, natural outcrops of stones that could be used as building material are rare in the Quaternary sandy area. The scarcity of stone walls reported in the archaeological records, except for the Mosque from Mahanara and some graves (Vérin, 1986), and the significant presence of post holes on archaeological excavations therefore suggests that wood was widely used as a building material, especially for dwellings. The few settlements that have been studied however do not give enough information to reconstruct the spatial organisation of the houses. In addition, wood was used to make charcoal, especially for cooking or as fuel for iron metallurgy, as well as to manufacture boats. Wood is probably a resource exploited only for local needs, since other regions more central in the Indian Ocean, such as Kenya, used to export wood massively.

On the contrary, it is very likely that copal, a scented resin, was exploited to be exported. Unfortunately, there is no archaeological or historical evidence of this trade.

The climate is also favourable for the production of spices, which are an important economic resource in this region of Madagascar today. However, there is no evidence that these spices grew naturally in this part of Madagascar. They were probably introduced later by Europeans (Seland, 2014).

3.3.3 Food production

The northeastern coast of Madagascar can be considered as a rich environment for small human communities. The dry climate of the northern part of our study area allows livestock breeding whereas the tropical southern part has ideal conditions for growing rice, some tubers such as cassava or other tropical plants such as coconuts.

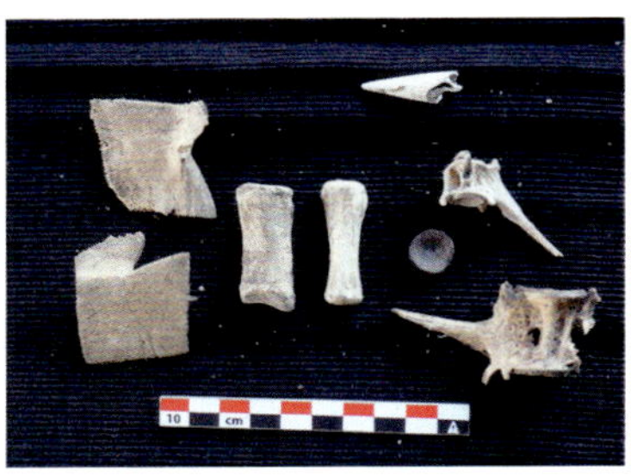

Figure 3.8: Example of an assemblage of bones found during a survey in the habitat of Andrangana. No zooarchaeological studies have been undertaken on this material at this time. From left to right: Sea turtle shell bones, zebu bones, a shell and fish bones.

The proximity to the sea is also ideal to access to marine resources such as fish, shellfish or sea turtles. A rich fauna must also have lived inland which is favourable to hunting. Although Madagascar is known for its giant fauna, such as the elephant-bird Aepyornis, it is not known whether humans cohabited with this fauna (Hansford et al., 2021).

The few test pits carried out on settlements give valuable information on the diet of these populations. Fish bones, zebu bones or turtle shells have been mainly uncovered (Figure 3.8). At Irodo we observed concentrations of shellfish that are the same as those eaten today by local people (Figure 3.9). There are no studies published on agriculture, pollen or seeds in this region of Madagascar, so the precise diet of these people is not known.

Figure 3.9: Left: concentration of shells at the Irodo site. The shells are pierced in a rather characteristic way that confirms that they were consumed by humans. Right: Current exploitation of these same shells by local populations at Irodo.

3.4 Occupation of the territory

Important survey campaigns have been carried out by several teams in the framework of this project in the whole studied area. The study of iron production, stone quarries and ceramics, combined with previous studies (Pearson, 2003; Vérin, 1986), provided a more detailed understanding of the occupation of the territory (Serneels et al., 2021 – Figure 3.10).

The human occupation is located mainly along the coast, apart from the chlorite schist quarries located in the hinterland of Vohémar (Figure 3.10). Before the beginning of our work, the picture of human occupation we had was a few important harbours oriented towards trade, such as Vohémar or Benavony, and little occupation of the rest of the territory. Our surveys lead to the discovery of more settlements, sometimes probably very small ones. It would thus appear that the coast was fully and relatively densely populated. Moreover, some of the chlorite schist quarries are difficult to access and far from the coast, attesting to a good knowledge of the hinterland and its resources. The *Rasikajy* were thus not only sea-oriented but, on the contrary, many villages were settled on a permanent basis and turned towards the exploitation of the hinterland.

It would seem that the settlements are located along navigable rivers and therefore close to estuaries and lakes. Protected bays seem to be favourable to more important occupations such as Vohémar or Benavony. Some settlements were located inland, even if no radiocarbon dating has been done on these sites. The settlements furthest from the coast are about fifteen kilometres inland (Andampy-Be, Vohotsarivo, Antoaka).

Overall, little is known about inland sites because they are difficult to reach. Current knowledge about the occupation of the territory is clearly biased by our survey approach. In particular, the northern half of the study area was much less surveyed and it would not be surprising to locate settlement close to every river mouth, including the Lokia River. It is likely that there are more settlement sites far from the coastline to be found, but it is unlikely to be a dense occupation otherwise it would have been reported by previous authors and by local people during our visits.

At the current state of research, it is not known whether another population cohabited with the *Rasikajy* inland. The northern part of Madagascar was occupied by a population before the *Rasikajy* arrival (Dewar et al., 2013) but there is no archaeological evidence that this population and the *Rasikajy* cohabited. It is therefore not known whether there was competition for the occupation of the territory or whether the *Rasikajy* settled on land that was not occupied by humans.

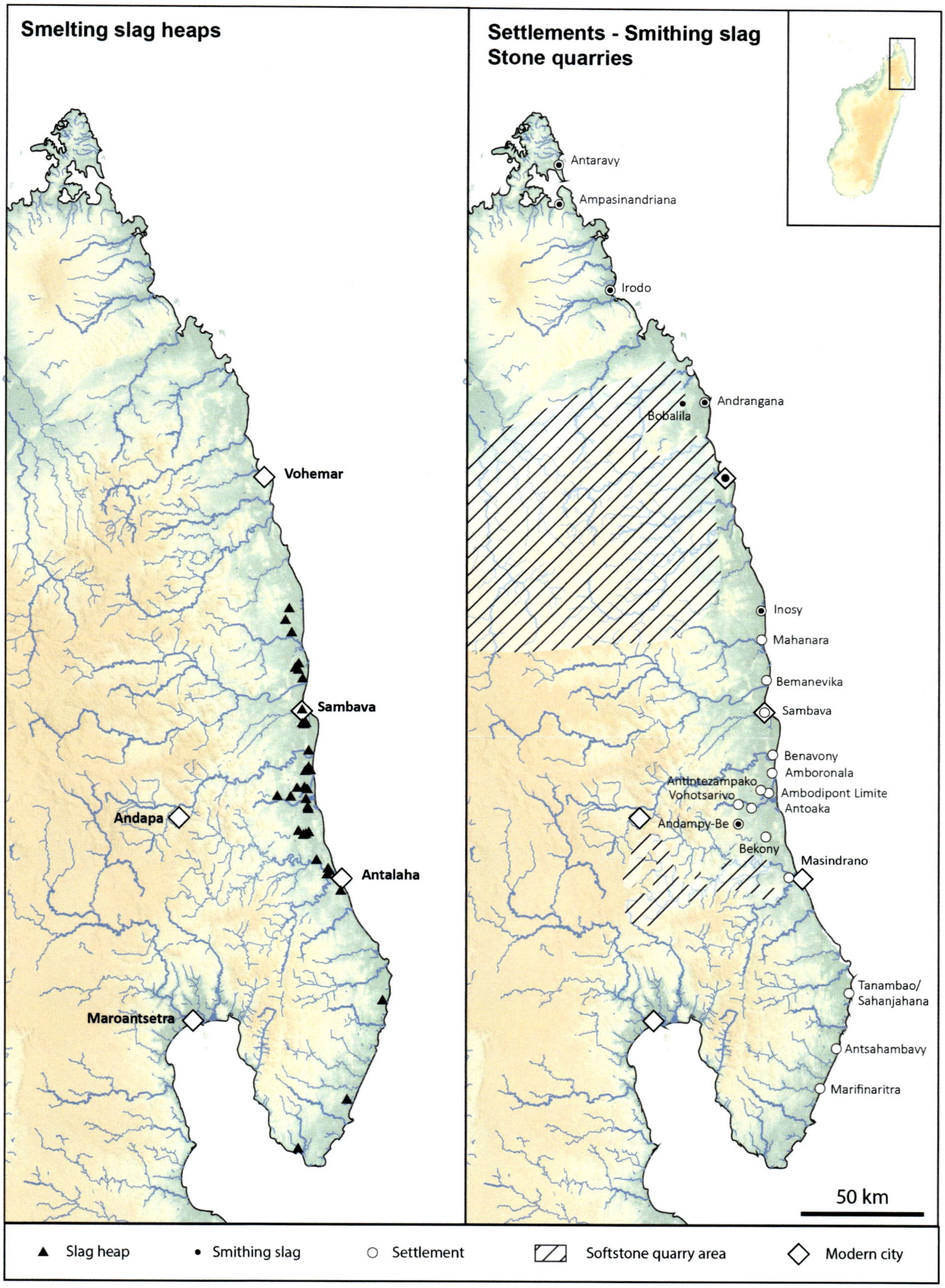

Figure 3.10: General maps of the *Rasikajy* occupation of the territory in northeastern Madagascar. **Left**: Map of the smelting slag heaps located during our surveys and of the metallurgical district. **Right**: Map of the settlements and of the area of chlorite schist extraction based on our surveys (sources: USGS GMTED2010 Elevation Data, OpenStreetMap).

Archaeological Excavations at Smelting Sites 4

Iron production remains in northeastern Madagascar, and in particular the presence of slag, have been documented since the beginning of the 20th century (Vérin, 1986). However, only four sites were reported in the archaeological documentation (Benavony, Matavy, Amboronala, Bemanevika), which gave the picture of a scattered and isolated production. However, the production of iron was widely spread since about 150 slag heaps, distributed over about 20 sites, have been located during our surveys (Figure 4.3). Given the intensity of our surveys in the study area, we know that we could not locate all the existing slag heaps. Some sites are very difficult to reach and it would take a long time to locate all the slag heaps in the region. Moreover, some of them are probably already destroyed by the intensification of agriculture, in particular palm monoculture. However, we covered a large area during our surveys and many discussions were held with the local population. It is therefore unlikely that any significant production area was missed and we are confident that we have a good representation of the distribution of metallurgical sites. We estimate that at least half of the slag heaps in the area were located. This means that there may have been between 150 and 300 slag heaps, which correspond to 500 to 1000 tons of slag.

Two excavation campaigns were conducted in 2017 and 2018 (for the complete and detailed studies, see Serneels et al., 2018; Serneels et al., 2019b; Serneels et al., 2020; Serneels et al., 2021). Three slag heaps on three different sites were excavated (Benavony BNY410, Matavy MTY11 and Amboronala MBR140) and twelve test pits were carried out on other slag heaps. These sites form the basis of our knowledge about iron metallurgy in northeastern Madagascar.

Figure 4.1: Picture of a weathering profil of basalts in Antanandava, close to Sambava.

All iron smelting sites are located in the southern part of the study area and seem to be absent in the north (Figure 4.2). The location of the studied metallurgical sites coincides with the presence of basalts and the warm, humid, tropical climate zone (Figure 3.1). The weathered basalts are transformed into lateritic rocks (Figure 4.1) which were used as iron ore (Chapter 6). North of Vohémar, no slag heaps have been found, probably because there is no good ore available.

It is possible that the metallurgical district extends further south of our study area, into the Antongil Bay. The toponymy of some places suggests that they are associated with ancient metallurgical production. The toponymy of places in Madagascar generally has a meaning and sometimes refers to metallurgical practices, which guide the surveys. Thus, places named *Antaimby* ("the place of slag") or *Betaimby* ("a lot of slag") are frequently encountered. Unfortunately, no survey was carried out in this area in the course of this thesis.

Figure 4.2: Map of the iron smelting district according to our surveys (Sources : USGS GMTED2010 Elevation Data, OpenStreetMap).

Sites	Location details	Number of heaps	Slag heap	GPS		Type of intervention	Size	Slag assemblage
				X	Y			
Inosy	Antaimby/Anjala	5	INY110	39L402480	8467504	Visit	Small	BFS?
			INY120	39L402468	8467491	Visit	Small	BFS?
			INY130	39L402463	8467483	Visit	Small	Mixed
			INY140	39L402469	8467481	Visit	Small	Mixed
			INY150	39L402451	8467479	Visit	Small	BFS?
	Masomamangy	2	INY210	39L400901	8462497	Visit	Unknown	Unknown
			INY220	39L400898	8462478	Visit	Unknown	Unknown
	Lac Inosy	1	INY310	39L406183	8466889	Visit	Unknown	Mixed?
Mahanara	Ambodimanga	6	MHN110	39L403566	8457063	Visit	Unknown	Unknown
			MHN120	39L403562	8457066	Visit	Small	Unknown
			MHN130	39L403555	8457062	Visit	Small	Unknown
			MHN140	39L403549	8457065	Visit	Unknown	Unknown
			MHN150	39L403546	8457058	Visit	Unknown	Unknown
			MHN160	39L403581	8457040	Visit	Unknown	Unknown
Bemanevika	Andranginalo	4	BMK510	39L406626	8444025	Survey	Small	BFS
			BMK520	39L406609	8444019	Survey	Small	BFS
			BMK530	39L406613	8444013	Survey	Small	BFS
			BMK540	39L406603	8444003	Survey	Medium	BFS
	Tsaratanana	4	BMK110	39L405694	8442228	Survey	Small	BFS
			BMK120	39L405700	8442213	Survey	Small	BFS
			BMK130	39L405699	8442204	Test pit	Small	BFS
			BMK140	39L405717	8442174	Survey	Small	BFS
	Tanambao-Lex	5	BMK210	39L406144	8441236	Survey	Small	BFS
			BMK220	39L406149	8441258	Survey	Small	BFS
			BMK230	39L406162	8441341	Test pit	Small	BFS
			BMK240	39L406184	8441317	Survey	Small	BFS
			BMK250	39L406194	8441299	Survey	Small	BFS
		2	BMK410	39L408160	8437505	Survey	Small	BFS
			BMK420	39L408154	8437346	Test pit	Small	Mixed
Sambava	Antaimby	5	ATB1	39L407968	8423790	Survey	Small	Mixed
			ATB2	39L407928	8423836	Survey	Small	Unknown
			ATB3	39L407892	8423884	Survey	Small	Unknown
			ATB4	39L407912	8423866	Survey	Small	Unknown
			ATB5	39L407921	8423840	Survey	Small	Mixed
Matavy		19	MTY1	39L408663	8417706	Survey	Small	BFS
			MTY2	39L408659	8417729	Survey	Small	BFS
			MTY3	39L408681	8417769	Survey	Small	BFS
			MTY4	39L408594	8417877	Survey	Small	BFS
			MTY5	39L408563	8418211	Survey	Small	BFS
			MTY6	39L408855	8418159	Survey	Small	BFS
			MTY7	39L408945	8418249	Survey	Small	BFS
			MTY8	39L408854	8418275	Survey	Small	BFS
			MTY9	39L408845	8418293	Survey	Small	BFS
			MTY10	39L408825	8418297	Survey	Small	BFS
			MTY11	39L408881	8418300	Excavations	Big	Mixed
			MTY12	39L408867	8418322	Test pit	Small	BFS
			MTY13	39L408899	8418311	Survey	Small	BFS
			MTY14	39L408904	8418326	Survey	Small	BFS
			MTY15	39L408916	8418334	Survey	Small	BFS
			MTY16	39L408929	8418337	Survey	Small	BFS
			MTY17	39L408912	8418361	Survey	Small	BFS
			MTY18	39L408928	8418353	Survey	Medium	BFS
			MTY19	39L408966	8418346	Survey	Small	BFS
Benavony		5	BNY410	39L410957	8406592	Excavations	Big	Mixed
			BNY420	39L410987	8406642	Survey	Big	Mixed
			BNY430	39L410816	8406526	Test pit	Big	Mixed
			BNY440	39L410975	8406693	Survey	Big	Mixed
			BNY450	39L410908	8406686	Survey	Big	Mixed
Amboronala		12	MBR110	39L410832	8398584	Test pit	Small	TS
			MBR120	39L410836	8398624	Test pit	Small	BFS
			MBR131	39L410856	8398609	Survey	Small	BFS
			MBR132	39L410856	8398609	Survey	Small	TS
			MBR141	39L410891	8398807	Excavations	Small	TS
			MBR142	39L410891	8398807	Excavations	Small	TS
			MBR143	39L410891	8398807	Excavations	Small	BFS
			MBR210	39L410738	8399857	Test pit	Medium	BFS
			MBR220	39L410734	8399868	Test pit	Small	BFS
			MBR240	39L410740	8399854	Test pit	Small	BFS
			MBR411	39L411516	8398162	Survey	Small	BFS
			MBR412	39L411516	8398162	Survey	Small	TS

Figure 4.3: Table listing from north to south the iron smelting slag heaps located in northeastern Madagascar (northern part of the district) and their general characteritics (for a more detailed table see Figure A.1).

Sites	Location details	Number of heaps	Slag heap	GPS		Type of intervention	Size	Slag assemblage
				X	Y			
Ambodimadiro		14	DMD810	39L409849	8397868	Survey	Small	TS
			DMD820	39L409874	8397865	Test pit	Small	TS
			DMD830	39L409852	8397868	Survey	Small	TS
			DMD840	39L409863	8397868	Survey	Small	BFS
			DMD850	39L409913	8397863	Test pit	Small	TSs
			DMD860	39L409928	8397858	Survey	Small	BFS
			DMD910	39L409807	8398073	Survey	Big	Mixed
			DMD920	39L409814	8398069	Survey	Big	Mixed
			DMD930	39L409868	8398075	Visit	Small	BFS
			DMD940	39L409886	8398103	Visit	Small	BFS
			DMD950	39L409873	8398112	Visit	Small	TS
			DMD960	39L409866	8398114	Visit	Small	TS
			DMD970	39L409859	8398112	Visit	Small	TS
			DMD980	39L409884	8398117	Visit	Small	Unknown
Ambodipont Limite	Hill	1	APL100	39L409789	8390892	Survey	Small	BFS
	Village	1	APL200	39L410360	8390220	Visit	Small	Unknown
	Andranovato	1	APL300	39L406243	8390699	Survey	Medium	Unknown
	Antintezampako	5	APL410	39L409452	8391121	Survey	Medium	Mixed
			APL420	39L409449	8391164	Survey	Small	BFS
			APL430	39L409452	8391160	Survey	Small	BFS
			APL440	39L409437	8391111	Survey	Small	Mixed
			APL450	39L409437	8391101	Survey	Big	Mixed
Ampanantova	Antaimby	5	PTV110	39L409782	8385848	Survey	Small	TS
			PTV120	39L409781	8385854	Survey	Small	TS
			PTV130	39L409753	8385824	Survey	Small	BFS
			PTV140	39L409688	8385981	Survey	Small	TS
			PTV150	39L409722	8385981	Survey	Small	TS
	Antoaka	3	PTV210	39L403386	8386852	Survey	Unknown	Unknown
			PTV220	39L403386	8386852	Survey	Unknown	Unknown
			PTV230	39L403386	8386852	Survey	Unknown	Unknown
	Vohotsarivo-Betaimby	2	PTY310	39L398144	8387156	Visit	Medium	BFS
			PTV320	39L398164	8387146	Visit	Medium	BFS
Ambodipont Isahana		3	ISH110	39L410565	8382954	Survey	Small	BFS
			ISH120	39L410815	8381974	Survey	Small	BFS
			ISH130	39L410717	8382355	Survey	Small	BFS
Ampaha	Ankarango	2	MPH110	39L408787	8371108	Visit	Unknown	Unknown
			MPH120	39L408780	8371110	Visit	Unknown	Unknown
		4	MPH210	39L408827	8370727	Visit	Unknown	BFS
			MPH220	39L408829	8370715	Visit	Unknown	BFS
			MPH230	39L408829	8370701	Visit	Unknown	BFS
			MPH240	39L408813	8370702	Visit	Unknown	BFS
		4	MPH310	39L408513	8370782	Visit	Small	BFS
			MPH320	39L408513	837078	Visit	Small	BFS
			MPH330	39L408512	8370788	Visit	Small	BFS
			MPH340	39L408505	8370812	Visit	Small	BFS
		2	MPH410	39L409339	8371209	Visit	Unknown	BFS
			MPH420	39L409376	8371183	Visit	Medium	Mixed
		1	MPH510	39L406798	8372725	Visit	Small	Mixed
		2	MPH610	39L411037	8372052	Visit	Unknown	Unknown
			MPH620	39L411030	8372080	Visit	Unknown	Unknown
Ankavanana	Andrika	3	TLH110	39L419461	8355975	Survey	Unknown	Mixed
			TLH120	39L419467	8355989	Survey	Unknown	Mixed
			TLH130	39L419469	8355982	Survey	Unknown	Mixed
	Andripitra	3	TLH210	39L419450	8355677	Survey	Unknown	BFS
			TLH220	39L419440	8355663	Survey	Unknown	BFS
			TLH230	39L419442	8355626	Survey	Unknown	BFS
	Masindrano	3	TLH310	39L419126	8355217	Survey	Unknown	Mixed
			TLH320	39L418992	8355191	Survey	Unknown	Mixed
			TLH330	39L418744	8354799	Survey	Unknown	Mixed
	Ankavanana	3	TLH410	39L414347	8360393	Visit	Unknown	Unknown
			TLH420	39L414449	8360386	Visit	Unknown	TS
			TLH430	39L414403	8360369	Visit	Unknown	Mixed
Ambodikakazo		8	KKZ110			Survey	Unknown	BFS
			KKZ120	39L424727	8347401	Survey	Unknown	BFS
			KKZ130			Survey	Unknown	BFS
			KKZ140			Survey	Unknown	BFS
			KKZ150			Survey	Unknown	BFS
			KKZ160	39L424531	8347255	Survey	Unknown	BFS
			KKZ170	39L424518	8347310	Survey	Unknown	BFS
			KKZ180	39L424516	8347310	Survey	Unknown	BFS
Cap Masoala	Antsiragnamatso / Ambohimahery	4	CPM110	39L442094	8300913	Visit	Unknown	Mixed
			CPM120	39L442105	8300910	Visit	Unknown	Mixed
			CPM130	39L442099	8300904	Visit	Unknown	Mixed
			CPM140	39L442092	8300911	Visit	Unknown	Mixed
	Marifinaritra/ Ampanavoana	2	CPM210	39L427256	8259733	Visit	Unknown	Unknown
			CPM220	39L427265	8259711	Visit	Unknown	Unknown
	Antsirabato	1	CPM310	39L408515	8237726	Visit	Small	Mixed
	TOTAL	147						

4.1 Methodological approach to excavating an iron smelting site

Once the metallurgical sites were located, either through descriptions in previous archaeological studies (Clist, 1995; Vérin, 1986) or through oral surveys with the local population, some sites were selected for archaeological excavations. Four main objectives were then sought to be achieved in order to better understand the implemented technique: to understand the spatial organisation or the workshop, to reconstruct the furnace architecture, to describe the assemblage of metallurgical waste and associated materials, and, finally, to collect charcoal for dating.

Generally, before excavations can begin, the slag heap and the surrounding area are completely cleaned of vegetation on the surface to reveal the archaeological structures. A detailed map, with elevation measurements, is also drawn. Only once there is a global vision of the sector, can the first test pits be implanted.

Workshop spatial organisation

The global study of the workshop spatial organisation requires the observation of the shape of the slag heaps, their number and their orientation. The number of furnaces can also be identified and the test pits can thus be efficiently placed.

A detailed study of the working area also gives the opportunity to observe the structures associated with the furnace. The expected remains and structures on a metallurgical site are, for example, ore stockpiles, ore preparation area (washing, roasting, crushing) or built structures with remains of masonry or post holes. In some cases, it is possible to find post-smelting hearth structures, identifiable in particular by the presence of hammerscales.

The repeated observation and systematic measurement of these criteria are necessary to compare different smelting workshops.

Furnace architecture

The excavations of the furnace remains were in our case the most delicate part, given the fragility of the poorly-preserved structure. A cross-sectional and plan excavation of the furnace is optimal when possible. In some cases, the remains of the furnaces are sufficiently well preserved to allow a fairly good reconstruction of how they operated. In our case, the study of the metallurgical remains (slag, tuyere, wall) was essential to reconstruct the furnace (Chapter 5). For this reason, all fragments of metallurgical waste observed during the excavation of the furnace, such as wall fragments or ceramics, must be recorded.

Detailed field description of the metallurgical waste assemblage

The first step is to identify as many slag categories as possible, based on macroscopic observations (eyes, hands to test the mass or the density of the slag pieces, magnet). The shape, the size and the kind of material have to be recorded on a large number of slag pieces. The greater the number of pieces measured, the greater the chance of identifying rare artefacts which are not visible solely by checking the surface of the heap, such as fragment of metallic iron or iron ore. Several slag morphologies or shapes can thus be defined. In our case, two slag morphologies were observed: internal bottom furnace slag and external tapped slag. These wastes will be described in details in the next chapter (Chapter 5).

The volume and the mass of metallurgical waste can then be estimated. To estimate the volume, a trench must be dug through the heap, at least to its centre, to estimate the maximum thickness of slag layer. The heap is mapped and the heights are recorded. An estimate of the volume of the heap, whose shape is approximated by simple geometric figures, is calculated. A test pit of fixed and measured dimensions is then excavated and all the artefacts are sorted according to the previously defined morphologies (slag - internal/external, ore, tuyere, wall, ceramic), and weighted. The mass of waste per unit of volume in the heap can be calculated. Knowing the volume of the heap, we can then approximate the mass of metallurgical waste, i.e. slag and associated materials, and their proportions (Figure 4.4). This data is essential for the further study and estimation of the quantities of iron produced (Chapter 6).

Site	Heap	Volume (m3)	Mass (kg)					
			Bottom furnace slag	Tapped slag	Small unsorted fragments	Total		
Benavony	BNY410	0,550	494,5	87,5	191,5	773,5		
Matavy	MTY11	0,375		243,5	2,5	246		
Amboronala	MBR120	0,665	622			622		
	MBR141	0,850		681		681		
	MBR220	0,555	476,5			476,5		
	MBR210	0,350	409,5					
	MBR240	0,300	274					
For 1m3							**Volume of the heap (m3)**	**Total mass (T)**
Benavony	BNY410	1	899	159	348	1400	22-30	30-42
Matavy	MTY11	1		649	6,5	660	46-50	30-33
Amboronala	MBR120	1	935			935	2,7	2,5
	MBR141	1		801		800	3,4	2,7
	MBR220	1	859			860	11,1	9,5
	MBR210	1	1170			1170	1,4	1,6
	MBR240	1	910			910	3,2	2,9

Figure 4.4: Table reporting the masses of slag weighed according to their morphology in the seven studied slag heaps, as well as the calculations of total volume and mass.

Dating of the metallurgical activity

The dating was carried out on charcoals by the radiocarbon method[1]. The charcoals were taken directly from the sediment, no charcoal taken from a slag was dated. Indeed, the accuracy of the datings carried out on charcoal entrapped in slag are still debated. Charcoals were sampled mainly at the base of the slag heaps to date the beginning of the production. For the heaps excavated in detail, several charcoal samples were taken along the stratigraphy.

1: The samples were dated at the ETH Zürich under the responsibility of Dr. Irka Hajdas.

4.2 Chronological framework of metallurgical production in northeastern Madagascar

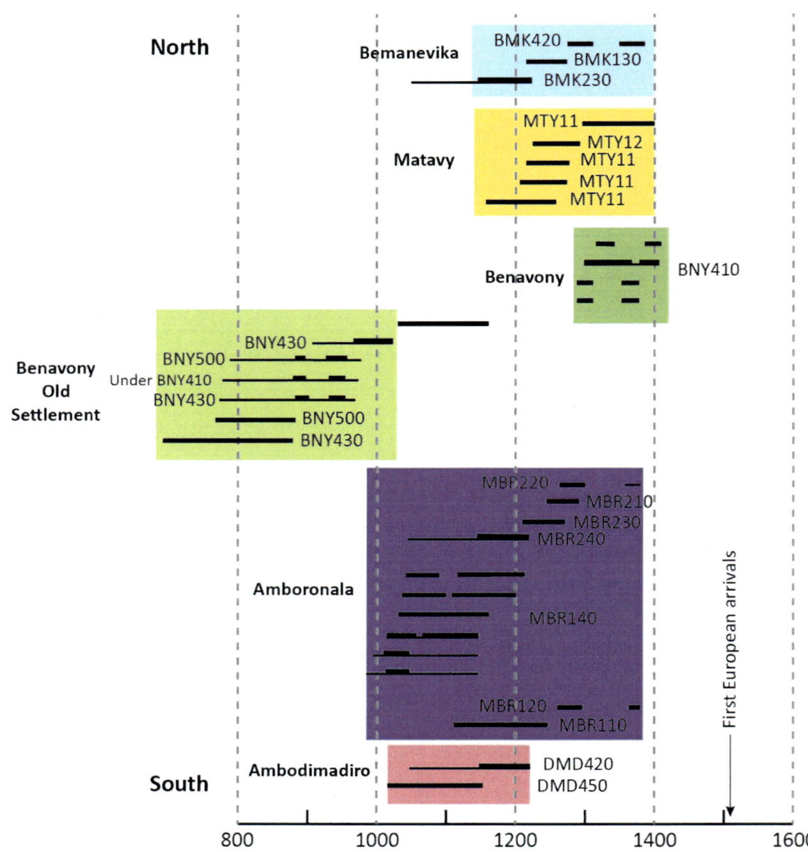

Figure 4.5: Representation of the chronology of the metallurgical activity in northeastern Madagascar. Charcoals pieces were collected at the bottom of the excavated slag heaps, and then dated by the radiocarbon technique (For detailed dating information, see Figure A.2.

A total of 30 datings were carried out on fifteen slag heaps from five different smelting sites (Figure 4.5 and Figure A.2). Until now, no trace of metallurgical production before the 11th century has been found. It appears that iron production began around the 11th century and was active until the 14th century. The sites of Amboronala and Ambodimadiro seem to operate throughout the period (11th-14th century), while Matavy and Bemanevika seem to have been operating from the second half of the period (13th-14th century).

At Benavony, the production was dated to the 13th-14th century. Only the BNY430 slag heap was dated to the 7th-9th century. This would be the oldest trace of metallurgical activity in Madagascar. We believe, however, that these charcoals come from an earlier occupation layer and that they do not come from the slag layer. A ancient human occupation, dating from the 9th-10th century, has indeed been identified in Benavony. During this period, people were already producing chlorite schist tripods. They needed iron tools to cut the stone in the quarries. They were not making their own iron yet, but they were already using it. They probably imported iron from the Indian Ocean trade.

A large number of dated samples would be needed to establish a detailed chronology and to study the period of activity of each site. The few dated samples can only confirm that a site was active at a specific period of time. This does not give any definite indication of the duration of the activity.

The Amboronala MBR140 and Benavony BNY410 slag heaps are the two excavated areas with the most numerous and reliable dates, and the best estimate of the total mass of slag. If we consider that one bottom furnace slag corresponds to one smelting operation, then we can estimate the number of operations carried out at each site, and in particular estimate the number of operations per year (Figure 4.6). If we consider the largest intervals, both of these heaps appear to have been operating for maximum two centuries. However, the annual production at BNY410 (40-50 operations per year) was much higher than at MBR140 (less than 5 operations per year). Benavony is probably an exception compared to the other visited sites of the district, Amboronala is probably more representative. Hence, a low scale iron production was taking place in northeastern Madagascar for about four centuries.

Slage heap	Total mass (T)	Total mass of BFS (T)	Mass of one BFS (kg)	Number of BFS	Estimated duration of activity (years)	Number of smelt/year
Benavony 410	30-40	27-36	3,5	8000-10000	200	40-50
Amboronala 140	4	0.4	1,5	400	100-200	5<

Figure 4.6: Estimation of the number of iron smelting operations per year for Benavony BNY410 and Amboronala MBR140, according to the total mass of slag and the average mass of a bottom furnace slag piece (BFS).

The reason for the interruption of metallurgical production is also still unknown. The first European travellers visited the island of Madagascar at the beginning of the 16th century, so it is not possible to put forward the hypothesis of a *Rasikajy* iron not being competitive with European iron. Some authors suggest that the *Rasikajy* left the Northeast around the 15th century to settle further south (Vérin, 1986). However, there is no solid archaeological evidence to support this hypothesis at the moment.

At the current state of research, we do not know what drove the

Rasikajy to produce their own metal. It may have been a change in the organisation of the trading network in the Indian Ocean or perhaps an increase in the need for iron.

4.3 Benavony (BNY): a large smelting site at the periphery of a major coastal settlement

4.3.1 General presentation of the site of Benavony

Benavony 19th century. M. Meurs reports in 1897 the existence of a treasure: a vase containing gold jewellery and coins (Vérin, 1986; Grandidier and Grandidier, 1908). These gold coins are imitations of gold dinars of the fifth and eighth Fatimid caliphs (Egypt, 10th and 11th centuries CE - Chauvicourt and Chauvicourt, 1968). In 1943, E. Vernier studied this site and excavated several tombs with artefacts similar to those found in the necropolis of Vohémar. P. Vérin compiled and published some artefacts from Vernier's excavations, but there are unfortunately no detailed reports of the excavations. The reputation of the Benavony treasure has brought in many treasure seekers since then and to this day. There are many pits dug all around the site and it is difficult to attribute them to ancient excavations or looting.

The site is located on a sandy strip parallel to the coast, north-south oriented, at the mouth of the Lokoho River. The substratum of the coastline is uniformly sandy. This ancient coastline, 250 metres wide and 1000 metres long, is connected to the dune deposits from the north. The northern part of the strip is covered by dense forest vegetation, while the south is covered by pastures and fields. On both sides of the dune belt are marshy lowlands, now used for rice cultivation (Figure 4.7.A-B). These are the old beds and mouths of the river. At the time of the medieval settlement, the river estuary may have been located at the foot of the settlement, separated from the ocean by a sandy strip (Figure 4.7.C). The estuary could thus provide a calm and protected anchorage for boats. Boats could also penetrate inland by sailing up the Lokoho River. The location of the site therefore offers a number of favourable conditions for the establishment of a permanent settlement.

Because of the dense forest cover, tall grass and scrub, it is very difficult to locate archaeological remains and to make systematic observations. The presence of vanilla plantations does not facilitate surveys and localization of archaeological remains. Surveys could neither be systematic nor exhaustive, but the site was intensively visited over its entire surface.

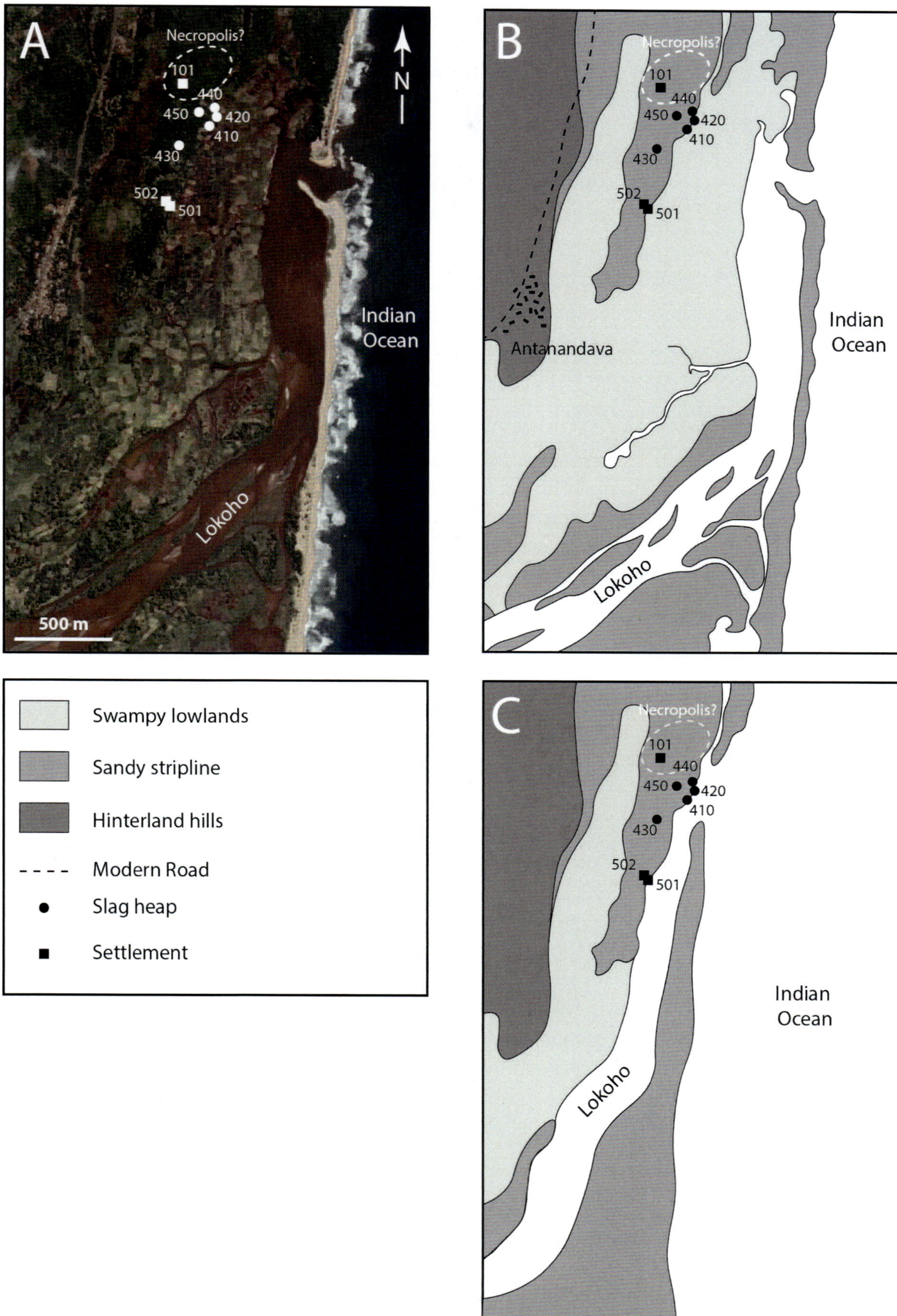

Figure 4.7: Map of the topography of the Benavony archaeological site (Google Earth 7.3, (2018) *14°24′37″S 50°10′25″E. [Accessed 5 June 2022])* . A. Satellite images of the archaeological site of Benavony. B. Diagram of the current topography of the site of Benavony. C. Hypothesis on the topography of the site of Benavony during its medieval occupation.

On the basis of these observations, the distribution of human occupation can be mapped. The necropolis is located in the northern part of the sandy strip. Basalt slabs were observed and may have been used to build the graves. Ceramic is rare but few concentrations have been observed (BNY101). Going south, five slag heaps, of significant size (at least 5 metres of diameter), were located (BNY410 to 450). Further south, several important concentrations of ceramics suggest the presence of significant human occupation (BNY500).

Three test pits were carried out in the settlement area, where the artefacts were most abundant at the surface. Some slag pieces were collected but they show no morphological or chemical difference from the slag sampled in the heaps. They are probably not smithing slags but rather smelting slags that have been transported. It is very likely that some smithing workshops existed in the medieval village but we could not locate them during our surveys. Four small pieces of iron, maximum 3 centimetres long, were found during the excavations but these are so corroded that their study is impossible.

The slag heap BNY410, located in the northeast of the site, could be studied in detail. The entire surface of the slag heap was cleaned and the working area was extensively excavated. A test pit was conducted on a second slag heap (BNY430) in order to identify the stratigraphy and to collect samples.

4.3.2 Excavations of the slag heap BNY410

The sector BNY410 is located on the eastern slope of the dune belt. The slag heap is a mound with an irregular crescent shape, open towards the south. The diameter of the structure is 12 to 14 metres (Figure 4.8 and Figure 4.9) and covers a surface of about 100 m^2. A large trench (15 x 1 m: 410-411-412-413) at the intersection of the slag heap allowed a better understanding of the stratigraphy of the slag accumulation. A second test pit (2.5 x 5 m: 414), located parallel to the trench, in the concavity of the crescent, revealed traces of a furnace and few other structures.

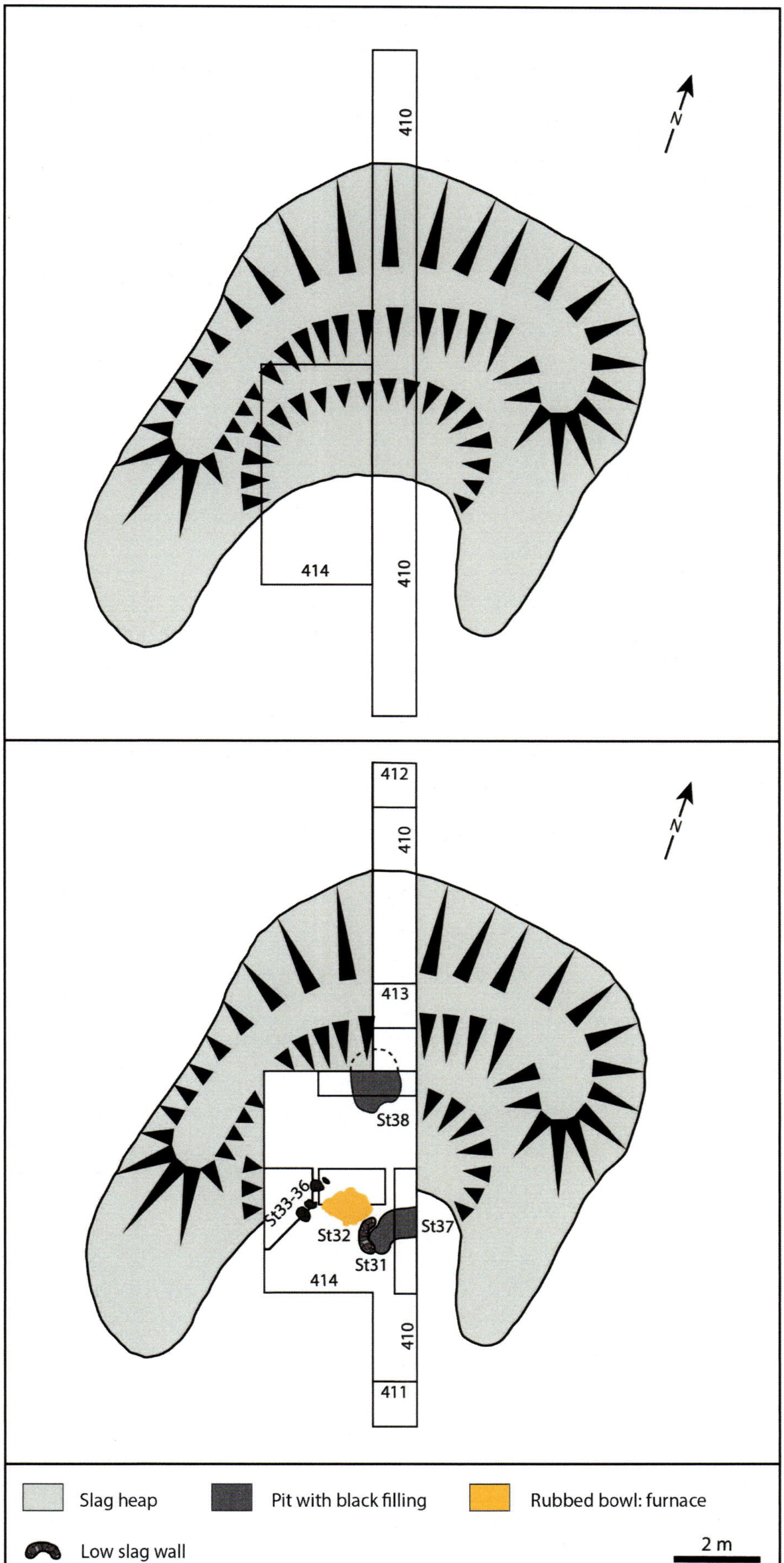

Figure 4.8: Map of the archaeological excavations of slag heap BNY410 (From Serneels et al., 2018).

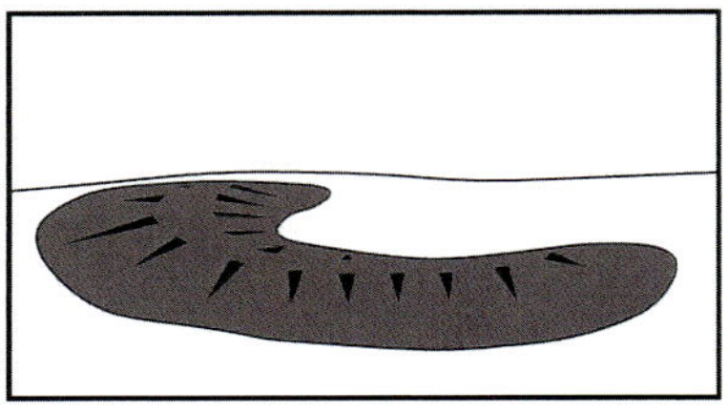

Figure 4.9: View from the west of the BNY410 slag heap before excavations (Benavony).

BNY 410: Lower occupation level without iron production remains

The natural substratum consists of a loose, very homogeneous, white-beige sand (C6, Figure 4.10). This layer was not disturbed at a depth of 80 centimetres. In the upper forty centimetres, this sand gradually becomes greyish hue due to the presence of disseminated organic materials (C5). This level has been disturbed by human activities. Artefacts, such as local ceramics, imported ceramics and chlorite schist, were collected in the two test pits (411-412). The boundary between the undisturbed white sands (C6) and the disturbed grey sands (C5) is sometimes progressive and rather horizontal, whereas elsewhere it is steep and with traces of digging and postholes.

This early occupation has been radiocarbon dated before the 13th century and is one of the earliest dated human occupations in this region (Figure 4.5 and Figure A.2). The populations settled in Benavony were already in contact with the Indian Ocean trade and the production of chlorite schist was also already active in the

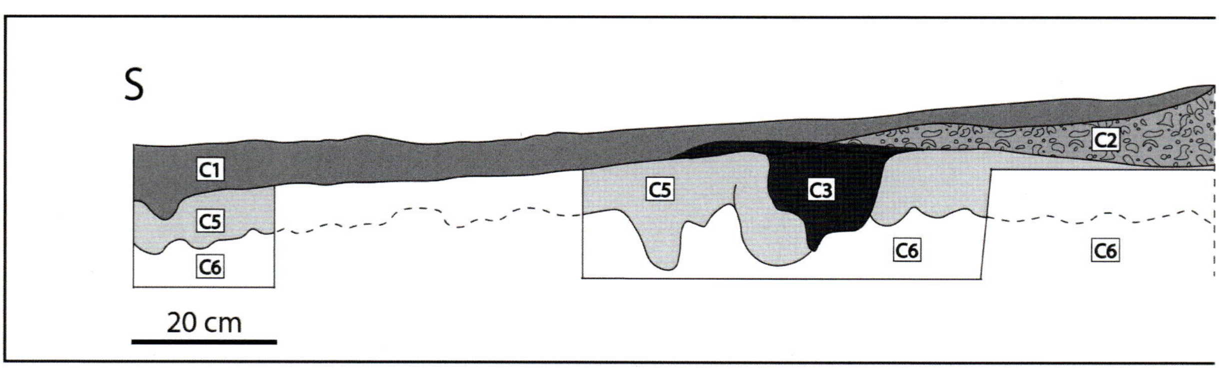

Figure 4.10: Stratigraphic section from north to south of the slag heap BNY410, Benavony (from Serneels et al., 2018).

region. However, no trace of metallurgical production was found in layer C5. This human occupation therefore predates the establishment of metallurgical production. It is nevertheless difficult to assess whether the occupation is continuous between this early phase and the later metallurgical production phase.

BNY 410: Upper occupation level with smelting remains

The BNY410 slag heap is made up of a disordered layer of iron slag and other metallurgical waste piled up with a small proportion of black sandy interstitial sediment. Both slag morphologies are found mixed in the heap, with the bottom furnace slag pieces in much greater proportion (85%) than the tapped slag (15%).

In the central part of the trench, the C2 slag layer overlies the C5 grey sands and the interface between the two units is regular and horizontal. The thickness of the slag layer reaches 55 centimetres in the central part and slopes gently towards the north. Towards the south, on the side of the working area, a break in the slope is clearly visible, indicating that the original flank of the heap must have been steeper initially (slope of about 45°). This slope collapsed and eroded after the cessation of metallurgical activity. The shape of the pile is relatively irregular but its volume can be approximated by calculating a half-ring. The volume of BNY410 can be approximated around 30 m^3. The total mass of slag can therefore be reasonably estimated between 30 and 40 tons (Figure 4.4).

In the working area, in the concavity of the crescent, the remains of at least one furnace were found. The sediment is much less sandy and very black (C3). The layer contains abundant archaeological material and a high proportion of metallurgical waste. Some ceramic fragments showing thermal impact and adhering slag were found. For the moment however, no function related to metallurgy has been found to explain the existence of these slagged ceramics. The C3 layer lies on top of the grey C5 sands and fills

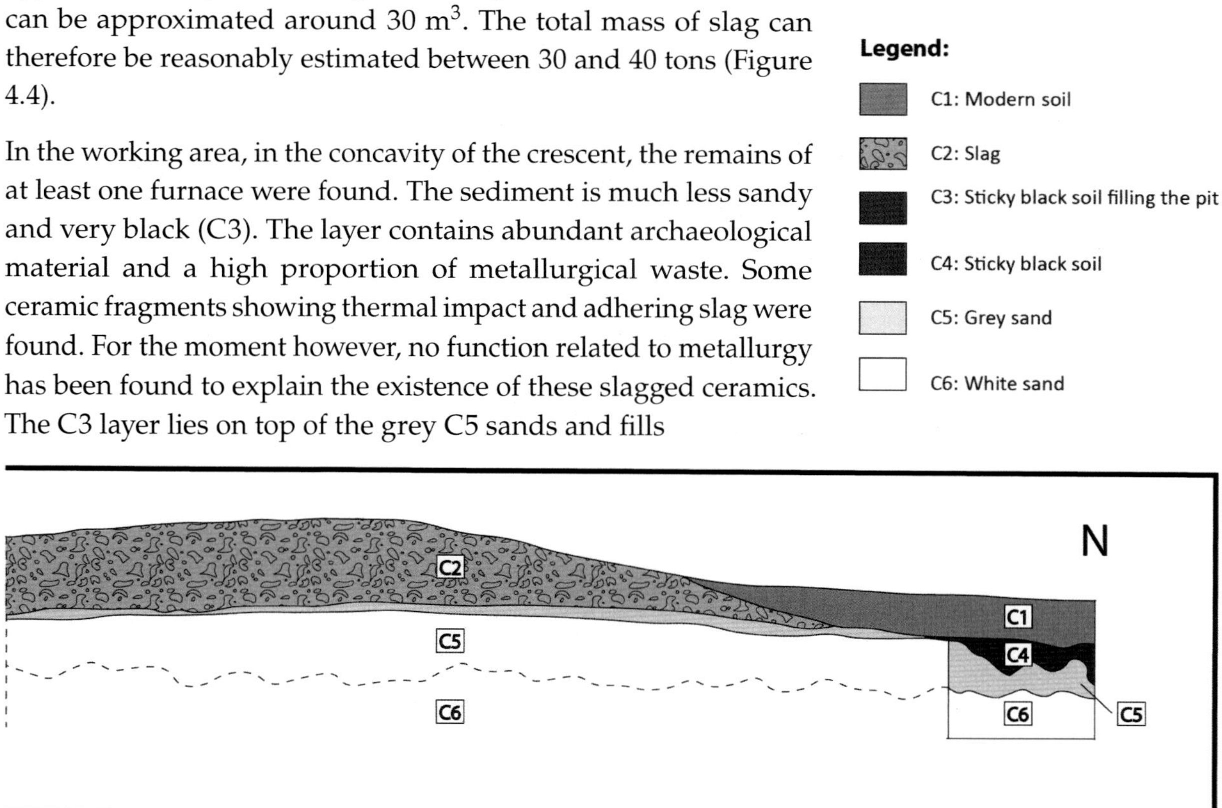

the structures excavated in this layer. In addition to the base of the central central furnace (St 32), there are several pits and postholes and an accumulation of unusually large slag blocks which form a kind of curved wall (St31, Figure 4.11).

BNY 410 : The smelting furnace

The archaeological remains of the furnace (St32) are fragile and poor. No superstructure or wall fragments, either in stone or clay, were found. Only an elliptical bowl (95 x 66 cm) dug directly into the grey sand (C5), without any clay lining, is left (Figure 4.12). The bowl is 20 centimetres deep. The bottom and walls of this basin have undergone a significant thermal impact, as evidenced by the strong orange rubbing of the sandy sediment. However, there is no trace of melting and the orange halo reaches a maximum thickness of 10 centimetres. The thermal impact seems to have been stronger on the eastern side of the bowl than on the western side, showing a non-homogeneous repartition of the temperatures in the furnace. The hottest temperatures in a furnace are close to the air supply, so to the tuyere. Bellows might thus have been installed on the eastern side of the furnace and slag could be tapped on the western elongated side.

Figure 4.11: View from the north-west of the smelting furnace (St32 - BNY410) and of the curved wall (St31 - BNY410).

The curved wall made of big pieces of slag (St31) is next to the pit of the furnace. But there is no trace of thermal impact even if the orange rubbed sand touches the base of the wall. The stratigraphic relationship is unclear and it was not possible to determine if the wall and the furnace were contemporary. If it was the case, the wall could be a structure linked with the bellows.

North of the furnace (St32), another elliptical pit, dug into the grey sand (C5), was partially excavated (St38, Figure 4.11). This bowl is 30 centimetres deep and the bottom also shows traces of rubbing.

These could be the remains of a second furnace. The stratigraphic relationship with the other furnace (St 38) is however unclear and it is not possible to decide whether the two furnaces were working together. According to the other excavated smelting workshops in our study area, it is more likely that they operated successively. The oldest furnace St38 might have been operating first and when the slag heap became too big, the smelters probably moved and built a new furnace (St32) few metres south.

Figure 4.12: Left: View from above of the remains of the smelting furnace in Benavony 410 (St31). Right: Cross-section of the smelting furnace in Benavony 410.

4.3.3 Test pit in the slag heap BNY 430

A test pit was conducted on the slag heap BNY430 which is located 150 metres southwest of BNY410. The area around BNY430 was not cleared of vegetation so we could only estimate of the size of the heap. It appears to be slightly smaller than BNY410 and about 5 metres in diameter.

The stratigraphy of BNY430 is very similar to BNY410. The white-beige sandy substratum (C4) is devoid of human disturbance and of ceramics. The C4 layer is overlain by a grey sand (C3) in which local ceramics, imported ceramics, chlorite schist fragments and a piece of coloured glass were found. No traces of metallurgical activity were found in this layer. Above this layer begins the slag heap (C2), which reaches a thickness of about 30 cm. The slag pieces are compact and mixed with a dark sediment rich in organic matter. As for BNY410, local ceramics, some with traces of thermal impact, were found in the C2 layer. The slag pieces from BNY430, both internal and external slag, are morphologically indistinguishable from the slag from BNY410.

Radiocarbon dating was carried out on charcoal samples from layer C3 under the slag layer (Figure 4.5). This dating (BNV08: 1281 ±21 BP, 689-878 cal AD) is consistent with that of BNY410 and corresponds to an early human occupation prior to any metallurgical production. Three other charcoals collected at the basis of the slag layer (C2) were dated. These three dates are ambiguous (BNV09: 1211 ±21 BP, 773-970 cal AD, BNV10: 594 ±21 BP, 1303-1409 cal AD and BNV11: 1049 ±21 BP, 908-1025 cal AD). BNV09 and BNV11 delivered abnormally old dates and are not compatible with BNV10. It is quite likely that these old charcoals were in fact either taken from the upper part of layer C3, or that they moved into the sandy layers because of pedogenetic effects. Metallurgical activity would therefore be operating in the 14th century.

4.4 Matavy River (MTY): a medium sized smelting site on an inland sandy plateau

4.4.1 General presentation of the site of the Matavy River

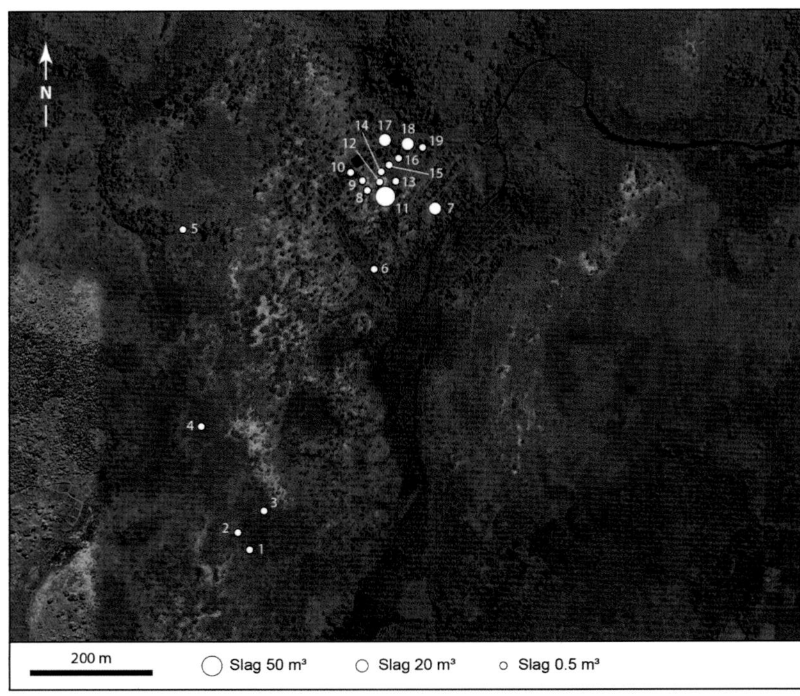

Figure 4.13: Satellite image and location of slag heaps from the site of Matavy (Google Earth 7.3, (2018), *14°18′24″S 50°09′15″E*. [Accessed 15 June 2022]).

Running along the coast from south to north, a river serves as an outlet for the lakes Andohabe and Andomoty. Perpendicular to this river, a tributary flows westwards to drain the hills of the hinterland: the Matavy river. One kilometre upstream, a plateau, covered by very white sand, extends over a kilometre[2]. This plateau is surrounded by marshy areas and by hills few dozens of metres high. The soil on these hills is red to brown and is covered by iron-rich alterite which could be a source for iron-rich concretions.

2: This place is named as Ankazofihatra by the local population.

This metallurgical site was already partially studied by B. Clist (1995). In 1993, he located 14 slag heaps and excavated one (MTY18). This heap was dated to the thirteenth century by radiocarbon dating. The trench from these excavations is still visible today.

The low vegetation and the very white sand allows the easy detection of 19 slag heaps (Figure 4.13). The plateau was intensively surveyed and the slag heaps described by Clist were located. A second cluster of heaps to the southwest of the plateau was also identified. No evidence of ancient settlement was found on this plateau. The largest slag heap, MTY11, was extensively excavated and a second very small slag heap was tested (MTY12).

4.4.2 Excavations of the slag heap MTY 11

Figure 4.14: View from the south-west of the slag heap MTY11 (Matavy) after cleaning the vegetation and before excavations.

The MTY11 slag heap has a stocky crescent shape and extends over an area of 125 m^2 and 13 metres in diameter. The maximum height reaches 90 centimetres above modern ground level (Figure 4.15 and Figure 4.14). The shape of the heap can be approximated by a regular spherical dome from which a quarter of the volume, corresponding to the central depression, must be removed. This calculation gives a volume of 50 m^3. A first test pit (MTY11-S1 - 5x5 m) was installed in the northern part of the heap to study its stratigraphy. A second test pit (MTY11-S2 - 3x4 m) in the hollow of the crescent was excavated to study the working area and the remains of the furnace. Contrary to Benavony, little ceramic or chlorite schist material was found during the excavations at Matavy.

MTY11: Occupation level with smelting remains

There is a spatial organisation of the metallurgical waste discharge in the MTY11 slag heap, contrasting with what was observed in Benavony. The northern part of the crescent, which forms a second small bump, is exclusively composed of small fragments of tapped slag. On the other hand, the parts surrounding the working area to the south of the crescent are made up almost exclusively of internal bottom furnace slag. It is estimated that tapped slag accounts for at least one third, if not one half, of the total slag assemblage. Because we weighed slag from the northern part of the heap (MTY11-S1), we cannot give an accurate value of the total mass of slag. Indeed,

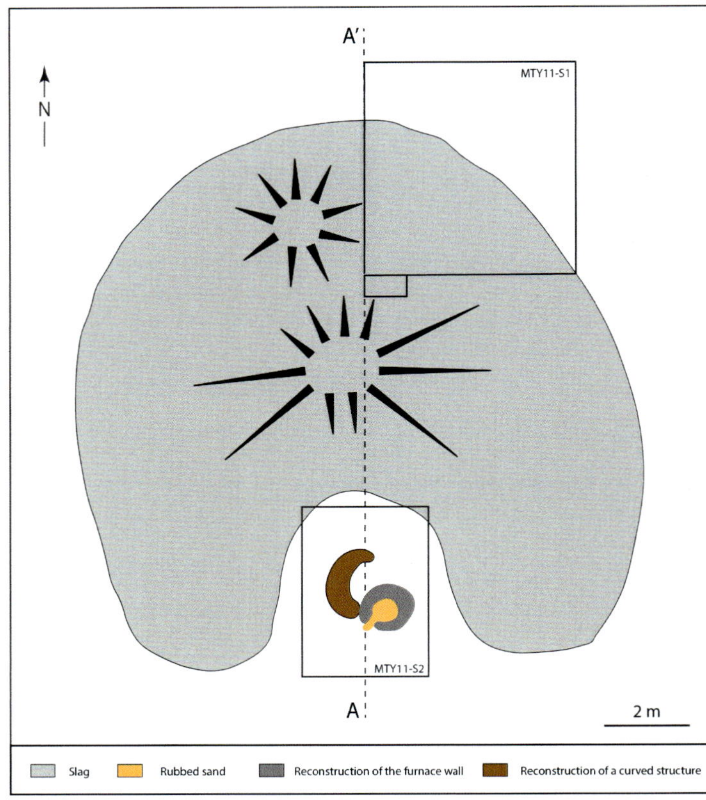

Figure 4.15: Map of the shape of the slag heap MTY11 and location of the test pits. Section AA' in Figure Figure 4.16 is also shown in dashed lines.

we cannot be sure if the density is similar in all parts of the slag heap. A total mass of 30-32 tons of metallurgical waste is estimated (Figure 4.4). The natural substratum of the plateau is a very white sand (C9), probably of aeolian origin. It is completely sterile and its surface is almost horizontal. The slag heap lies directly on the sandy substratum C6 (Figure 4.16 and Figure 4.17). There is no period of occupation of the site prior to the smelting activities. The site was only used for metallurgical production.

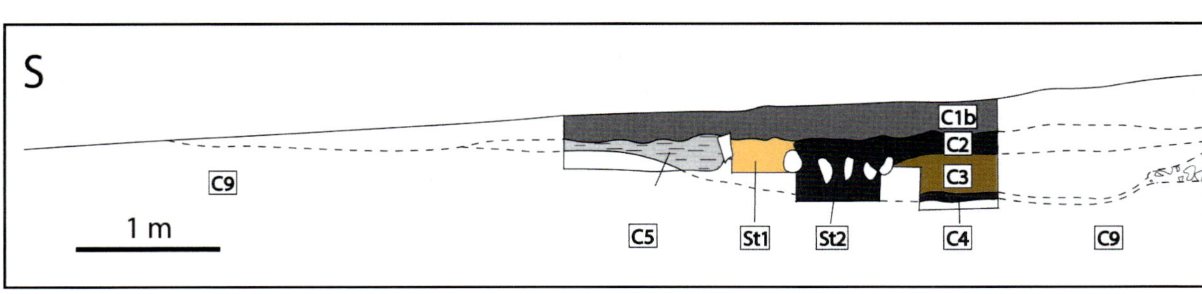

Figure 4.16: Stratigraphic section from north to south of the slag heap MTY11, Matavy (from Serneels et al., 2018).

Figure 4.17: Stratigraphy of the northern part of section AA′ of the slag heap MTY11, Matavy (Figure 4.16). The white sterile sandy substratum at the base can be observed as well as the slag heap that lies directly on it.

The stratigraphy of the excavation area MTY11-S2 is complex. In the middle of the pit there is an elliptical shallow depression dug directly into the white sandy substratum (C9), oriented north-south (3x2.5 m) and about 15 centimetres deep. It is filled by a compact brown sandy sediment (C3), containing small amounts of metallurgical waste. The bottom of the depression is covered by a 5 cm thick layer of very black and sticky soil (C4), enriched in charcoal micro-fragments. Above C3, in the central part of the basin, there is a similar layer of black, sticky earth (C2), which reaches a thickness of 20 centimetres. The upper layer is a grey-brown sandy sediment (C1b), poor in metallurgical material and rich in roots. On the edges of the basin, to the west, south and east, a complex layer of grey sands with a laminar structure (C5) was observed. The lamination is underlined by the variable presence of heated and rubbed sandy-clay elements (orange shades), charcoal particles

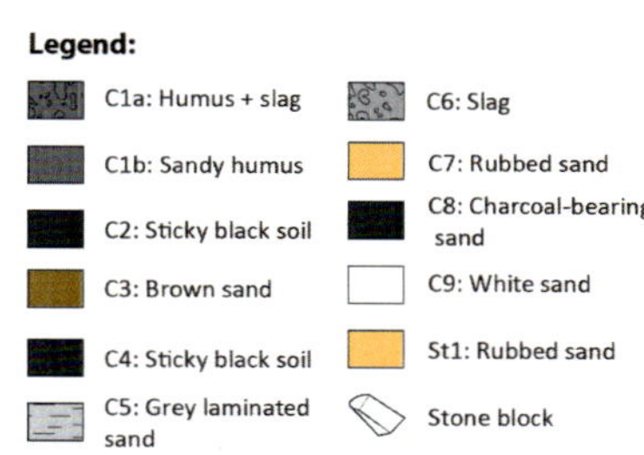

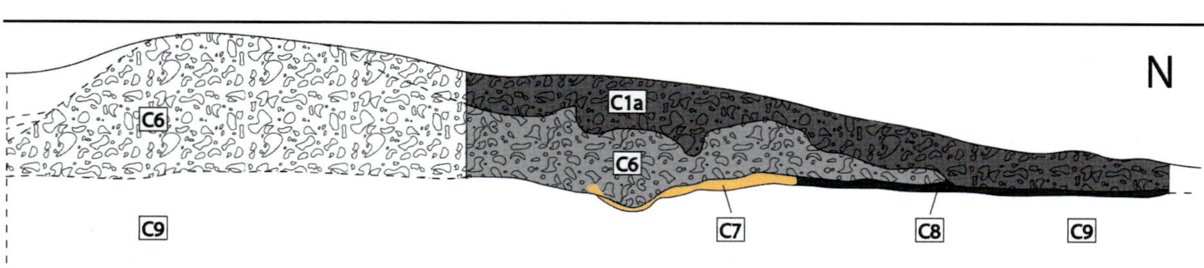

(black shades), ashes (grey shades) or ferruginous concretions (dark red shades). These laminations have little lateral continuity and have to be linked to metallurgical activities. Stratigraphically, this C5 layer is intercalated above the white sandy substratum (C9) and below the filling of the depression (C3).

The excavation also revealed some structures dug into the substrate C9. These are probably post holes, but no real organisation of these structures emerges. They could be traces of a non permanent shelter for protection from the weather or the sun.

MTY11: The smelting furnace

In the centre of the bowl there are several angular stone blocks whose sides seem to have been carved out. These stones were brought on the site by the smelters. Some of them are laid on their side, which is not a natural arrangement but, but rather an intentional one. The general organisation of these stone blocks is not very clear.

Five of these stones are laid around an elliptical area of rubbed sediment (St 1). This area is about 80 centimetres in diameter and is interpreted as the remains of a furnace (Figure 4.18). The stones could have been used to reinforce the wall. A strip of rubbed sediment enters between two stones to the southwest of the furnace. This could be the area meant for the tapping of the slag.

Further north, there are other stones, some of them juxtaposed, which form an arc in a slightly lower stratigraphic level than the furnace (St 1). No traces of rubbed sediment or of any thermal impact were observed around these stones. The function of this structure is not understood. It could be an earlier abandoned smelting furnace, as observed during the excavation of BNY410 (Benavony). There is however no evidence to support this hypothesis.

Figure 4.18: Top: Reconstruction of the organisation of the working area south of the MTY11 heap (Matavy). **Bottom:** View from the south of the excavation of the working area and of pit MTY11-S2. (From Serneels et al., 2018).

4.4.3 Test pit in the slag heap MTY12

The slag heap MTY11 is by far the biggest heap on the plateau. Most of the other slag heaps are very small and cover a few square metres and their thickness is only about ten centimetres maximum. Three medium-sized heaps reach 8 to 10 metres in diameter and a thickness of 50 centimetres of slag, but these are not predominant.

Cluster MTY12 (Figure 4.19) is typical of small slag heaps and measures barely 5 metres in diameter. It covers an area of approximately 20 m^2 and the thickness of the slag layer does not exceed 5 cm. The volume of slag is estimated to be no more than 1m^3 and the total mass of waste less than 1 ton.

Figure 4.19: Left: Photo of the MTY12 (Matavy) slag heap before excavation. **Right**: Photo of the excavation of the MTY12 cluster. The slag layer lies directly on the white sandy substratum.

4.5 Amboronala (MBR): a small smelting site on the coastal sand dunes

4.5.1 General presentation of the site of Amboronala

The site of Amboronala is located along the coast 25 km south of Sambava, south of the estuary of the Lokoho river. The archaeological remains are located on an big dune belt, which can reach ten metres high. This sandy strip is wedged between the hinterland hills to the west, where another metallurgical site (Ambodimadiro - DMD) was found, and marshy lowlands to the east (Figure 4.20). Nine slag heaps were identified during the surveys, as well as

Figure 4.20: Satellite view of the sites of Amboronala (MBR) and Ambodimadiro (DMD). (Google Earth 7.3, (2018), 1429′05″*S*5010′16″*E*. [Accessed 15 June 2022]).

traces of settlement. Based on the previous observations of Battistini in 1971, Vérin (1986) already mentions the presence of an additional slag heap in this area, but it could never be located. The area where Vérin locates the heap is now covered by palm cultivation, the archaeological layers have been extensively reworked and destroyed.

The pair of slag heaps MBR140 was intensively excavated as well as the associated working area. The volume and mass of waste from other heaps (MBR110, MBR120, MBR210, MBR220 and MBR240) were also measured and test pits were conducted.

4.5.2 Excavation of the smelting site MBR 140

The workshop MBR140 is located at the bottom of the dune belt, towards the east, on a dry platform but barely overhanging the marshes. The top soil and grass cover was removed from the entire surface (Figure 4.21 and Figure 4.22). Two small elliptical heaps of tapped slag fragments are set side by side (MBR141 to the east - 6x5 m and MBR142 to the west - 4x4 m). The slag layer is about twenty centimetres thick in both cases. The slag heap MBR141 and

Figure 4.21: General view from the west of the workshop MBR140 (Amboronala) before excavation. In the foreground we see the slag heap MBR142 and in the background the slag heap MBR141.

its surroundings have been fully excavated and no furnace was uncovered. To the south of MBR141, we could observe scattered iron rich pisolitic lateritic concretions. The slag heap MBR142 was only partially excavated. A few metres to the south of this heap, an extended layer of grey sediment can be seen on the surface. A test pit (5x4 m) was set on this grey layer and the working pit MBR143 was located on the darkest area. Further south, there is a large quadrangular block of stone with percussion traces on the wider faces. Fragments of broken pisolites were found in the sediment around it. It was probably an anvil used to crush iron ore before the smelt.

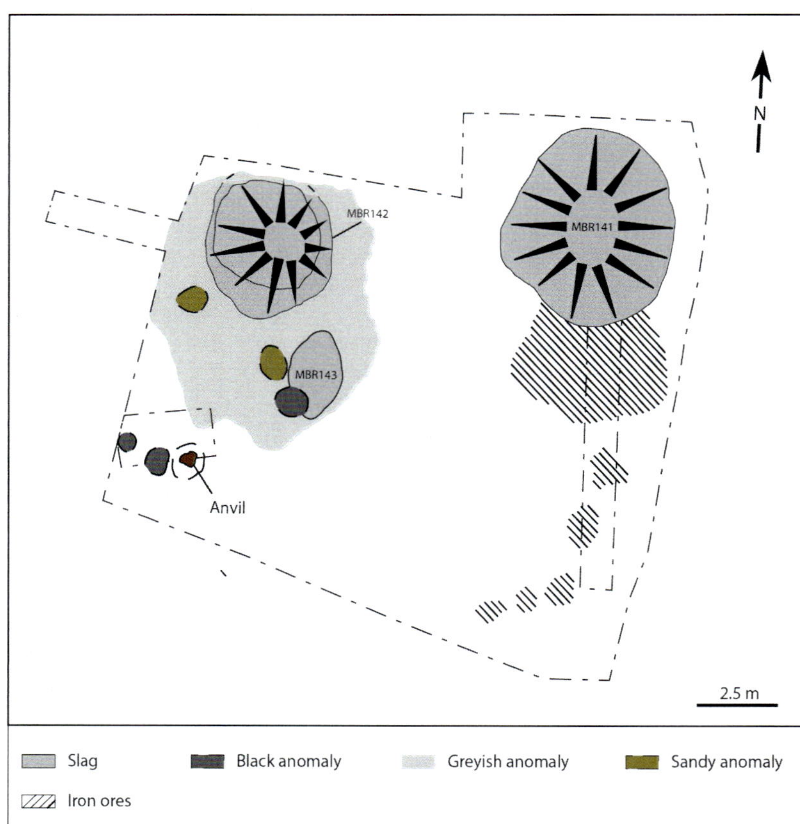

Figure 4.22: Map of the main archaeological features of the excavated workshop MBR140 (Amboronala).

MBR 140: Spatial organisation

The slag heaps lie directly on a barren, orange sandy substratum. There is no trace of previous human occupation. The excavation of the entire metallurgical site did not reveal any ceramic shards. The site is therefore disconnected from any settlement context.

To the north-west of the MBR 141 cluster, the stripping of the sterile sand layer revealed the presence of about fifteen more or less circular black disturbances, arranged in a disordered manner over an area of about 4 m^2. The diameters are variable, from 5 to 30 cm, as are the depths (2 to 25 cm). Some disturbances are more extensive with irregular shapes (Figure 4.23). Fragments of slag were collected in the fills. These structures probably correspond to post holes contemporary with the metallurgical activity. It is difficult to reconstruct an organisation.

Figure 4.23: View from the west of post holes uncovered under the slag layer of MBR141 (Amboronala).

The metallurgical waste from the MBR141 slag heap was weighed (Figure 4.4). The mass of metallurgical waste for MBR141 and 142 is therefore estimated at 2,700 kg and 1,200 kg respectively. The slag pieces in pit MBR143 have not been weighed but the mass of slag is estimated to be 400 kg. The total mass of slag for MBR140 is in the range of 4 to 4.5 tons.

MBR 140: The working pit and the smelting furnace

The large pit MBR143 is 2.5 m long from north to south and 1.8 m wide (Figure 4.24 and Figure 4.25). It is dug directly into the sandy substratum. In the northern part, the pit is deeper and the excavation reaches 35 cm below the current ground level. To the north and east, the edges of the pit are almost vertical. They are reinforced by a pile of large slag fragments. The pieces are joined together, practically embedded against each other. Almost exclusively bottom furnace slags are used. In the middle of the pit, on the bottom, the slag pieces are smaller and less compactly arranged. On the western side, the edge of the pit slopes gently upwards and the slag pieces are scarcer and smaller. The southern part of pit MBR 143 forms a shallower, trapezoidal depression, adjacent to the northern part. The bottom of this southern part lies 10 centimetres above the lower level of the northern part. The junction between the two parts of the pit is marked by bumps on either side, leaving only a narrow central passage. A shallow depression occupies the centre of the southern part. In and around this central depression the traces of thermal impact are the strongest. The filling of the whole is made up of a black sticky sediment, very rich in organic matter and charcoal.

These remains are rather poorly preserved but the southern part of the pit, with the thermal impacts, can be interpreted as the location of a hearth for the smelting of the ore. The very simple structure

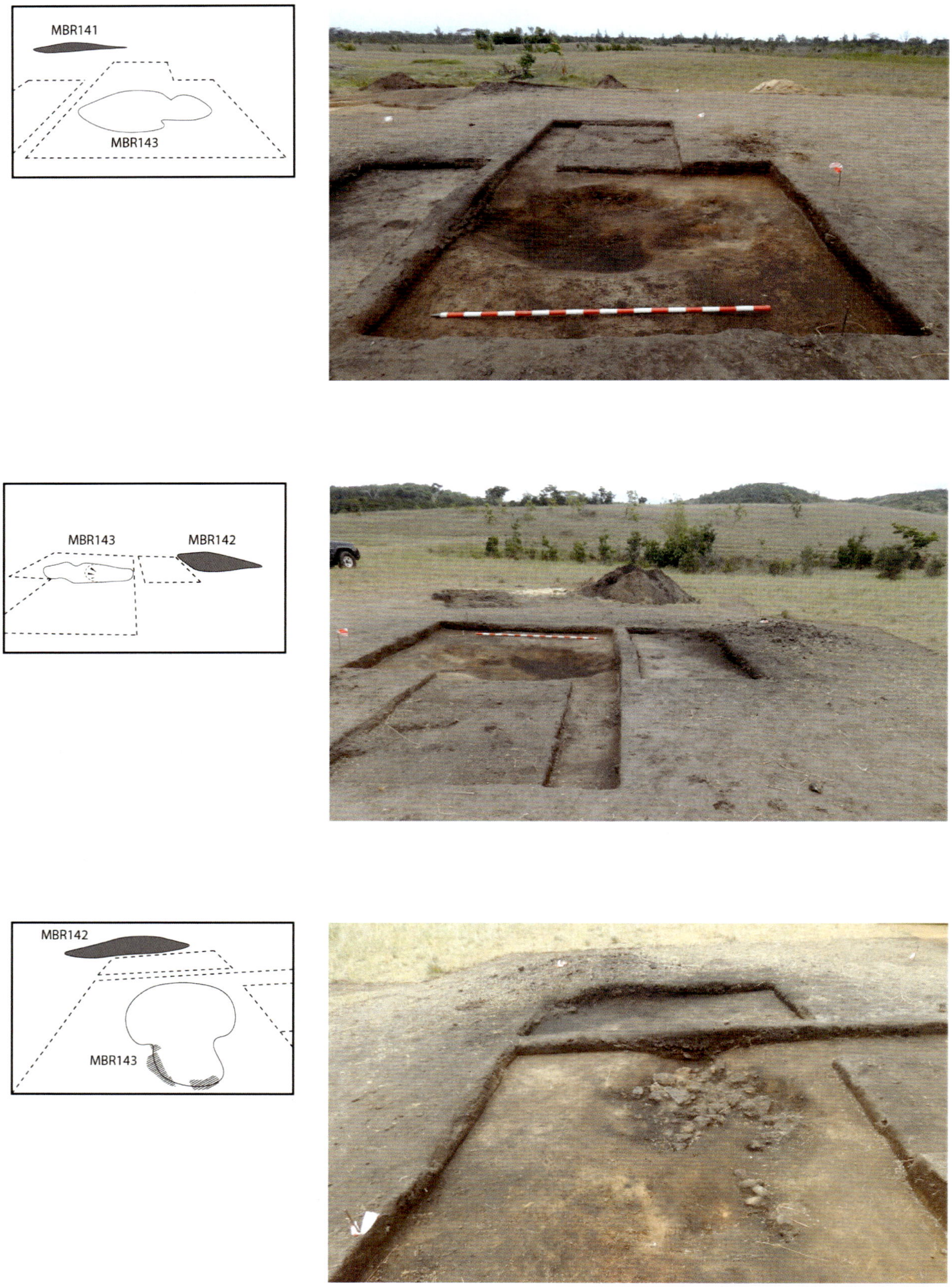

Figure 4.24: MBR143 working pit (Amboronala). **Above**: View from the west of the working pit after excavation. **Middle**: View from the east of the MBR143 working pit after excavation. **Bottom**: View from the south of the MBR143 working pit before complete excavation. The bottom furnace slag pieces that lined the bottom of the pit are still visible.

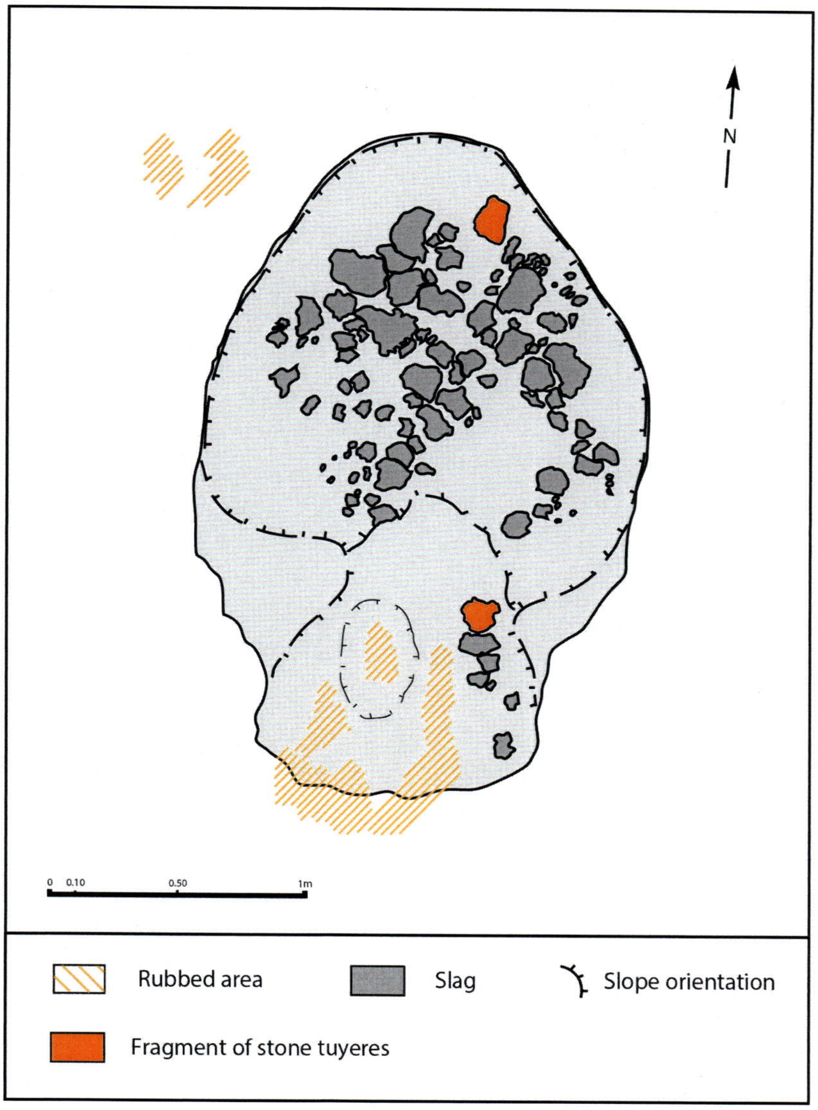

Figure 4.25: Map of the MBR143 working pit (Amboronala). To the south are the rubbed areas corresponding to the smelting furnace. The northern part of the pit is the actual working area and the slag tapping area.

of the furnace was probably re-dug frequently, perhaps even after each smelt. As the heat impacts are slightly more pronounced to the south of the pit, the bellows and the tuyere should have been installed in this position. The northern part of the pit is probably a working area giving access to the furnace. This is where tapped slag is evacuated during the smelt.

4.5.3 The other slag heaps on the dune ridge near Amboronala

Seven other concentrations of slag have been located nearby on the sandy dune ridge. These are systematically small and with an elliptical shape. The mass of metallurgical waste is estimated for each of these clusters to be between 1 and 3 tons. The only exception is the larger cluster MBR210, which is estimated to be composed of 9.5 tons of metallurgical waste (Figure 4.4).

The bottom furnace slags predominate in the metallurgical assemblages (see Appendix - Figure A.1). The majority of the slag heaps contain only bottom furnace slag pieces. In the case of the sectors with tapped slag (MBR110, MBR130 and MBR410), the disposal of metallurgical waste is always organised according to the morphology of the slag (internal/external). A heap of tapped slag will be juxtaposed with a heap of internal slag. There is no heap composed of a mix of the two morphologies. For example, MBR130 and MBR410 are each composed of two mounds with a spatial distinction according to the morphologies.

The test pits around the MBR200 sector unearthed ceramic shards associated with the slag. Ceramics were also found in the soil from modern charcoal burning pits. However, the exact location of the former settlement could not be found.

4.5.4 Ambodimadiro (DMD): a small smelting site on the altered basement

Ambodimadiro is located two kilometres inland from Amboronala, in the transition zone between the Quaternary sandy soils and the hills covered with red alterites. These two sites are most likely a single production centre. The Amboronala site is close to the shoreline and potential settlement, whereas Ambodimadiro could have been closer to the ore source. The slag heaps are divided into two clusters distributed among two low sandy hills, separated by a wet lowland oriented West-East (Figure 4.20 and Figure 4.26). Today, the site is planted with coconut trees. Test pits were conducted on two slag heaps to sample charcoal (DMD820 and DMD850).

Eight slag concentrations were identified on the top of the northern hill (DMD900). DMD910 and 920 are two larger heaps which reach a diameter of more than ten metres. These two heaps are made up of tapped slag associated with pieces of bottom furnace slag. Tapped slag seems to be the most abundant. The other heaps are smaller and are grouped close together, but no organisation emerges from their arrangement. Some heaps appear to be only made up of internal bottom furnace slag, others of tapped slag.

On the southern hill (DMD800), six slag concentrations are visible. They form a rectilinear East-West alignment, more or less in the axis of the dune belt. All the clusters are small (tableau annexes). The organisation of the discharge of metallurgical waste is also organized according to the morphology of the slag. Some of them contain mostly hemispherical slag, the others mostly tapped slag.

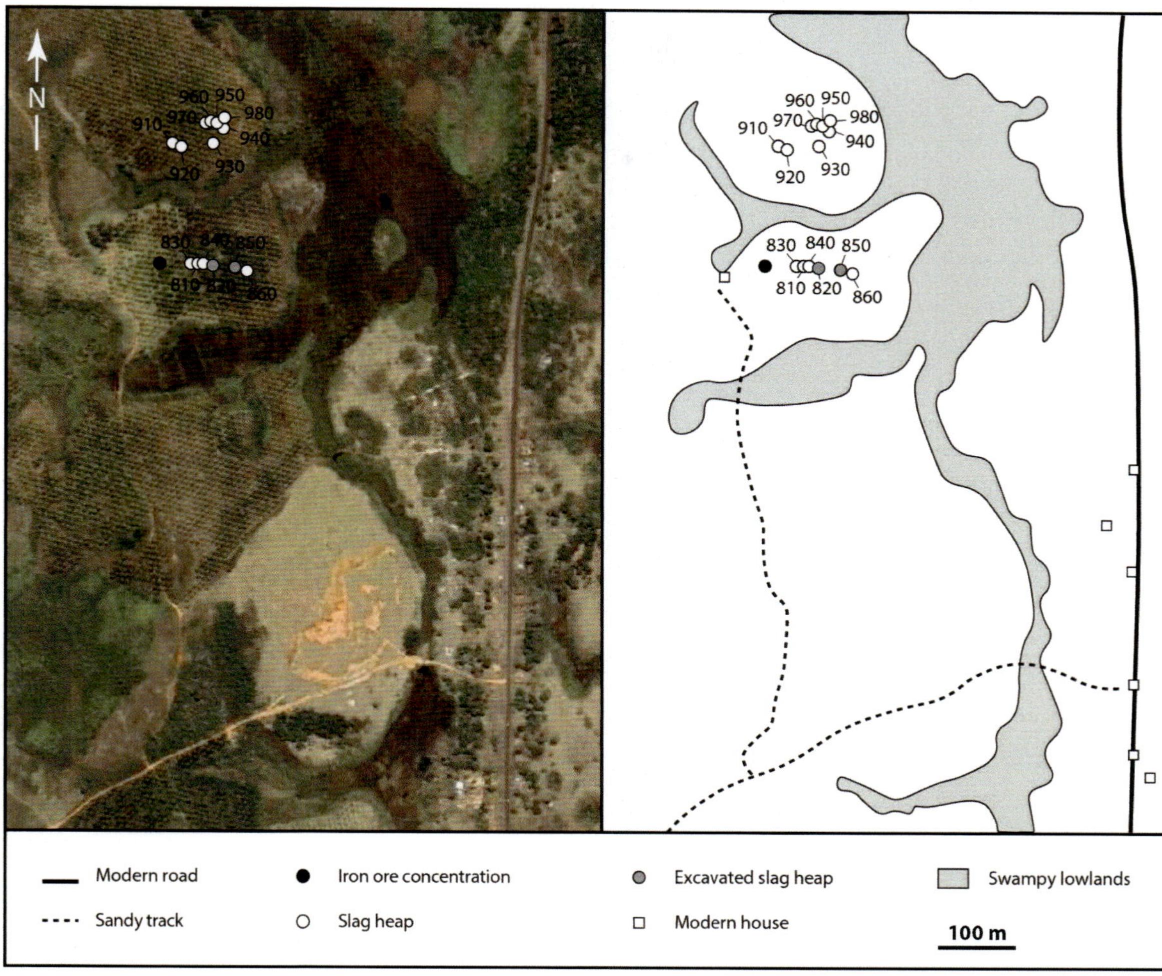

Figure 4.26: General map of Ambodimadiro (DMD) and location of the excavated slag heaps (Google Earth 7.3, (2018), 14°29′22″S 50°09′46″E. [Accessed 15 June 2022]).

4.6 Bemanevika (BMK) : several groups of slag heaps near a major coastal settlement

Bemanevika is located on the northern bank of the Bemarivo river mouth, on the Quaternary sandy strip, at the junction with the low hills of alterites. Vérin already mentioned this site (1986) and excavated an area of early settlement (BMK400), which is located on the sandy strip 1 km from the coast and which also shows evidence of metallurgical production.

A total of at least eleven slag heaps in four metallurgical sectors were identified in the direct vicinity of the village (Figure 4.27). The BMK300 sector is currently on the road. Slag fragments outcrop on the surface but it has not been possible to investigate these slag concentrations. The slag heaps are systematically located on the sandy strips and never on any other bedrock type. Test pits were conducted in three areas (BMK100, 200, 400) to collect slag and charcoal samples.

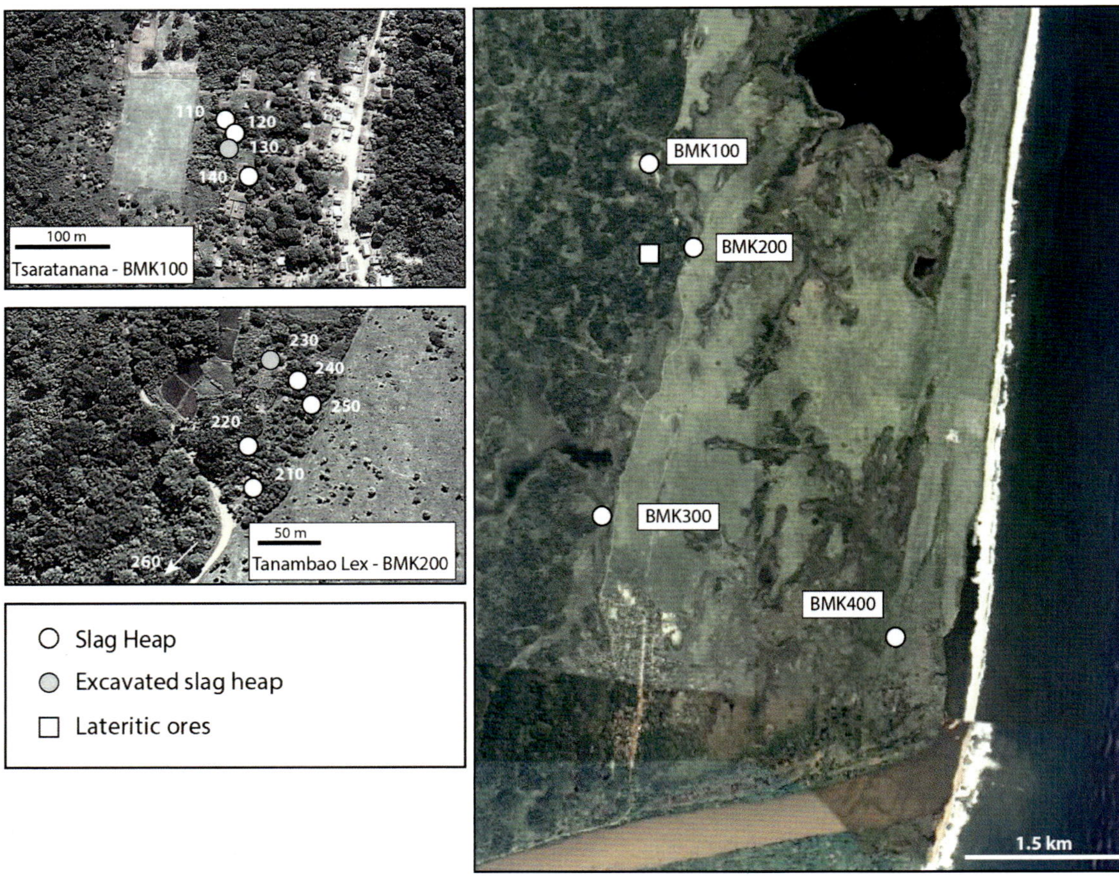

Figure 4.27: Satellite view of Bemanevika (BMK) and location of the excavated slag heaps (Google Earth 7.3, (2018), 14°06′16″S 50°08′29″E. [Accessed 15 June 2022]).

The slag heaps in the sectors BMK100 and 200 are small and with an elliptical shape. The maximum thickness of the slag layer is between 20 and 30 centimetres depending on the heap. The mass of metallurgical waste is estimated to be no more than 4 tons per heap. The vast majority of the slag pieces are bottom furnace slag, mixed with rare tapped slag fragments. Very few ceramic sherds were found in these test pits. These two sectors were probably chosen to be situated within a reasonable walking distance of the early settlement (BMK400) but closer to the ore. There is indeed a low hill called *Antaimby* ("the place of the slag") located between BMK100 and 200. No slag has been observed on this hill, but it is covered with iron-rich rocks, either in the form of pisolithic ferruginous concretions or massive blocks of more than 1 metre in diameter. This is probably the area where the ore was collected.

The slag heap excavated in the settlement sector (BMK420) is larger (10.5x4.5m). The slag layer reaches a maximum thickness of 40 cm. The metallurgical waste consists exclusively of furnace bottom slag. In contrast to the other areas, many ceramic sherds were uncovered during the excavation of BMK420. This is consistent with the proximity of an ancient settlement.

4.7 Global description of the metallurgical district

4.7.1 Distribution of smelting sites along the coast

During our surveys we located 150 slag heaps spread from Lake Inosy in the north, which is about 50 km south of Vohémar, to Cape Masoala in the south (Figure 4.2). Not all the sites listed are documented with the same level of accuracy (See the column 'Type of intervention' in Figure 4.3). An important part of the surveys were carried out by two students from the University of Antsiranana, Walker Chrisoël JAONY and Mangany Hervé TOTOBEMAHEFA, who located and quickly described most of the slag heaps. Other heaps were visited by myself, some were the subject of in-depth surveys that led to test pits to collect charcoal samples for radiocarbon dating, and finally, some were fully excavated and dated. Unfortunately, the development of agriculture and in particular the acceleration of the development of palm tree monoculture have significantly destroyed some archaeological sites. It is still possible to attest the presence of smelting slag concentrations but without being able to quantify them.

As for the settlement sites, iron production sites are concentrated along the sea coast. As mentioned previously, it is likely that human occupation was concentrated along the coast and following the major navigable river axes. Many more archaeological surveys away from the coast would be needed to confirm this hypothesis. It can however be a survey bias since the hinterland is much less accessible, leading to slow and difficult surveys.

The surveys conducted further north than the Lake Inosy did not reveal any traces of iron production, only traces of settlements. Only few surveys have been carried out directly around Vohémar, but this area has been intensively prospected in the past (Vérin, 1986; Vernier and Millot, 1971; Gaudebout and Vernier, 1941). If slag heaps existed, they would certainly be mentioned in the literature. It can therefore be stated that the northern limit of the metallurgical district is certainly located close to Lake Inosy.

The northern part of the district: Lake Inosy and the Mahanara River

Lake Inosy is the northernmost part where we found slag heaps during our surveys. This area had not been described in the previous literature.

The Inosy lake is a 2 km long body of water drained towards the north through a small meandering river. It is a favourable location for human settlement, as evidenced by the presence of local ceramics found on the shore of the lake but also on an island in the middle of the lake. Three scattered metallurgical sites have been located in the vicinity of the lake (**INY100**, **INY200** and **INY300**), i.e. a total of 8 slag heaps. Lateritic ore outcrops have also been located in the surroundings.

The mouth of the Mahanara river is located about fifteen kilometres south of the Inosy lake. The **Mahanara** sector was already listed in the archaeological records as it was described by Vérin (1986) and by older authors:

> "Mr. Jully, head of the civil buildings department, currently on mission, has made a most interesting discovery on the east coast of Madagascar, which will certainly attract the attention of archaeologists and which may be the starting point for new historical research on the island.
>
> M. Jully has uncovered, on the Mahanara river, north of the village of Sahambavany [Sambava], ruins presenting the character of Arab constructions, masoned with lime and which seem to prove by their importance, that a great city once existed at this place.".
>
> Translated from Journal officiel de Officiel Madagascar et Dépendances, 5 mai 1898:1812 (Vérin, 1986)

It would seem that the site was then briefly studied by Jully (BAM I (1):16), and then by Guillaume Grandidier, who would have conducted excavations in 1899. Other authors were interested in the site and Vérin (1986:845-853) gives a good account of the successive reports concerning the site of Mahanara. The main discoveries at that time are the presence of a lime wall, of which little remains today and which has been interpreted as a mosque, as well as massive cylindrical nozzles carved from a single block of stone. Local ceramics, gold dinars and Chinese ceramics (from the 12th century, from Vérin, 1986) are also mentioned. This set of imported goods attests that it was a *Rasikajy* settlement.

However, at the mouth of the river, no evidence for iron smelting was attested. Upstream, three kilometres from the coast, near the village of **Ambodimanga**, five concentrations of slag and tuyere were found during our surveys (**MHN100**). Outcrops of ferruginous concretions can also be seen in several places, as well as the presence of large blocks of laterite, which could have been used as iron ore.

From the Bemarivo River to Antalaha

The mouth of the Bemarivo river is located about twenty kilometres south of the Mahanara river, where the modern village of Bemanevika is established. The slag heaps of **Bemanevika** have already been detailed above (**BMK100, 200, 300, 400**). However, north of the Bemanevika slag heaps, in the modern village of **Andranginalo**, slag concentrations could also be observed (**BMK500**). Unfortunately, the sandy coastal strip where the heaps are located is covered with palm monoculture, which has destroyed the archaeological layers. Only one preserved slag heap could be observed and three other concentrations of scattered bottom furnace slag and tuyere fragments.

According to the local population, more slag heaps and ceramic sherds can be found on the shore of a lake in the vicinity of the village, the Ampasimbato lake. However, this place is considered as sacred so we could not go and confirm these testimonies.

Continuing southwards is the town of **Sambava**. A small necropolis was excavated in 1899 and attests to the existence of a *Rasikajy* settlement on the site of the present-day town of Sambava (Vérin, 1986:267-268).

Figure 4.28: Satellite view of the Antaimby district in the modern city of Sambava and location of slag and ceramic concentrations (Google Earth 7.3, (2018), 14°15'22"S 50°08'48"E. [Accessed 15 June 2022].

The oldest district is called Antaimby ("*the place of slag*") and is indeed linked to ancient metallurgical activities. In this densely populated area, our surveys revealed the presence of scattered slag over an area of about 300 m^2 and at least five significant concentrations of slag (**ATB**, Figure 4.28). For two heaps, the slag layer is about 20 cm thick. The slag pieces are very fragmented, but it is possible to identify the BFS morphology for most of them. In the same area, there is a large amount of ceramic debris belonging to the *Rasikajy* material culture. In addition to the common ceramics, we note the presence of chlorite schist fragments and *sgraffiato* ceramics imported from the Middle East.

Along the next 30 kilometres to the south, along the quaternary coastal sandy strip, are Matavy (**MTY**), Benavony (**BNY**), Amboronala (**MBR**) and Ambodimadiro (**DMD**), which have already been detailed.

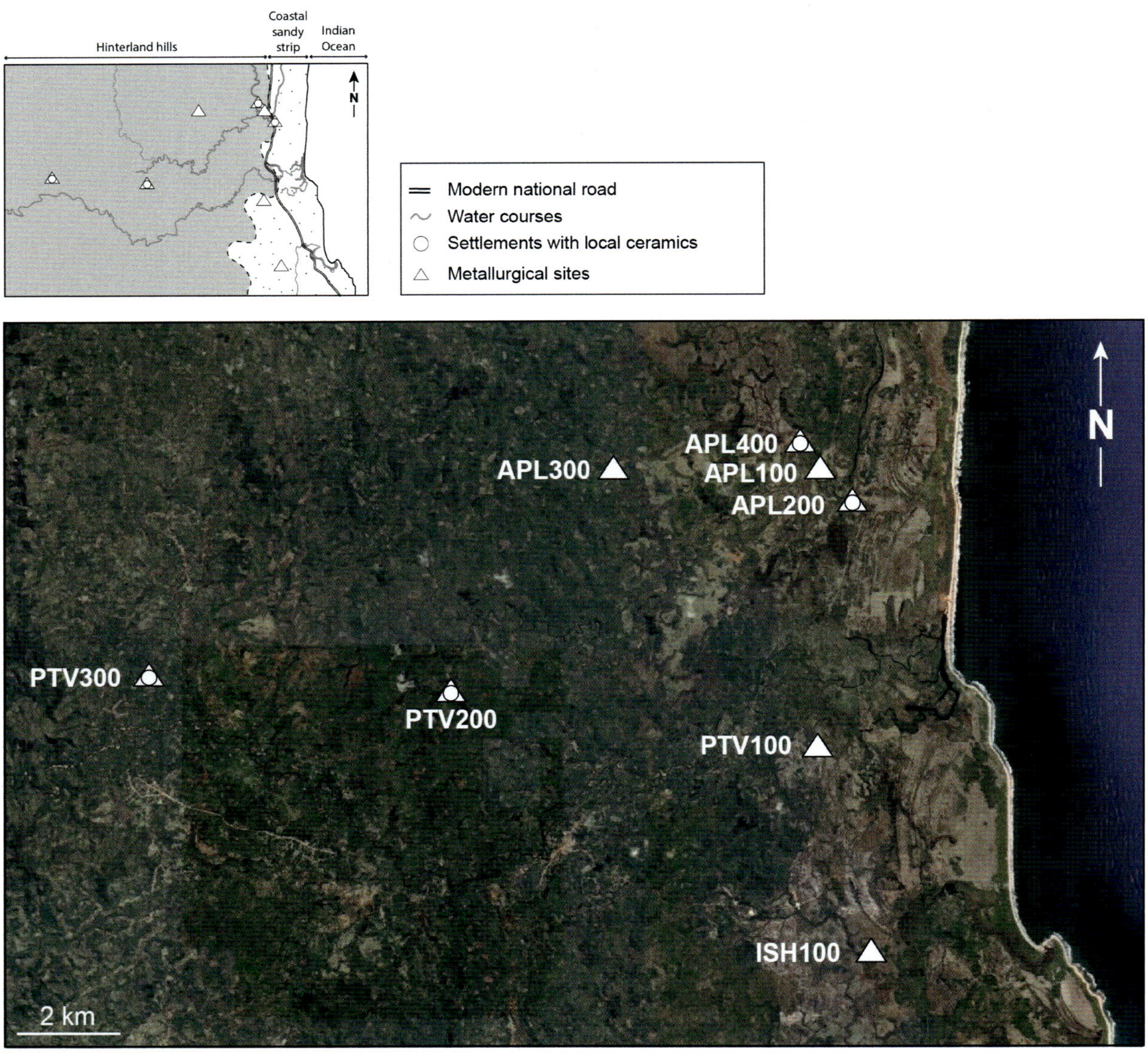

Figure 4.29: Satellite view and general map of the sectors of Ambodipont Limite (APL), Ampanantova (PTV) and Ambodipont Isahana (ISH). (Google Earth 7.3, (2018), 14°34′58″S 50°08′38″E. [Accessed 15 June 2022]).

South of the Amboronala-Ambodimadiro metallurgical complex stands a village called **Ambodipont-Limite** (**APL**, Figure 4.29). This village lies along the meander of a river and on the coastal sandy strip. Local people report that slag pieces are regularly excavated during the building of houses (**APL200**), as well as local ceramics and chlorite schist fragments in the rice fields, testifying to the presence of a *Rasikajy* settlement. Directly to the west of the modern national road, on the top of a hill, slag concentrations were observed associated with a stone block, interpreted as an anvil for crushing the ore (**APL100**).

To the west of this hill, 5 slag heaps are located on a plateau covered with white aeolian sand (**Antintezampako - APL400**). These heaps are ovoidal and elongated. Some of them are installed along the river on a steep slope. The slag fragments are therefore strongly spread along the slope and the slag layer must be relatively thin. Surface observation of the slag pieces shows that BFS are systematically present, but TS were only observed in two heaps.

The village **Andranovato** is located five kilometres inland and is part of the locality of Ampamipihamby (literally "*where iron is worked*"). This village is settled on a red clay substratum, with local accumulation of ferruginous concretions at the bottom of runoff slopes. Lateritic blocks also outcrop on the surface. The substratum is therefore radically different from what was observed at all the previous smelting sites. A blacksmith is still working in this village and has set up his workshop on an ancient slag heap (**APL300**). This heap covers 20 m^2 and the slag pieces are particularly fragmented. The blacksmith still uses archaeological stone tuyeres. Complete pieces of tuyere made of micaschist could be observed (Figure 4.30).

Figure 4.30: Complete stone tuyere made of micaschiste found during surveys at Andranovato (APL300). **a.** End of the cylinder on the side of the bellows. **b.** Side view of the tuyere. **c.** End of the cylinder which was positioned inside the furnace.

Five more slag heaps were located further South near the commune of **Ampanantova (PTV100)**, on a sandy hill called *Antaimby* (Figure 4.31). The site is located in the transition zone between sands and alterites. Of the five slag heaps, four appear to be composed exclusively of TS and only one of BFS. The clusters are small and the thickness of the slag layer does not exceed 40 cm. Another site was located 8 km inland from the coast at Antoaka (**PTV200**), where three separate mounds were observed in 2018 but in 2019 only scattered slag remains after the land was cultivated.

Further inland, other concentrations of slag were found associ-

Figure 4.31: View of two slag heaps from Ampanantova (PTV100 and PTV200).

Figure 4.32: Picture of a stone tuyere from MPH200 (Ampaha). The picture was taken from the tuyere side which is connected to the bellows. On the right-hand side a partially melted brown material can be seen, and corresponds to the furnace wall. No sand imprints are visible, which allows us to hypothesise that the wall of the furnace was made of clay.

ated with coarse local ceramics in the commune of Vohotsarivo (**PTV300**), in a place called Betaimby ("*Many slags*"). These slag heaps appear on the surface to be made up of BFS only. Fragments of tuyere made of stone (rubbed chlorite schist or gneiss), and others of ceramic material, were also observed. This is the furthest metallurgical site from the coast identified during our surveys.

Moving southwards, close to the village of **Ampodipont Isahana**, three further slag concentrations were observed (**ISH100**). These small slag heaps are found on dune strips covered by tall grass and are mainly composed of bottom furnace slags. Near one of the heaps (**ISH120**), a concentration of pisolites was found in the sandy substratum. This is probably a stockpile of ore brought by the metallurgists (see analyses Appendix, Figure B.6).

Twenty kilometres to the south, runoff accumulates and forms the **Ampaha** lake, the overflow of which flows directly into the ocean (Figure 4.33). To the northwest of this lake, several slag heaps were located during our surveys. The presence of lateritic bedrock containing ferruginous concretions is also noticeable everywhere. Two slag concentrations are present in the village of **Ankarango** (**MPH100**). The pieces are very fragmented, but both BFS and TS could be identified. On the other hand, many complete tuyere pieces, drilled into a metamorphic rock, were found on this site.

About 500 metres to the south, there are numerous local ceramic sherds and another group of slag heaps (**MPH200**). The metallurgical waste found on this site is extremely fragmented, which is certainly due to the impact of agricultural practices, since these heaps are located in rice fields. Nevertheless, a few pieces confirm the use of the same iron ore smelting technology as elsewhere in the region. The short cylindrical tuyeres are mostly made of stone, but there are also some ceramic pieces. One of the fragments shows a strong slagging on one side, which corresponds to the imprint of the wall in which the tuyere was inserted. In this case, it is a clay wall because, contrary to what was observed on other coastal smelting sites, neither imprint of sand nor indurated sand is conserved (Figure 4.32). It is consistent with the nature of the local bedrock.

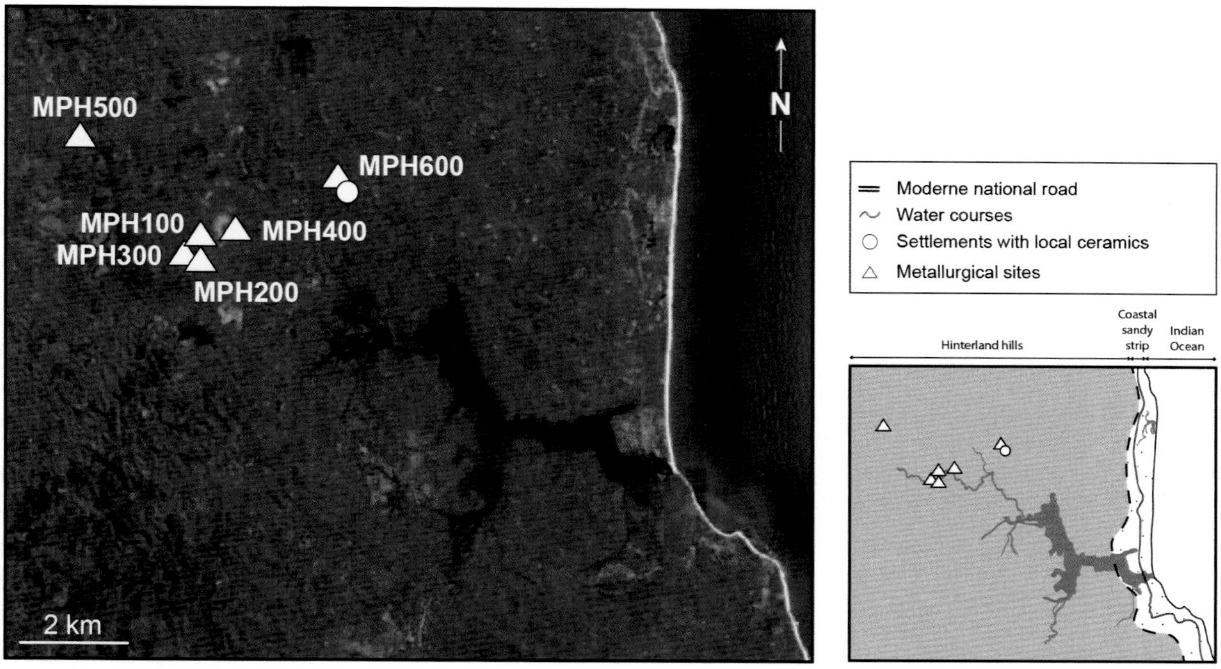

Figure 4.33: Satellite view and general map of the sector of Ampaha (MPH). (Google Earth 7.3, (2018), 14°44′27″S 50°10′20″E. [Accessed 15 June 2022]).

Nearby, four small slag heaps were observed (**MPH300**) with the same archaeological assemblage characteristics, i.e. stone tuyere (chlorite schist or gneiss) and FBS. A fourth concentration of slag is found a few hundred metres further north (**MPH400**), associated with local ceramic sherds. On this site, the slag pieces are much more massive although they present the usual morphological characteristics.

To the north (**MPH500**), not far from a village called Betaimby ("*A lot of slag*"), slag concentrations made both of tapped and bottom furnace slag pieces have also been located. Unfortunately, the archaeological layers have been extensively reworked by agricultural practices and it is difficult to delimit the extension of the heaps.

Finally, east of the village of **Bekony**, along the Ampahana River, a last concentration of slag was observed on the top of a hill (**MPH600**). The slag pieces are very fragmented but seem to be bottom furnace slags. Also here, there are stone tuyere and local ceramics.

The area around the Ampaha lake would deserve a much more thorough archaeological study. It is a rather large metallurgical complex, associated with settlements and it is located far from the shoreline. The substratum is different from what we excavated and it would thus be interesting to study the slag chemical compositions to make comparisons with slag formed in a sandy substratum. The question of the variability of the furnace architecture also arises.

The surroundings of the modern town of Antalaha and the Masoala peninsula

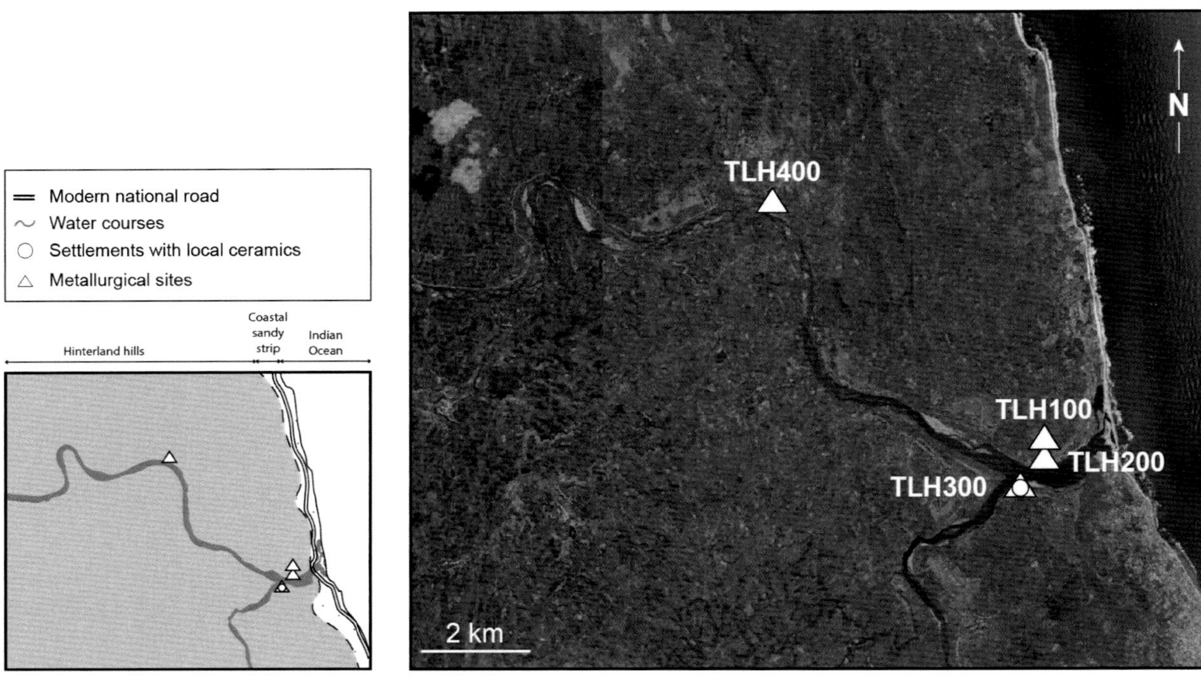

Figure 4.34: Satellite view and general map of the sector of Ankavanana, close to the modern town of Antalaha (THL). (Google Earth 7.3, (2018), 14°52′25″S 50°14′52″E. [Accessed 15 June 2022]).

The shared estuary of the Andampy and Ankavia rivers is situated right north of the modern town of Antalaha. One kilometre inland, several slag concentrations are located on the north bank of the rivers, near **Andripitra** (**TLH200**, Figure 4.34), as well as sherds attesting to ancient occupation. On the south bank, there is a similar situation near the modern village of **Masindrano** (**TLH300**). In the agglomeration itself we collected fragments of local ceramics, but also imported *sgraffiato* and sherds of chlorite schist. At the western exit of the village, at the top of a hill, a very large concentration of ceramics probably corresponds to an important settlement.

We tried to go as far up the Ankavanana river as possible. About 10 kilometres inland, three new slag heaps were discovered in the village of **Androranga** (**TLH400**, Figure 4.34). These slags are located in the middle of modern dwellings. It is interesting to note the dominant presence of tapped slags. Further upstream, for a few kilometres, only laterites with ferruginous concretions are observed, but no slag.

Five kilometres south of Antalaha, there are numerous concentrations of slag in the vicinity of the village of **Ambodikakazo** (**KKZ**). However, 5 kilometres further on, no slag could be found in the villages of Antsirabato and Namoana.

From this area, fewer surveys were carried out. Travelling in this area of Madagascar is especially difficult, so surveys had to be concentrated along the coast. During the oral surveys, local population repeatedly mentioned the existence of slag heaps, but we could not verify whether these accounts were well-founded. Nevertheless, some settlements and metallurgical workshops were visited. Ferruginous lateritic concretions were regularly observed all along the coast, even in the extreme south of the peninsula.

In the village of **Tanambao**, there are local ceramics and a stone tuyere. Oral surveys also mention the existence of slag, although none was observed. Continuing south to the village of **Antsiragnamatso** (**CPM100**), concentrations of slag are associated with local ceramics and stone tuyere. Finally, on the track between **Marifinaritra** and **Anjagnazagna**, slag was located (**CPM200**). They are unfortunately very fragmentary and it is therefore difficult to characterise their morphology.

The southernmost site observed is located at **Antsirabato** (**CPM300**), at the southern tip of Cape Masaola. The slag concentrations were mainly observed in the embankments of the houses or in the vanilla fields. It is therefore difficult to estimate the number and surface area of these heaps. Both slag morphologies, tapped and bottom furnace slag, were identified, associated with coarse local ceramic sherds.

4.7.2 Spatial organisation of the smelting workshops

The detailed study of three smelting workshops, the dozen test pits carried out and extensive surveys throughout the area show that the majority of the technical and spatial characteristics are shared by all sites. However, in the details each site and each smelting workshop is organized slightly differently and reveals some local variability.

Organisation of a slag heap

The slag heaps show a circular or an elliptical shape with an average diameter between 5 and 7 metres. Most of the heaps are solitary but some operate in pairs, as in Amboronala (MBR140 or MBR410). The maximum thickness of the slag layer rarely exceeds 30 cm. Benavony is an exception, as well as some other heaps from other sites (Matavy - MTY11, Ambodimadiro - DMD910 and DMD920), which are more massive and crescent-shaped. The thickness of the slag layer can reach 90 centimetres in these cases and the heap is about 15 metres large. The largest slag heaps do not exceed 40 tons of metallurgical waste.

The discharge of metallurgical waste can be selective according to the type of slag (internal bottom furnace (BFS) slag or external tapped slag (TS) or not. In the case of Matavy MTY11, it can be seen that one part of the heap is made up exclusively of tapped slag and the other part only of BFS. In other cases, such as at Amboronala or Ambodimadiro, a heap composed exclusively of BFS will be juxtaposed with a heap composed exclusively of tapped slag. It is likely that, at the start of metallurgical activity on a site, there are distinct separate disposal areas, one for TS and the other one for BFS. Then, as the amount of waste increases, the two piles merge. This leads to situations like in Matavy MTY11. On the contrary, in other cases, such as in Benavony, there is no sorting of waste according to its morphology.

The proportions of each type of slag are extremely variable from one heap to another, although bottom furnace slag pieces are largely predominant (Chapter 5).

Organisation of a smelting workshop

All the sites studied are slightly different from each other. Nevertheless, similar structures associated to metallurgical activities were recurrently observed on the excavated smelting workshops. Associated with the slag disposal areas, a furnace, ore preparation area, post holes and sometimes low walls were indeed found.

The smelting furnaces are poorly preserved structures, simply dug into the sandy substratum (Chapter 5). A single furnace is installed at the foot of the slag heap, or pair of slag heaps. No furnace battery organisation was excavated. For the three working areas excavated in detail, the furnaces are located to the south of the metallurgical waste disposal area. This may be a coincidence, however, and more excavations would be required to check this hypothesis. The other crescent-shaped heaps that could be observed during our surveys are also oriented with their entrance to the south. It is possible that there is a recurrent pattern to the installation and spatial organisation of the workshops. This is probably not a systematic pattern, however, since at Bemanevika the orientation of the BMK420 cluster and the presence of a slope make it unlikely that the furnace was installed to the south.

Moreover, the metallurgical production sites are systematically installed near a watercourse, probably to wash the ore (Chapter 6). Sometimes, there is an anvil to crush the ore (Amboronala - MBR140 and MBR200, Ambodipont Limite - APL100). Other structures must have been implemented, as shown by the presence of numerous post holes. However, no organisation could be deduced from these remains. It is likely that these holes were dug and rearranged over the years and production campaigns.

In all the smelting sites studied and surveyed, we found several slag heaps a few tens of metres apart. There is no site with a single slag heap but rather the juxtaposition of several small slag concentrations.

Relationship between a smelting site, the settlement and ore sources

Some smelting sites are associated to settlements while no settlement traces have been found for others. As explained above, settlements, and associated slag heaps, are either directly near the seashore, such as Benavony (BNY), Amboronala (MBR), Bemanevika (BMK400) or Sambava (ATB), or very far from the coast, such as Bekony (MPH) or Antoaka (PTV200). The choice of a village location is not based on the proximity to iron ore, but on water or food resources, or the search for protected areas. The smelters therefore had to fetch the iron ore a few kilometres inland and bring it back for smelting.

We could notice that smelting sites without any trace of settlement are found in the vicinity of ore deposits, within a reasonable walking distance from settlements. These sites were probably established to limit the transportation of the ore.

In most cases, a smelting site close to a settlement is associated with an inland smelting site, as if these sites were functioning as a pair. Matavy (MTY) would for example be the inland site associated with Sambava (ATB), or Ambodimadiro (DMD) could be paired with Amboronala (MBR). The radiocarbon dating carried out (Figure 4.5 and Figure A.2) does not allow us to see an evolution over time of the use of one type of site rather than another. It is even more likely that the two types of sites operated in parallel throughout the period of metallurgical activity. On the other hand, sites that are more than ten kilometres from the coast are again associated with evidence of ancient settlements (Antoaka, Vohotsarivo for example). The relationship of these sites, which are difficult to access, to the rest of the territory is still poorly understood.

It has to be noted that only workshops located on dune strips were excavated and studied. The smelting sites further away from the coast, and not installed on a sandy substratum, such as Ampaha (MPH) for example, could not be studied or sampled. They were only visited during the surveys conducted by two students. The organisation of these sites is therefore unknown, even if we do know that the slag heaps are also small and the slag pieces are similar in shape.

Study of Metallurgical Waste 5

The three excavations on smelting workshops (Benavony, Matavy, Amboronala) revealed poorly preserved bowl furnaces, built in sand with a non-permanent superstructure (Serneels et al., 2019b, Serneels et al., 2018). These vanishing remains are too scarce to reconstruct the architecture of the furnace. A detailed study of the morphology of the metallurgical wastes, both slag and tuyere, in close connection with the study of archaeological structures, was necessary to understand the furnace implementation. The morphology of the slag pieces reflects the shape of the furnace, the presence of some specific features, such as the air supply, or how the furnace was operated, for example if the slag was tapped. Detailed observations of the metallurgical wastes also allows us to estimate their variability and similarity. In our case, the slag samples examined in the laboratory and during fieldwork show a limited variability, which tends to demonstrate that the *Rasikajy* used the same smelting technique throughout the chronological period and in the whole study area ($11^{th} - 14^{th}$).

5.1 Slag morphology

Two types of slag morphology were identified on iron smelting sites: internal bottom furnace slag (**BFS**) and external tapped slag (**TS**). All of them are fayalitic slag with high iron content (> 60% Fe_2O_3tot). Metallic iron more or less rusted is frequently observed in the slag fragments, mostly in BFS but also in some TS pieces.

5.1.1 Bottom Furnace Slag (BFS)

Bottom furnace slag pieces (**BFS**) have a hemispherical shape which corresponds to the footprint of the bottom of the furnace (Figure 5.1). Indeed, during the smelt the slag melts and accumulates on the bottom floor of the furnace below the tuyere. At the end of the smelt, the slag solidifies in the shape of a roughly hemispherical bowl. The BFS regularly has large prismatic porosities that are charcoal impressions. This confirms that these slag pieces are formed by accumulation at the bottom of the furnace pit.

Frequently the bowl has one straight edge (Figure 5.1 [1]). Moreover, on several pieces tuyere material was observed attached to this straight edge. A central depression (Figure 5.1 [2]) with slightly raised rim (Figure 5.1 [3]), touching the straight edge, was observed on a significant amount of BFS. This depression is formed because of the air flow coming out of the tuyere and deforming the viscous slag.

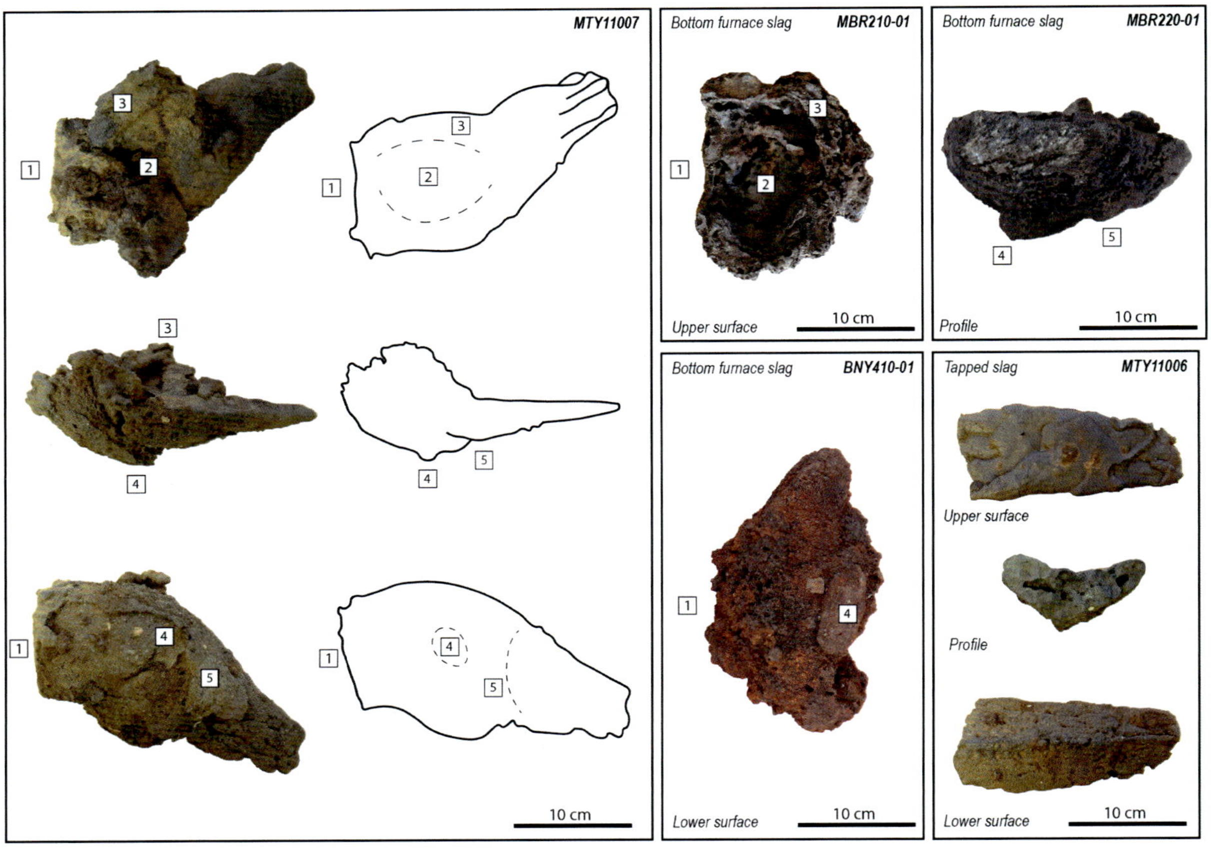

Figure 5.1: General morphology of smelting slag from northeastern Madagascar. **1**. Straight edge indicated the position of the tuyere. **2**. Central depression on the upper surface of BFS, formed because of the deformation of the still viscous slag by the air flow coming from the tuyere. **3**. Slightly raised rim on the edge of the BFS upper surface. **4**. Small bump observed on the BFS lower surface, and interpreted as an evidence for a ritual deposition in the bottom of the furnace before smelting. **5**. Angular protrusion reminiscent of the shape of a digging tool.

The lower surface is convex and very often sand grains are caught in the slag. It can form a real crust, up to several millimetres thick. This shows that the slag cools down directly on the sand and thus that the base of the furnace is dug directly into the sand. There is no clay lining. In the center of the lower part of the slag there is frequently a small prominence of a few centimetres, like a bump (Figure 5.1 [4]). The shape is generally rounded. The molten slag fills a small hollow produced by pressing the sandy floor of the furnace with the hand or a tool. The intentional deformation of the floor has no technical purpose. This repeated action could be related to a ritual of deposition or offering before the smelt. There is some archaeological and ethnographical evidence of such ritual reported from East African contexts (Brown, 1995; Schmidt and Childs, 1985).

Even if the BFS pieces from the largest slag heaps, in Benavony

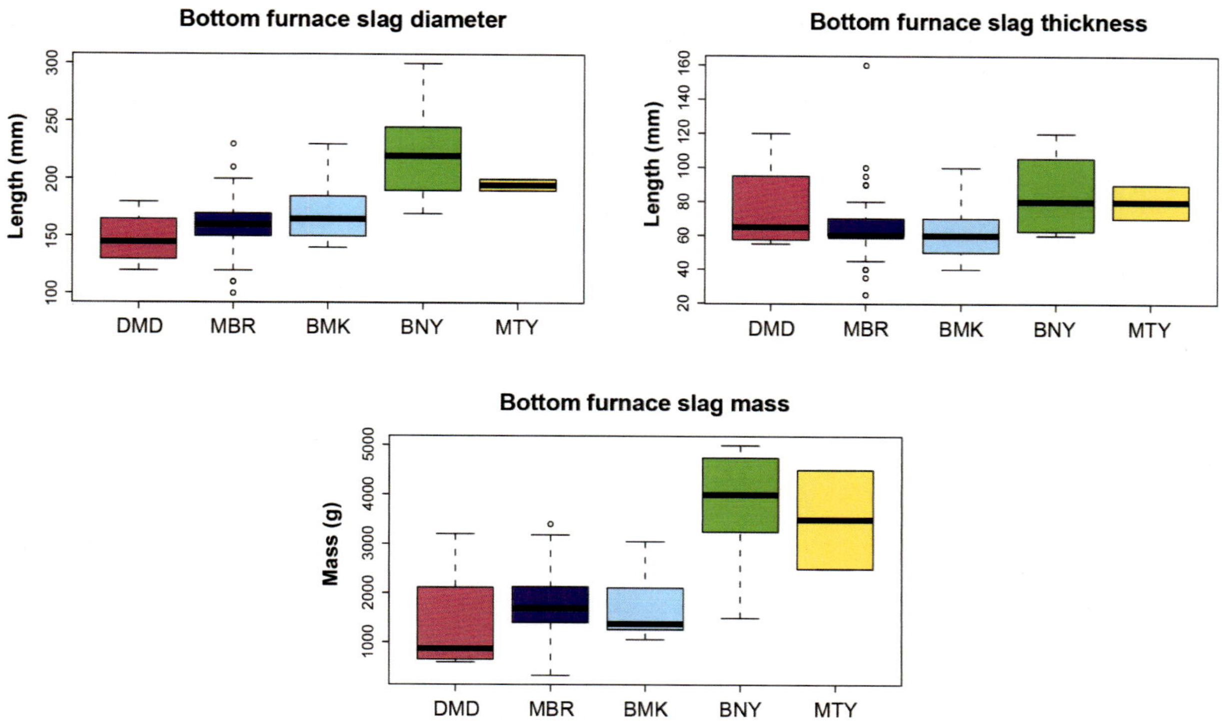

Figure 5.2: Comparison of the dimensions and masses of bottom furnace slag pieces from the five most studied sites. 136 complete BFS samples were measured and are plotted here. DMD: Ambodimadiro (n = 4), MBR: Amboronala (n = 89), BMK: Bemanevika (n = 14), BNY: Benavony (n = 16), MTY: Matavy (n = 13). Data from Figure A.3.

(BNY410) and Matavy (MTY11), tend to be a bit bigger (Figure 5.2 and Figure A.3), the slag dimensions are relatively homogeneous from one smelting site to another. For most of the slag pieces, their diameters are between 10 and 20 centimetres and they are 4 to 10 centimetres thick. Their weight is between 1 and 2 kilograms. Slag pieces from Benavony and Matavy have a diameter greater than 20 centimetres and their mass can reach 4 to 5 kilograms. Their thickness however, is similar to all the other measured slag pieces.

Figure 5.3: Example of three tapped slag fragments from Amboronala. **A.** Large slab of tapped slag overflowing out of the channel intended for the tapping of the slag. **B.** Typical tapped slag for northeastern Madagascar with overlapping ropy flows. **C.** Single flow of tapped slag.

5.1.2 Tapped Slag (TS)

Tapped slags (**TS**) are made up of small juxtaposed ropy flows stacked in a V-shaped profile (Figure 5.1). The slag solidifies in a channel dug into the sand, 4 to 6 centimetres wide and 3 to 5 centimetres thick. Tapped slags are mainly found as small fragments but a few pieces can be as long as 30 centimetres (Figure 5.3). The largest fragment is a 30 centimetres large slab, overflowing outside the original channel. Its weight is about 5 kilograms. The smaller fragments commonly found measure only a few centimetres and weigh only about ten grams.

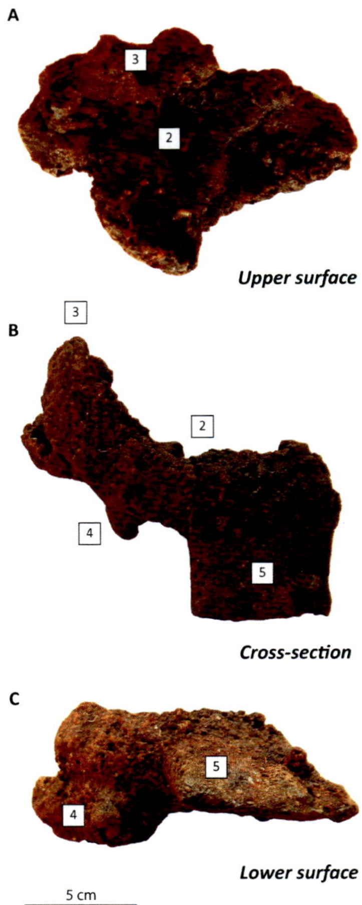

Figure 5.4: Profile view of a BFS (BNY41027 - Benavony) with a protrusion with the shape of a spade on the lower surface (see **B** and **C**). The slag has the same general morphological characteristics as those described in Figure 5.1.

5.1.3 Mixed morphology: TS flowing from a BFS

Some slag pieces show a mixed morphology with a flow of tapped slag still attached to a BFS (Figure 5.1-left). Sometimes this flow of slag from the BFS is significant, sometimes only the beginning of the flow is observed. This starting point of the flowing part is located on the opposite edge from the tuyere, usually slightly off axis. The flow is horizontal, at the same high than the circulation level.

Some BFS and the slag with a mixed morphology show an angular protrusion reminiscent of the shape of a digging tool (Figure 5.1 [5]). Some pieces are even shaped like a spade or a tool for digging the ground (Figure 5.4). In order to drain the slag by flowing, the metallurgist dug a V-shaped channel directly into the loose sand with such a tool.

The channel was dug during the smelting process and not beforehand because the slag has these very angular features and not smooth ones. It suggests that the tool was stuck in the sand and that the liquid slag immediately filled the cavity. If the channel had been dug beforehand, because the substratum is dry sand, the shapes would have been more rounded.

5.1.4 Proportions of each slag morphology in the excavated slag heaps

Bottom furnace slags are observed on every smelting area whereas tapped slag are not (Chapter 4). The proportion between the two types of slag varies radically from one heap to another. At least half of the slag heaps are only made of BFS such as the three MBR200 slag heaps (Amboronala - MBR210, MBR220 and MBR240).

Figure 5.5: Table showing the proportions of BFS and TS in each slag heap where an excavation or test pit was carried out. Assuming that one piece of BFS corresponds to one smelting operation, associated with a given amount of TS, it is possible to estimate the amount of slag produced for one operation for each slag heap.

Sites	Slag heaps	Proportions		Mass of slag for one smelting operation (kg)		
		TS	BFS	TS	BFS	Total
Benavony	BNY410	10	90	0,4	3,5	4
Matavy	MTY11	30-50	70-50			
Amboronala	MBR140	90	10	20	1,5	21,5
	MBR210	0	100	0	1,6	1,6
	MBR220	0	100	0	2,1	2,1
	MBR240	0	100	0	1,8	1,8

The other heaps are made up of a fluctuating proportion between tapped slag and furnace bottom slag (Figure 5.5). A slag heap composed mostly of BFS can be located on the same site only a few metres apart from a heap composed mostly of TS. For example, in Ambodimadiro (DMD), each slag heap is composed mainly of one slag morphology: DMD850 is mostly made of TS whereas DMD860 of BFS even if they are located only few metres apart. It is likely that these heaps work in pairs with a specific discharge area for BFS and one for TS. When the accumulation of waste becomes important, these heaps join and form a single heap with differentiated discharge areas, like MTY11 (Matavy). The furnace would then be located somewhere between the two heaps. However, no excavations have been carried out to confirm this hypothesis.

It would thus appear that slag tapping is not a systematic practice. It is probably implemented only when the evacuation of an excess of slag is necessary. If needed, the smelters could easily dig a V-shaped channel in the sand and pierce the floor, or maybe wall, of the furnace.

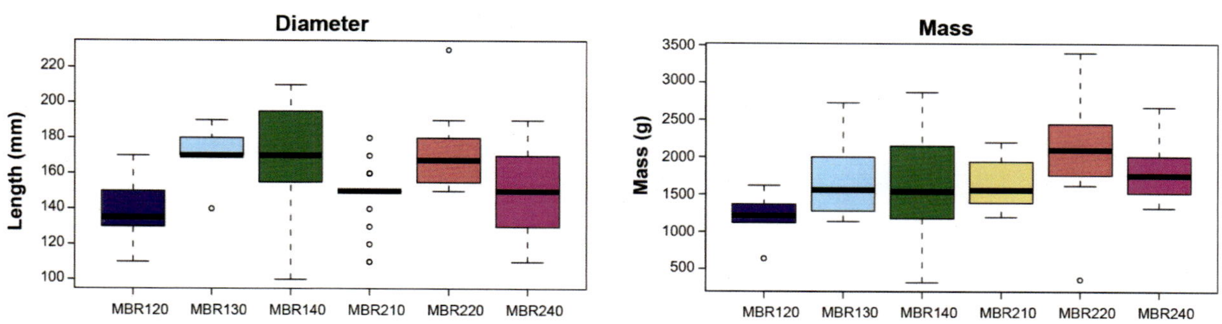

Figure 5.6: Comparison of the dimensions and masses of bottom furnace slag from several slag heap in Amboronala (MBR). A total of 89 complete pieces were measured (MBR120: n=6, MBR130: n=5, MBR140: n=17, MBR210: n=21, MBR220: n=20, MBR240: n=20).

The Amboronala sector (MBR) was studied in detail, with MBR140 mainly composed of TS and the three MBR200 slag heaps composed exclusively of BFS. There is no significant difference of mass between the BFS of the heaps composed only of BFS and those with TS (Figure 5.6). Thus, the size of the furnace bottom does not vary, but rather the amount of ore reduced. For the same mass of BFS produced in one smelting operation, about 20 additional kilograms of TS are produced at MBR140 than at MBR200. To explain such a difference, more ore had to be introduced into the furnace and the excess slag drained off.

The changes in the proportion between the two kinds of slags from one heap to another reflect the variability of the smelting operations, not only from one working area to another, but probably also from one smelt to the next. This variability should not be interpreted as the presence of two distinct techniques because the metallurgical wastes are very similar on all visited sites. It is rather a difference in practice, probably because in some cases a larger quantity of ore was introduced into the furnace. This reflects an extreme variability of the technical gesture and thus a low standardization of the metallurgical practice.

5.2 Tuyere morphology

The tuyeres represent a relatively small proportion of the metallurgical waste (1 – 2 %), especially in comparison with other African metallurgical sites where tuyeres are found in very large quantities (Serneels et al., 2015; Robion-Brunner, 2008).

The tuyeres are short, about 10 centimetres long. They have an internal funnel shaped pipe and the diameter of the outlet is always small, around 1.5 to 3 centimetres. All other characteristics are variable and apparently related to the type of material used. Some of the tuyeres are made from plastic ceramic paste whereas the others are made of rigid stone (Figure 5.8). In a single slag heap, most of the tuyeres are made of one single material, but a few pieces are made of other raw materials (Figure 5.7). During the excavations of the Benavony BNY410 slag heap for example, most of the recovered tuyeres are made of micaschiste but some pieces made of rhyolite, gneiss or orange ceramic were observed, whereas in Amboronala MBR140, most of the recovered pieces were made of white clay associated with few stone tuyeres.

Site	Dominant Materials
Bemanevika	Red ceramic
Matavy	Orange ceramic
Benavony	Rhyolite
	Gneiss
	Ceramic
Amboronala	White clay
Ambodipont Limite	Para Gneiss /Micaschiste
Ambodipont Isahana	Para Gneiss /Micaschiste

Site	Sample	Materials
Bemanevika	BMK13701	Rhyolite
	BMK13702	Ceramic
	BMK23702	ParaGneiss /Micaschiste
Matavy	MTY11702	Ceramic
Benavony	BNY41701	Rhyolite
	BNY41702	Rhyolite
	BNY41703	Gneiss
	BNY41705	Micaschiste /Gneiss
Amboronala	MBR14702	Ceramic
Ambodipont Limite	APL30701	Para Gneiss /Micaschiste
Ambodipont Isahana	ISH10701	Micaschiste
	ISH20701	Gneiss

Figure 5.7: Left: Summary table of the main materials used for the tuyere manufacturing in the studied sites. **Right**: Table summarising the tuyere materials identified during petrographic observations. Petrographic observations were conducted on thin section. The aim of these observations was not to have a representative sample of the main materials for each site, but rather to assess the diversity of materials at the metallurgical district level.

Ceramic tuyeres are cylindrical with wall 3-4 centimetres thick. The inlet diameter can be large, up to 6 centimetres. They are shaped by plastic deformation of a wet paste. The discovery of a batch of pure white clay during excavation at Amboronala supports the hypothesis that the clay tuyeres were produced at the smelting site. The clay tuyeres were probably not pre-fired before use as, in most cases, the part in contact with the fire is much better preserved than the rear part. Different ceramic pastes have been used on

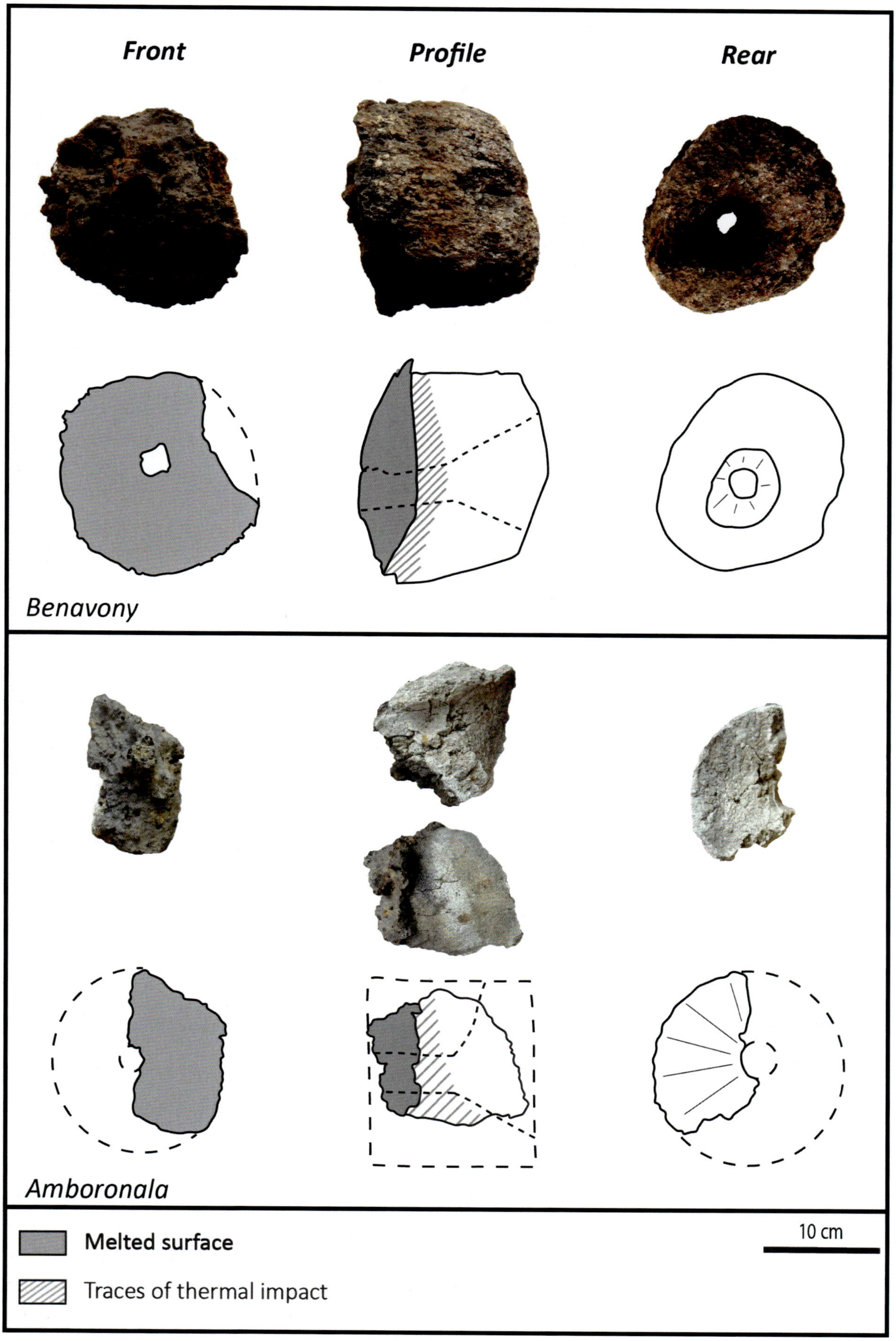

Figure 5.8: General morphology of stone (above) and ceramic (below) tuyere from northeastern Madagascar.

different smelting sites, possibly according to the availability of the raw materials. In the coastal range, the substratum is exclusively made of sand. Clays are really difficult to find. On the contrary, the inland hills are covered with a thick layer of alterites rich in clays.

Stone tuyeres are mainly cylindrical but can be brick shaped. The wall can be thicker, up to 7 centimetres, and the inlet is smaller, around 4 to 5 centimetres. Petrographic observations were made on thin sections (Figure 5.7). Several types of stone were identified: gneiss or mica schist,and rhyolite. However, although chlorite schist was exploited to a large extent for the production of tripod cooking pots (Serneels et al., 2020; Vérin, 1986; Gaudebout and Vernier, 1941), and one tuyere piece made of chlorite schiste is reported in the literature (Vernier and Millot, 1971), no petrographic observation in thin section has confirmed its use to manufacture tuyere. One reason may be that few chlorite schite deposits exist in the metallurgical production area and local stones are used to be drilled into tuyere. Indeed, the tuyeres appear to have been drilled by a rotary tool. Traces of this tool can be seen on some of the tuyeres, the most visible traces being preserved on those made of rhyolite (Figure 5.9).

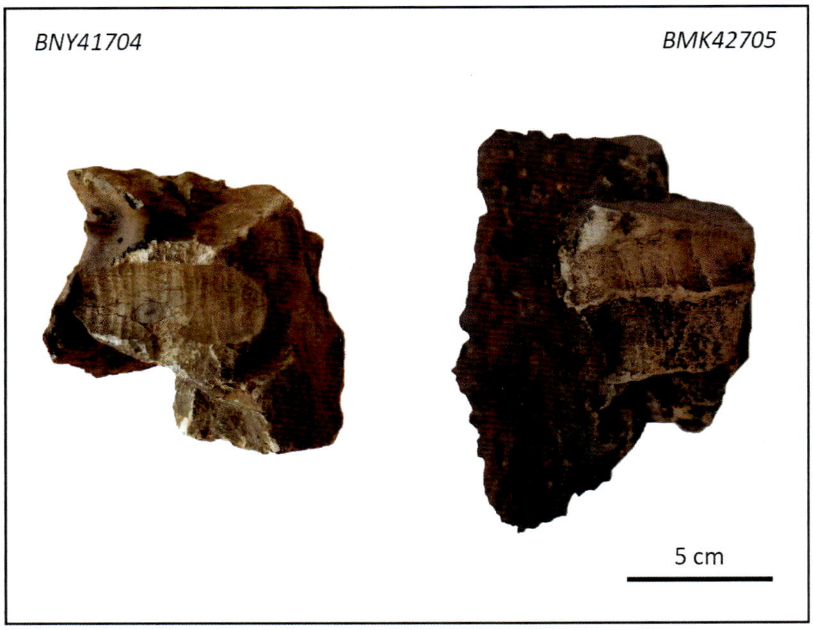

Figure 5.9: Example of two tuyeres samples made of rhyolite from two different slag heap (Benavony: BNY410 and Bemanevika: BMK420). Traces of a rotary tool used to drill the tuyere are visible on the duct.

Rich in biotite, gneiss and micaschiste are metamorphic rocks frequent in the Precambrian basement. It is also easy to collect boulders in the river beds. These rocks are hard, because they contain quartz, but are still workable due to the large amount of micas. On gneiss and micaschist tuyere, no trace of drill or chisel can be identified. Rhyolites are Cretaceous volcanic rocks present at several locations at the limit between the Quaternary coastal deposits and the Precambrian basement. The blocks used were probably collected at natural outcrops. This kind of rock is hard but homogeneous. Indeed, traces of drilling are clearly visible on the tuyere fragments made of rhyolite. Moreover, these fragments have undergone significant alteration due to fluid infiltration. This probably makes the rock less hard and easier to drill.

There is a surprisingly huge variability concerning the materials used to make the tuyeres and the associated manufacturing processes, even at the level of a single slag heap. On the other hand, their shape and size show a small variability. It is interesting to note that the two most important technical aspects, the outlet diameter and the length, show the lowest variability (Figure 5.8).

One fragment of white clay tuyere found at Amboronala MBR140 shows unusual details that have not been noticed on other fragments. These are small holes aligned around the tuyere along a notch. These holes are spaced about 2 cm apart and are 8 mm wide and 5 mm deep (Figure 5.10). These cavities are located at the edge of the thermal impact trace. They were therefore located inside of the wall. These holes are not deep enough to be used as a fixing points for the tuyere in the wall, using small sticks as dowels for example. For the moment, we could not find any satisfactory explanation for the existence of these holes on this exceptional fragment.

Figure 5.10: Picture of two clay tuyere fragments from Amboronala with unusual holes aligned on the external part of the pieces.

Many of the recovered tuyere fragments or complete pieces had heat impact traces, indicating which side of the tuyere was in the furnace. These heat traces are visible because of the rubefaction of the material and a slagging of the extremity of the cylinder. This means that the tuyere partially melts during smelting. Each material has a specific heat resistance. We did not systematically record the importance of the heat impact on tuyere fragment during fieldwork, but we could notice that tuyeres made of gneiss and micaschist probably have a better long term thermal resistance than those made of rhyolite, and even better than those made of clay or ceramic. We did not find any complete clay tuyeres whereas it was frequent to find, even during surveys, complete pieces made of gneiss or micaschist. Clay is a more fragile material, but we could observe that a bigger part of the extremity of clay samples melted.

The position of the heat impact on the tuyeres indicates that it was inserted more or less horizontally through a wall and was not protruding inside the furnace (Figure 5.8). The shape of the furnace bottom slags and especially the straight edge shows that a single tuyere was used. The small diameter of the outlet supports the use of bellows. The type of bellows that comes immediately to mind concerning Malagasy iron metallurgy is the Austronesian piston bellows. It is correct to make the connection between Indonesian piston bellows (Marschall, 1968) and the bellows described in the Malagasy Highlands by the first observers between the 17th and 20th centuries (Ellis, 1858; Lacombe, 1840; de Flacourt, 1661). Nevertheless, there is no historical source describing any Malagasy smelting process, and thus the associated bellows, before the 16th century. There are no sources describing the bellows used in northeastern Madagascar during early European visits in the 16th century either. There is today no archaeological or historical argument to describe the type of bellows used by the *Rasikajy*, even if the comparison with Austronesian technology is tempting.

5.3 Furnace architecture

On the three excavated sites the remains of the furnace are simple shallow pits dug in the sand without any internal clay lining. The bottom of the pit only shows traces of limited heat impact. The pits were also filled with charcoal rich sediment. Only in Benavony can the size of the shallow pit be clearly determined: it was an elliptic bowl of 100 by 70 centimetres and 20 centimetres deep (Figure 5.12). These dimensions are consistent with the size of the recovered BFS.

Figure 5.11: Rubbed sand wall around a clay tuyere, recovered in Amboronala. From top to bottom: Face on the hearth side; profile section of the wall; back face not in direct contact with the fire.

Digging a furnace directly in the sand involves fragility and low durability of the structure. The bowl of the furnace probably had to be regrooved after each smelting operation. It explains the great variability in the shape of the lower surface of BFS within a single slag heap.

The short single tuyere would be placed horizontally at the level of the external ground, approximatively 10 cm above the floor of the pit. The wall surrounding the tuyere was thus made of sand as can be seen on a few preserved pieces (Figure 5.11). The sand wall was at least 30 cm high so that there is about 10 cm of wall above the tuyere. At Matavy, a few non-contiguous blocks of stone, some of them standing vertically on one edge, were found around the firepit. They are interpreted as rigid elements strengthening the wall made of sand. No stone block with significant heat impact nor any fragment of burnt clay allows for the reconstruction of a superstructure built with another material other than sand. However, even if there is no evidence of it, it is not possible to rule out the existence of an external superstructure made of organic materials. Some populations use plants to build the walls of their furnaces, such as the Kuys people in Cambodia who use bamboo (Dupaigne, 1987) or the Musigati people in Burundi who use banana tree trunks (Celis, 1989).

This reconstruction of the furnace involves the construction of a more or less vertical and stable sand wall (Figure 5.12). It is unclear how a vertical sand wall could be built. No impressions of wood planks or other organic material, which could be used to stabilize the dry sand, were found on the wall fragments. Wet sand could have been used, which would make the furnace significantly less efficient.

Some slight variations in the furnace building could be observed on the three excavated sites. As seen above, at Matavy, stone blocks are set around the firepit. In Amboronala a second pit is installed on a lower level than the firepit, probably in order to facilitate the evacuation of the tapped slag during the smelting process. As for the tuyeres, despite the high variability of the furnaces, the main technical aspects are identical.

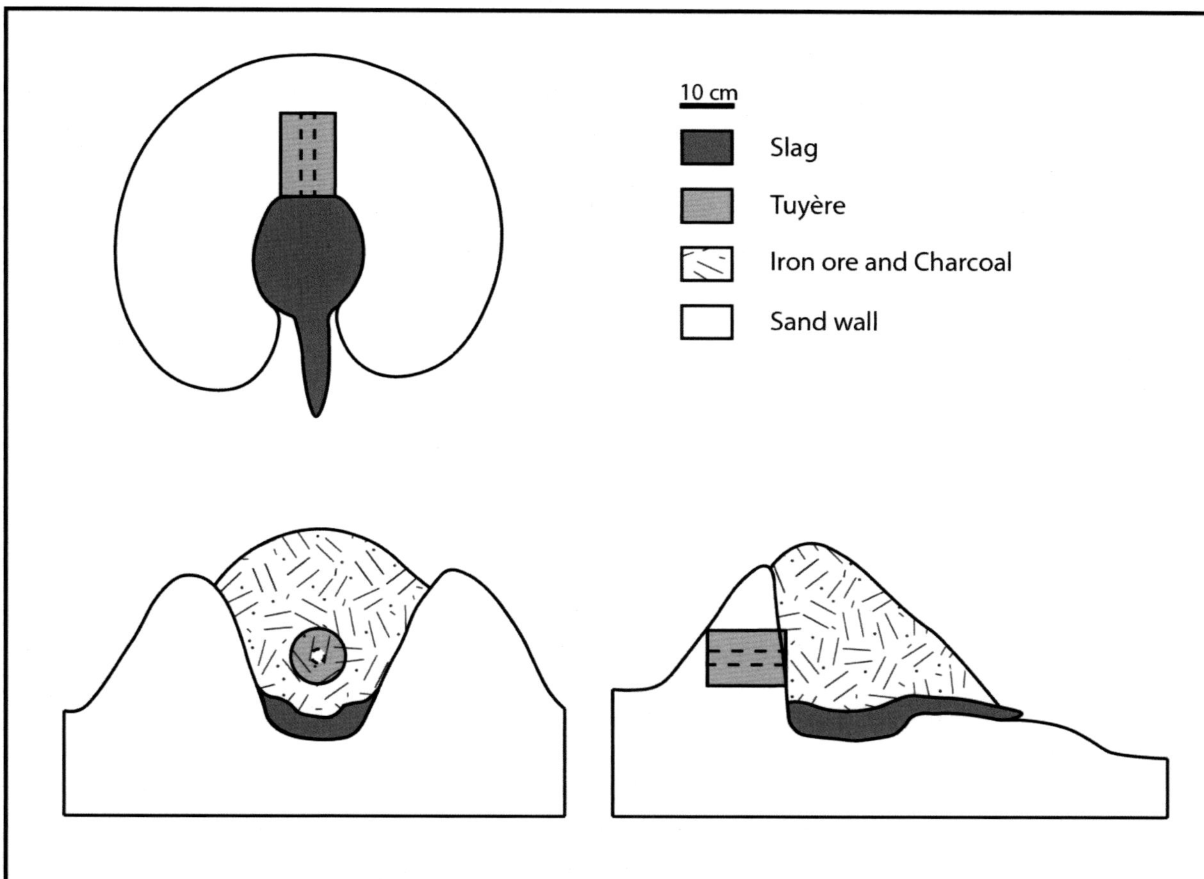

Figure 5.12: *Rasikajy* smelting furnace reconstruction from the study of metallurgical wastes and structure remains.

5.4 A single smelting technical tradition with local variations

An iron smelting technical tradition is defined as a given way to process iron ore into metal, following the same *chaîne opératoire* (Chapter 2). From an archaeological point of view, two sites with identical remains belong to the same technical tradition. The similarities of the remains must be based on the comparison of criteria as numerous as possible and as accurate as possible. A fully investigated site can be characterized by the spatial organisation of the remains, the architecture of the furnace and the assemblage of wastes and related materials. Laboratory examination of the materials can provide additional criteria. In some cases, the remains are indeed identical, but most of the time some variations can be observed. In the latter situation, it is always difficult to evaluate whether the differences are significant or not. Of course, when only a simple surface survey has been carried out, the number and the accuracy of criteria is limited and the interpretation is much weaker.

In the case of iron metallurgy in northeastern Madagascar, the question of the uniqueness of the studied technical tradition arose.

The spatial organisation of the smelting sites show common shared characteristics. The slag heaps are small and have a circular or an elliptical shape. The bigger ones are horseshoe shaped. It is likely that the small circular heaps become progressively elliptical and then horseshoe shaped as the number of wastes becomes larger. Recurrently, but not systematically, TS and BFS are not discharged in the same area.

Various observations on different smelting sites lead to a common reconstruction of the furnace (Figure 5.12). Every excavation revealed a bowl furnace without any well built superstructure or clay lining. Few fragments of indurated sand wall were uncovered but no clay wall. Few stones used as reinforcement in the sand wall were found only at Matavy.

Observations on tuyeres and slag pieces show systematically that only one air inlet was implemented. The single tuyere was linked to bellows. Even if the studied tuyere pieces show a significant variability in the material and the manufacturing process, they are all similar in size and shape.

On all the visited smelting sites, BFS are always found, and TS are also observed in most cases but in variable proportions. Slag pieces from each morphology are similar in shape and aspect and show little variability in size. They are all rich in Fe_2O_{3tot}. The internal variability of the slag is much smaller compared to the variability of slag in the archaeological record.

Hence, the investigated remains show a significant variability as well as numerous common shared criteria. These shared criteria define a unique technical tradition with local variations. It would thus seem that we are dealing with a smelting tradition that is extremely adjustable, not only to the production needs at a given time or in a given village, but also to the raw materials available in the surroundings.

Chemical Variability of Iron Production

6

By studying archaeological remains and metallurgical wastes, we could reconstruct the structure of the furnace. The smelting process can be better understood with an in-depth study of the chemistry of the slag, ores and building material of the furnace. Lateritic ores, in the form of centimetric ferruginous concretions, were used. The slag chemical compositions are highly variable and heavily contaminated by the sandy walls. The sandy substratum can therefore not be neglected and deserves special attention.

Moreover, the important variability of the chemical compositions implies a low standardization of the technical gesture and thus the work of non-specialized workers. The aim of this chapter is also to quantify the production, to assess whether enough iron was produced to meet the needs of the population.

6.1 Materials and Methods

6.1.1 Sampling strategy

Three types of materials will be studied and detailed in this chapter: slag, ore and the sandy building material of the furnaces (Figure 6.1). Samples were collected during the three fieldwork campaigns (2017, 2018, 2019). No samples were recovered from museum collections or previous archaeological excavations.

During the archaeological excavations of a smelting site, the proportion of the different types of waste were recorded (Chapter 4) and detailed macroscopic observations of the metallurgical waste assemblage were carried out. For the bottom furnace slag (BFS), a selection of at least 30 complete pieces were recorded (dimensions, weight, pictures, drawings). For the tapped slag (TS) pieces, only fragments could be observed. These fragments were recorded in the same way as the BFS, although it is not possible to know their original size. The most representative slag pieces were selected and then broken into two halves. One was taken to Switzerland for analysis while the other is stored in the Archaeological Museum Repository of Antananarivo. For the smelting sites where only test pits were carried out, only a few pieces were selected in order to estimate the variability of chemical compositions throughout the district. Most of the sites were only visited and no sampling were done, with the exception of three locations where slag samples were collected (Ambodipont Limite, Ambodipont Isahana, Sambava - See Figure 6.1). We are aware that the sampling in these cases is not representative but rather indicative of a general trend.

		LOI	XRF	XRD	OM	SEM
Amboronala (MBR) *Excavation 2018*	Slag	51	51	51	9	7
	Ore	10	10	6		
	Substratum	7	7	2		
Benavony (BNY) *Excavation 2017*	Slag	26	26	26	9	10
	Ore					
	Substratum	1	1	1		
Matavy (MTY) *Excavation 2017*	Slag	13	13	13	9	6
	Ore	13	13	8		
	Substratum	1	1	1		
Bemanevika (BMK) *Test pits 2018*	Slag	16	16	16	9	2
	Ore	6	6	6		
	Substratum					
Ambodimadiro (DMD) *Test pits 2018*	Slag	8	8	8	4	3
	Ore	3	3	2		
	Substratum					
Ambodipont Limite (APL) *Surveys 2018*	Slag	7	7			1
	Ore	6	6	6		
	Substratum					
Ambodipont Isahana (ISH) *Surveys 2019*	Slag	6	6			2
	Ore	3	3			
	Substratum					
Sambava (ATB) *Surveys 2017*	Slag	2	2	2	1	1
	Ore					
	Substratum					
TOTAL	Slag	149				
	Ore	41				
	Substratum	9				

Figure 6.1: Table with the number of samples studied for each smelting site and the analyses carried out (**LOI**: Loss Of Ignition, **XRF**: X-Ray Fluorescence, **XRD**: X-YRay Diffraction, **OM**: Optical Microscopy, **SEM**: Scanning Electron Microscope).

At Amboronala (MBR140 - 2018), Matavy (MTY11 - 2017) and Ambodipont Isahana (ISH120 - 2019), stocks of iron ore have been identified near the smelting area. These ferruginous concretions were collected. Due to the difficulties to identify the mining sites, the study of ores was mostly done by collecting iron-rich rocks in the surroundings of the archaeological sites in a non-systematic way. In the case of the ferruginous concretions, each sample consists of a batch of 50 grams of pisolites. Samples a and b are from the same large batch separated randomly using a splitter. The aim of the analysis is to obtain a representative mean composition of the ore batch which was loaded in the furnace.

As the study progressed, it became clear that the sandy building materials of the furnace played an important role in the final chemical composition of the slag. Unfortunately, too little substratum was sampled during fieldwork. Sand samples were collected in the immediate vicinity of the excavated furnaces. Some substratum samples were also collected directly on indurated furnace wall fragments. As the wall fragments are several centimetres thick, it is unlikely that chemical elements from inside the hearth contaminated the samples. Because only workshops installed on sandy substratum were excavated, the case of the workshops located on the earthy alterite of the hinterland cannot be investigated.

At the Department of Geosciences of the University of Fribourg, the samples were processed for geochemical, mineralogical and petrographic investigations. After cutting and cleaning, to remove all sediment from the porosity, an amount of 50 to 100 grams of material was powdered with an oscillating grinder in a tungsten carbide vessel to obtain a homogeneous powder with a grain size of 75 μm. In the case of slag pieces, only the internal parts were selected for crushing in order to avoid contamination by the sand which was stuck on the external surfaces.

6.1.2 Global chemistry analytical approach

A global approach for studying the chemistry of slag has been conducted. The successive steps follow a routine protocol of the Geosciences Department of the University of Fribourg, developed by Vincent Serneels over the years. Both bulk chemical compositions (XRF) and mineralogical compositions (XRD) were measured, thermogravimetric measurements (LOI) were conducted. Optical and electronic microscopy studies were also carried out. The processing of the data presented in this chapter was naturally conducted in accordance with the microscopy observations. However, the microscopy data are presented in Chapter 7.

Thermogravimetric analysis: Loss of ignition (LOI)

Loss of ignition (LOI) is a thermogravimetric analysis to measure the mass variation during heating of a powdered sample in an oxidising atmosphere. During the heating process, there is a loss of mass resulting from the release of the volatile substances present in the sample (H_2O, CO_2, S, etc.). At the same time, there is a gain of mass resulting from the oxidation of the reduced elements. In particular, Fe^{2+} is oxidized to Fe^{3+}. If present, metallic iron Fe^0 is also oxidized to Fe^{3+}. Indeed, slag can contain several iron-rich minerals, with different oxidation states. However, inside the furnace, the conditions must be reductive and most of the iron should be normally turned to Fe^{2+} and Fe^{0}.

- During smelting, some of the iron is turned to metal (Fe^{0}). Some of it can be trapped in the slag. Then, during cooling of the slag, the contact between the surface of the liquid and reducing gas or pieces of charcoal can lead to the reduction of iron oxides and the formation of small droplets or small filament of metallic iron.

- Most of the iron is turned to FeO during smelting (Fe^{2+}). This FeO is involved in the formation of the liquid slag. During cooling of the slag, FeO participates in the formation of the main mineralogical phases of the solidified slag: wustite (FeO), fayalite (Fe_2SiO_4) and spinels ($Fe_{1+x}Al_{2-x}O-4$).

- During smelting, all the Fe_2O_3 (Fe^{3+}) should have been reduced to Fe^{2+} and Fe^{0}. This transformation can however be uncompleted if the smelting conditions (temperature and fugacity of CO) are not strong enough or in the presence of more refractory minerals in the ore.

- During cooling of the slag, in oxidising conditions, Fe^{2+} in the liquid slag can oxidise to Fe^{3+}, but the reaction is limited and only takes place at the surface. Solidification then leads to the formation of hematite (Fe_2O_3), magnetite (Fe_3O_4) or Fe^{3+}-rich spinel (ferro-hercynite: $Fe^{2+}Fe^{3+}AlO4$. Fe^{3+} can be incorporated into fayalite but only at a very low amount. Other silicate minerals containing Fe^{3+} can also form in these conditions, in particular minerals from the pyroxene family.
- During cooling, at the solid state but still at elevated temperature, under oxidising conditions, Fe^{0} and Fe^{2+} can be oxidised:

$$Fe^{0} + \frac{1}{2} O_2 \longrightarrow FeO \tag{6.1}$$

$$2\,Fe^{0} + \frac{3}{2} O_2 \longrightarrow Fe_2O_3 \tag{6.2}$$

$$2\,FeO + \frac{1}{2} O_2 \longrightarrow Fe_2O_3 \tag{6.3}$$

Wustite (FeO) in particular can oxidise. The transformation can be partial (incorporation of limited amount of Fe^{3+} in the structure of the wustite), or it can lead to a structural transformation (change into magnetite Fe_3O_4, maghemite $\gamma - Fe_2O_3$ or hematite Fe_2O_3). This transformation is frequent in reheated slag. Slag fragments can be reheated for many reasons, but in particular during bloomsmithing.

- Finally during deposition in the sediment, in an aerobic and humid context, the metallic iron Fe^0 is rapidly altered into rust. A suite of minerals of the iron oxi-hydroxides family forms (Cornell, 2006). Goethite ($FeO(OH)$) is one of the typical alteration products. They frequently replace the previous metal particles but the oxi-hydroxides can also crystallise forming veins in cracks, or filling pores.

$$Fe_2O_3 + H_2O \longrightarrow 2FeO.OH \tag{6.4}$$

Wustite, hematite and magnetite are in general stable under deposition conditions. Fayalite and spinels are also quite stable. It seems however that Fe^{3+} content rises at the scale of many centuries, especially in tropical conditions (*Unpublished preliminary data*, Serneels).

Total elementary iron is measured by XRF but the two oxidation states cannot be differentiated with XRF analysis. However, it is possible to estimate the proportions of Fe^{2+} and Fe^{3+} from the LOI analyses. These estimates are indicative and not quantitative. All samples analysed underwent both XRF and LOI measurements. The thermogravimetric analyses were performed on a LecoTGA701 instrument, using the TGA701 Software (Leco Europe, 2005). The mass of the sample is monitored during the whole process (Figure 6.2).

- The first step is a one hour heat treatment at 110°C to remove residual water (moisture). The sample mass decreases (Figure 6.2- Phase 1).

- The temperature then progressively increases to 900°C during an hour and a half. During this stage, the CO_2 and the water present in the crystalline structure of the minerals are removed. Goethite is completely dehydrated at 600°C and gives Fe_2O_3. The mass loss observed during this stage gives an estimate of the amount of goethite in the sample, and thus an estimate of the amount of Fe^{3+} related to post-depositional alteration (Figure 6.2- Phase 2).

- During the last step, the samples are exposed to temperatures reaching 900°C for 2 hours. The temperature does not rise above 900°C because the fritting of the sample is common at

1000°C. An air flow circulates through the furnace, oxidising the iron. When the samples are exposed to temperatures of 900°C in an oxidising environment, metallic iron and Fe^{2+} are oxidised to Fe^{3+} following reactions 6.1, 6.2 and 6.3. The mass gain during this step corresponds to the oxygen added to the system. The quantity of metal and Fe^{2+} present in the sample can then be deduced (Figure 6.2- Phase 3).

- The loss of mass is also linked to the evacuation of other gases such as CO_2. Moreover, as the temperature rises, charcoal particles entrapped in the slag burn off, which also releases CO_2 and leads to a loss of mass.

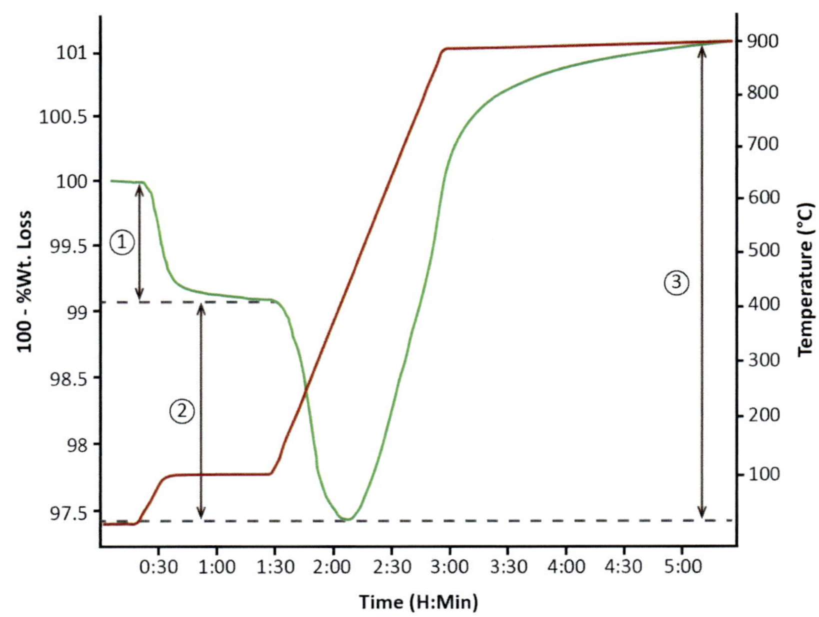

Figure 6.2: Example of a curve obtained from thermogravimetric analysis (sample BMK13001). The red curve represents the variation in temperature and the green curve corresponds to the variation in mass of the sample. The mass loss (1) observed is related to the evaporation of volatile elements and in particular of humidity. The loss of mass (2) is explained by the structural dehydration of some minerals such as goethite ($FeO.OH$). Finally, the gain in mass (3) is due to the oxidation of metallic iron Fe^0 and of FeO (Fe^{2+}) to Fe_2O_3 (Fe^{3+}) and therefore to the incorporation of oxygen into the sample.

Unfortunately, it is not possible to distinguish metal from Fe^{2+} and the quantities of metallic iron must be neglected in the calculation. Although macroscopic observations show that there is a significant amount of metal in the slag, metallic iron was sorted out during milling. Only small quantities of metal should be remaining as XRD data show for very few samples the pattern of ferrite (see Appendix, Figure B.4). The metallic phase must thus be less than 10% of the sample. It can therefore be approximated that:

$$Fe_2O_{3\ total\ XRF\ calculated} = Fe_2O_3 + Fe_2O_{3\ ox-FeO} + Fe_2O_{3\ ox-Fe} \tag{6.5}$$

An example of the full calculation is given in the Appendix (see Appendix B.1), as well as the results for each sample (Figure B.1).

Bulk chemical compositions: X-Rays Fluorescence (XRF)

X-Ray Fluorescence (XRF) analysis provides measurements of the bulk chemical composition of a samples, both for trace elements and for major and minor elements. XRF measurements were performed with a Zetium X-ray wavelength dispersive spectrometer (Malvern-PANalytical, Malvern, UK). Major components (Si, Ti, Al, Fe, Mg, Ca, Na, K and P) were measured using the routine program on fused glass beads made of 0.7 g of calcined powder, 0.35 g of lithium fluoride (LiF) and 6.65 g of lithium tetraborate ($Li_2B_4O_7$). The analytical error is below 1% relative. The program was calibrated using 40 international reference materials (Pfeifer et al., 1992). Trace elements were measured on pressed powder pellets (12 g) using the Pro-Trace analytical software of Malvern-PANalytical (version 6.1., Malvern-PANalytical, Malvern, UK). Out of 40 trace elements measured, we present only the 19 ones with significant results (Ba, Ce, Co, Cr, Cu, Hf, La, Mn, Nb, Nd, Ni, Rb, Sc, Sm, Sr, Th, V, Y, Zn, Zr). The analytical error is below 10% relative. Measurements below 2 ppm are not considered.

Mineralogical compositions: X-Ray Diffraction (XRD)

The main minerals present in a sample can be identified with X-Ray Diffraction (XRD) measurements. Only the crystallised phases are identifiable and therefore not the amorphous glassy phases. XRD measurements were made using an Ultima IV diffractometer (Rigaku, Tokyo, Japan) and the identification software X'Pert Highscore Plus (Malvern-PANalytical, Malvern, UK). Components below 5% are difficult to detect especially since the presence of a glassy phase increase the background effect. No quantitative approach of the mineral phases, with the Rietveld method for example, was undertaken. It is only a qualitative estimation based on the relative size of the peaks.

6.2 Relationship between slag, ore and sandy building material of the furnace

6.2.1 Slag

The slags from northeastern Madagascar are fayalitic slag and the other main minerals are aluminous spinels, titaniferous spinels and wustites. Pyroxenoids and feldspathoids such as leucite ($KAlSi_2O_6$) are also found in smaller quantities. Slag chemical composition is extremely variable. However, there is no significant difference between the chemical compositions of tapped slag and the bottom furnace slag pieces (Figure 6.4). Both morphologies show the same range of compositions. This observation applies to the whole study area but also to each smelting site.

Iron content in slag

Iron oxides content is systematically high, with an average value of 64 wt.% Fe_2O_{3tot} for all the studied slag samples (Figure 6.5). The average values for Matavy are slightly higher (68 wt.% Fe_2O_{3tot}) while those for Benavony (62 wt.% Fe_2O_{3tot}) and Bemanevika (61 wt.% Fe_2O_{3tot}) are slightly lower. The variability between the smelting sites is therefore negligible. On the other hand, the iron oxides content ranges from 55 to 75 wt.% Fe_2O_{3tot}, if outliers (4 samples) are not taken into account (Figure 6.4). There is therefore a very high variability in the quality of the smelting process from one smelt to the other.

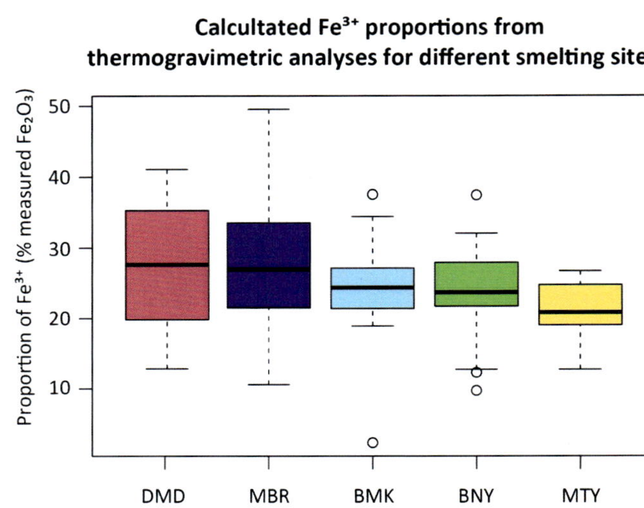

Figure 6.3: Boxplot illustrating the variability of Fe^{3+} proportions in slag from northeastern Madagascar.

The heterogeneity of the process is also supported by the study of iron oxidation states through thermogravimetric analyses (Figure 6.3 and Figure B.1). Iron located in the alteration products is detected in variable proportions in almost all the samples. It represents between 0 and 48% of the measured iron. This is consistent with the X-Rays spectra on which goethite ($\alpha - FeO(OH)$), lepidocrocite ($\gamma - FeO(OH)$) and akaganeite ($Fe^{3+}O(OH,Cl)$) peaks have been identified. The high proportion of iron from post-depositional weathering is not surprising in a warm and humid tropical climate. But this mainly demonstrates the high proportion of metallic iron originally in the slag, goethite being mainly the alteration product of metallic iron (6.4).

Iron under the Fe^{3+} oxidation state represents on average 25% of the measured iron. The proportion of Fe^{3+} in the slag from the north east of Madagascar is very high. There is no significant difference between samples from different smelting sites nor between the two slag morphologies. Moreover, the variability is very important since the proportion of Fe^{3+} varies between 2 and 50% of the measured iron (Figure 6.3). The presence of such high proportions of iron in its most oxidised (Fe^{3+}) state suggests that the reducing conditions in the furnace were not optimal. They were also very

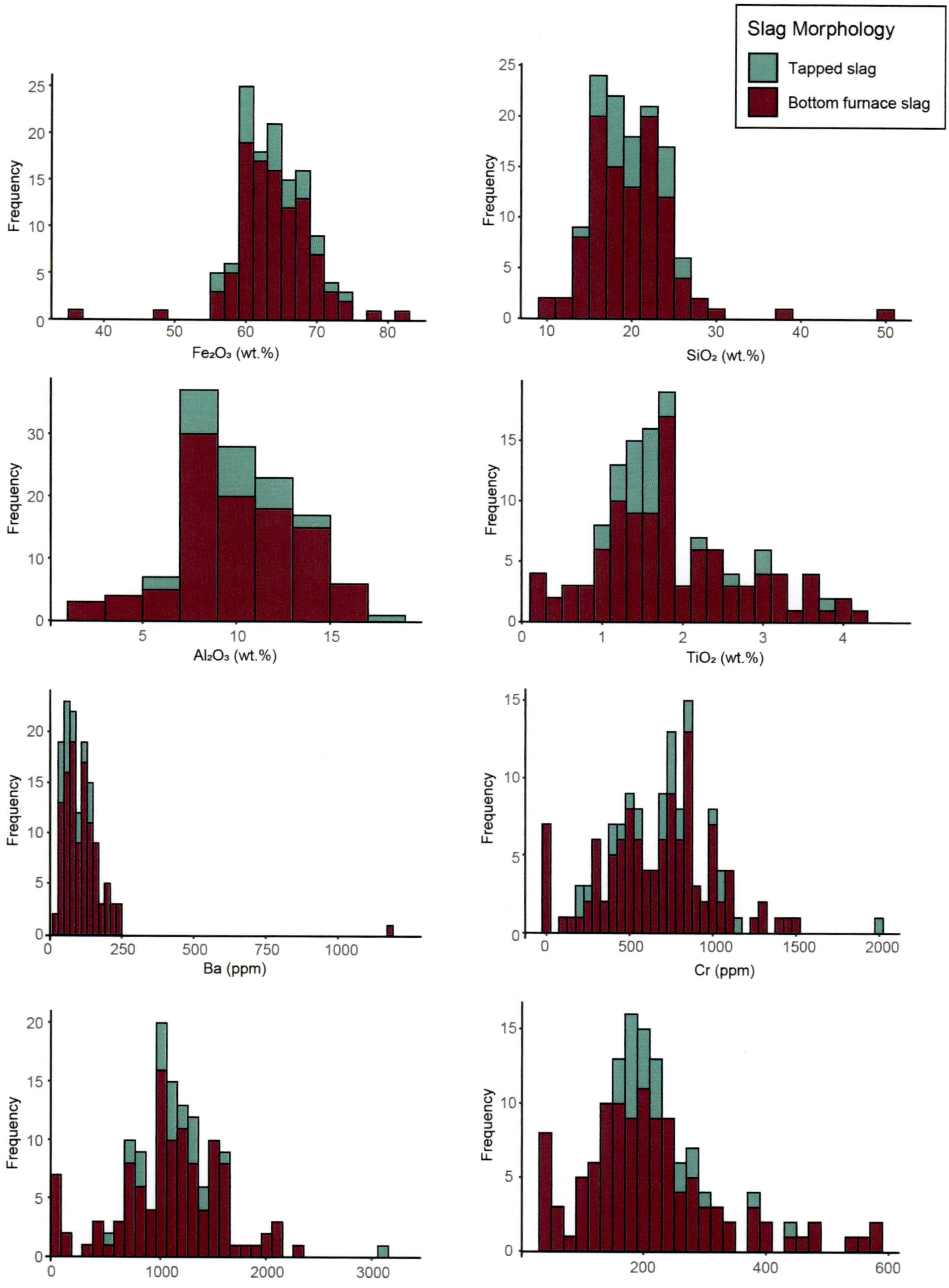

Figure 6.4: Histograms illustrating the chemical variability of all the studied slag pieces according to their morphology (n=129).

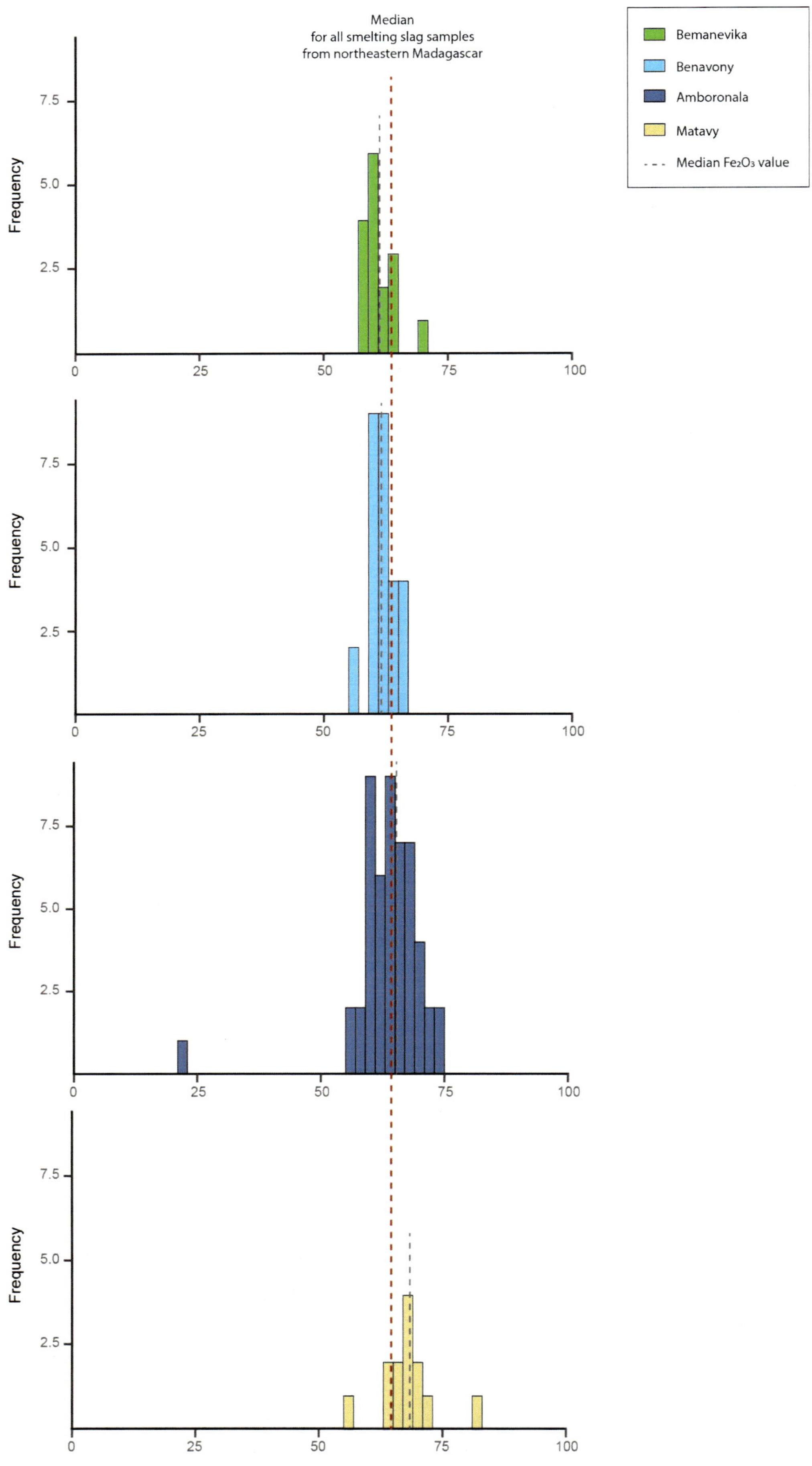

Figure 6.5: Histograms illustrating the variability of the total iron content in slag from smelting sites (Bemanevika, Benavony, Amboronala and Matavy). The red dotted line represents the median iron content for all slag samples analysed. The grey dotted lines represent the median iron content for each smelting site.

variable from one smelt to the other. This variability is found throughout the study area.

Chemical variability of slag

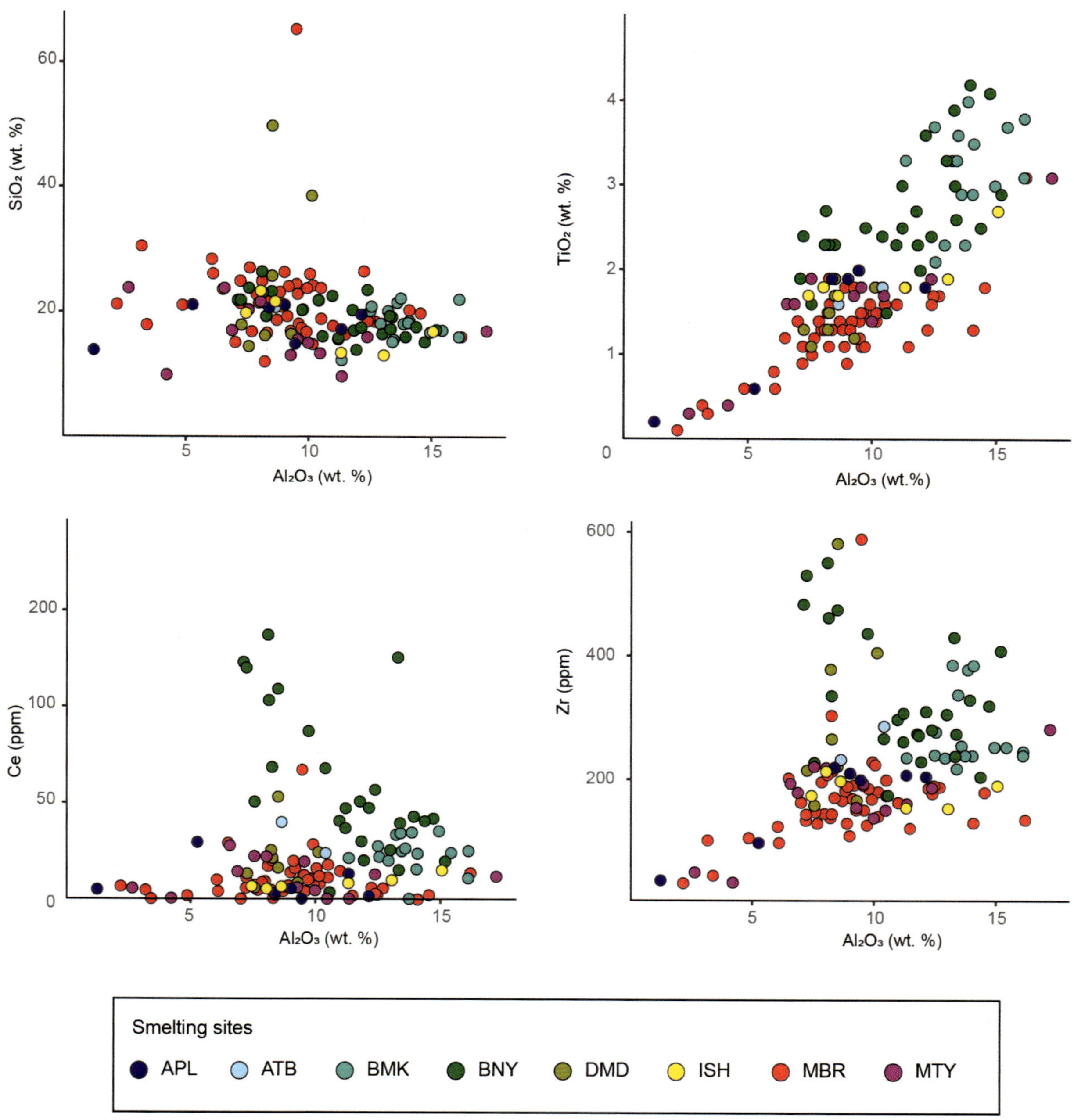

Figure 6.6: Bivariate diagrams $Al_2O_3 - SiO_2$, $Al_2O_3 - TiO_2$, $Al_2O_3 - Ce$ and $Al_2O_3 - Zr$ for the bulk compositions of slags from northeastern Madagascar.

Because of the high variability of the iron oxides content, the slag from northeastern Madagascar also has the characteristic of having extremely variable chemical compositions (Figure 6.4), especially the silica content (on average 20wt.% SiO_2) which varies between 10 and 30 wt.% SiO_2 if we exclude the outliers. The ratio between alumina (1.2–17.2 wt.% Al_2O_3) and silica is highly variable from one sample to another, even coming from the same slag heap. This is due to the highly variable quantity of silica found in the slag.

Titanium oxide content is remarkably high, between 0.1 and 4.9 wt.% TiO_2. Oxides of calcium (0.2–3.0 wt.% CaO) and potassium (0–1.5 wt.% K_2O) are important minor components. No correlation can be seen between these two elements. The most abundant trace elements are vanadium (V: 600–2200 ppm) and chromium (Cr: 250–1200 ppm). Manganese shows a strong variability (Mn: 100– 3000 ppm). There are small quantities of barium, rubidium, strontium and zirconium.

Despite the wide variability in composition across the district, trends can be seen at the scale of each smelting site. The variability of chemical compositions between smelting sites reflects both the variability of the ores and of the sandy substratum used as building materials. The furnaces of Benavony, for example, are built on substratum enriched in heavy minerals, such as ilmenite ($TiFeO_3$), monazite ($(REE)PO_4$) or zircon ($ZrSiO_4$) (Schreurs and Rakotoarisoa, 2011). As a consequence, the Benavony slags tend to have high TiO_2, REE and Zr content (Figure 6.6). The study of sandy substratum can therefore not be neglected.

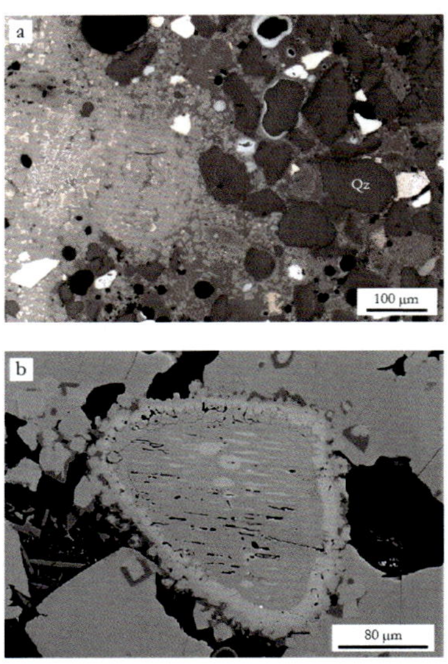

Figure 6.7: (a) Metallographic microscope image of quartz grains trapped in a slag (BNY44004). **(b)** Backscattered electron SEM image of an ilmenite grain trapped in a slag (BNY41005).

6.2.2 The sandy substratum as a building material for the furnace

Iron smelting techniques documented in other archaeological contexts rarely show such an important contamination of the slag from the building materials. The wall usually only slightly affects the chemical composition of the slag. Clay walls melt very little and their chemical contribution is, most of the time, insignificant. On the contrary, in the case of the *Rasikajy* technique, the sandy substratum impacts significantly the chemical composition of the slag, both for major and traces elements. As described previously (Chapter 3), the furnaces were dug directly into the sandy substratum and a 30 cm high sand wall was built to support the tuyere. No solid wall fragments were discovered. The walls are weakly indurated by the firing process and the loose sand can mix easily with the slag. Macroscopic and microscopic observations of slag cross-sections show that the slag contains many unfused quartz grains (Figure 6.7.a). In the case of slag heaps located on a heavy mineral rich substratum, as in Benavony, unmelted ilmenite grains can be entrapped into the slag (Figure 6.7.b). Depending on the location along the coast as well as on the distance from the coast, the substratum composition can vary. The variability in the chemical composition of the slag reflects the chemical variability of the furnace construction materials.

6.2.3 Iron ore

Black sands

Sample		SBV610	SBV611	SBV620	SBV621	SBV630	SBV631	BMK610	BMK611	SBV650	SBV651	SBV652
		Before panning	After panning	Before panning	After panning	Before panning	After panning	Before panning	After panning	Before panning	After panning	Sables noirs
SiO_2	(wt.%)	46,02	17,88	52,73	21,04	37,91	27,61	52,35	30,40	38,27	26,43	15,60
TiO_2	(wt.%)	11,60	17,90	9,40	18,80	13,80	17,00	5,90	9,50	14,80	18,30	22,20
Al_2O_3	(wt.%)	4,97	3,67	5,33	4,84	4,80	4,34	4,76	4,63	4,83	4,58	3,72
Fe_2O_3	(wt.%)	29,10	42,90	24,20	45,50	34,30	42,00	30,10	48,60	34,90	42,50	50,30
MgO	(wt.%)	1,77	1,78	1,85	2,51	1,98	2,03	1,47	2,28	1,73	1,93	1,69
MnO	(wt.%)	0,38	0,55	0,33	0,62	0,46	0,55	0,35	0,59	0,46	0,56	0,64
CaO	(wt.%)	1,70	1,40	1,90	2,00	1,90	1,80	1,70	2,40	1,30	1,40	1,10
Na_2O	(wt.%)	0,93	0,54	0,91	0,58	1,06	0,69	0,86	0,53	0,78	0,52	0,70
K_2O	(wt.%)	0,83	0,23	0,97	0,26	0,67	0,42	1,11	0,52	0,68	0,40	0,21
P_2O_5	(wt.%)	0,28	0,40	0,23	0,39	0,31	0,36	0,07	0,09	0,34	0,41	0,49
SUM	(wt.%)	98,64	89,00	98,80	98,16	98,55	98,38	99,06	99,94	99,23	98,44	98,26
Ba	(ppm)	346	335	351	310	322	324	367	224	350	330	366
Ce	(ppm)	1564	2895	1217	2369	1739	2112	211	254	2133	2704	3291
Co	(ppm)	133	117	157	118	126	99	167	115	111	129	99
Cr	(ppm)	236	401	212	417	297	361	105	158	285	370	436
Cu	(ppm)	6	10	5	11	8	10	3	8	7	7	11
Hf	(ppm)	125	218	110	215	195	236	24	29	113	138	164
La	(ppm)	788	1487	633	1206	884	1067	113	139	1092	1393	1641
Nb	(ppm)	189	327	160	316	231	290	60	93	235	296	357
Nd	(ppm)	600	1106	470	909	676	816	87	104	810	1028	1245
Ni	(ppm)	27	36	25	38	29	33	22	31	27	33	33
Rb	(ppm)	19	6	21	6	15	10	24	9	15	9	5
Sc	(ppm)	23	38	22	42	27	34	15	28	28	36	38
Se	(ppm)	1	0	0	1	1	1	0	0	1	1	1
Sm	(ppm)	85	163	69	137	107	124	9	15	116	146	175
Sn	(ppm)	11	20	8	17	13	17	7	9	13	15	17
Sr	(ppm)	75	31	79	35	69	51	74	42	45	33	19
Th	(ppm)	478	817	363	686	508	610	63	76	645	789	958
V	(ppm)	439	790	364	734	524	650	457	760	584	749	897
W	(ppm)	863	621	956	627	707	528	1072	626	664	747	526
Y	(ppm)	162	261	143	269	217	257	51	73	170	214	238
Zn	(ppm)	174	305	148	304	215	269	120	210	211	282	330
Zr	(ppm)	4461	7865	4099	7789	7030	8249	984	1166	4046	5076	5837
$SiO_2:Al_2O_3$		9,26	4,87	9,89	4,35	7,90	6,36	10,99	6,57	7,92	5,77	4,20
$SiO_2:TiO_2$		3,97	1,00	5,61	1,12	2,75	1,62	8,87	3,20	2,59	1,44	0,70
V:Cr		1,9	2,0	1,7	1,8	1,8	1,8	4,4	4,8	2,0	2,0	2,1

Figure 6.8: Bulk chemical compositions of the black sands collected on the shore of Benavony, Sambava and Bemanevika (see Appendix Figure B.5 for more detailed data).

At the beginning of our study, the use of black sands as iron ore was the privileged hypothesis. The former literature (Vérin, 1986) assumed that black sands, which are available all along the coast, were smelted. There is indeed a similar spatial distribution of the black sands reported on geological maps (Schreurs and Rakotoarisoa, 2011) and of the formerly recorded metallurgical waste. The sands of the beaches on the northeastern coast of Madagascar derive from the erosion of the granites and gneisses of the Precambrian basement of the hinterland. The numerous coastal rivers carry sediments that are deposited along the coast. During the Quaternary era, the sediments are reworked by waves, tides, longshore currents, and wind, which are effective mechanisms for sorting the mineral grains on the basis of differences in their mass, that is the combination of size and density (Van Gosen et al., 2014). Gradually, the heavy minerals are concentrated. Magnetite (Fe_3O_4) and ilmenite ($TiFeO_3$) accumulated together with other heavy minerals such as zircon ($ZrSiO_4$), garnet ($Ca_3[Cr, Al, Fe]_2(SiO_4)_3$) or monazite ($(Ce, La, Y, Th)PO_4$). Locally, these minerals form laminated or lens-shaped, heavy-mineral-rich sedimentary layers. In favourable topographic situations, these layers are superposed over a thickness of a few metres and form bands of a few hundred or thousands of metres in length (Figure 6.9). These black sands are very easy to extract and wash out. With a basic panning method, a concentrate containing all the heavy minerals can be produced quickly. However, this simple washing does not separate the iron-rich minerals from the other heavy grains. In the concentrate, magnetite is still abundant but it is linked with a high proportion of other minerals such as ilmenite. The iron content of the concentrate is therefore variable. Ethnographical and archaeological records mention the use of black sands as iron ores in some other smelting traditions such as in Africa (Killick and Miller, 2014; Ige and Rehren, 2003; David et al., 1989), in India (Buchanan, 1807), or in southeastern Asia (Park and Rehren, 2011; Chen, 2000). However, in most of these cases, black sands are alluvial or colluvial sands and not beach sands.

Figure 6.9: (a) Stratigraphy of the beach sands near Benavony: successive layers of black sands can be seen. **(b)** Black sands strips on the beach close to Benavony.

Black sand samples were thus collected at several locations along the coast. For each sampling location, a non-washed sample was analysed, as well as several samples washed with panning techniques, until only black minerals remained visible to the naked eye (Figure 6.10). The evolution of chemical compositions before and after panning could then be measured, in particular the enrichment in iron. The black sands samples before panning have extremely variable content depending on whether the sample was taken from a strip of sand already concentrated in heavy minerals or not (Figure 6.8 and Appendix, Figure B.2).

Figure 6.10: Black sands samples (SBV640) before **(a)** and after **(b)** the panning process. An enrichment in dark heavy minerals can be observed.

The iron oxides content of the black sands before panning varies between 7 and 49 wt.% Fe_2O_{3tot}. It is difficult to enrich these black sands to more than 60 wt.% Fe_2O_{3tot}. They are thus relatively poor potential ores with low iron oxides content. Titanium oxides grades are between 2 and 24 wt.% TiO_2. There is a clear enrichment of iron oxide and titanium oxide content after the panning stage, which corresponds respectively to an enrichment of magnetite or hematite, and ilmenite or rutile (TiO_2). MgO content increases as well during the panning process, which is due to the presence of ferro-magnesian minerals such as garnets and amphiboles. The content of some trace elements also increases during the panning process. Zirconium (Zr) content can reach 8500 ppm and hafnium (Hf) content slightly increases due to the presence of zircons ($(Zr,Hf,U)SiO_4$). Rare earth elements (Ce, La, Nd, Sm, Y) and P_2O_5 content also increases, which confirms the presence of monazite ($(REE(PO_4))$. Chromium (Cr) content also increases because chromium can be a substitute element in magnetite. Finally, niobium (Nb) content increases as well because it can substitute titanium in minerals such as ilmenite or rutile.

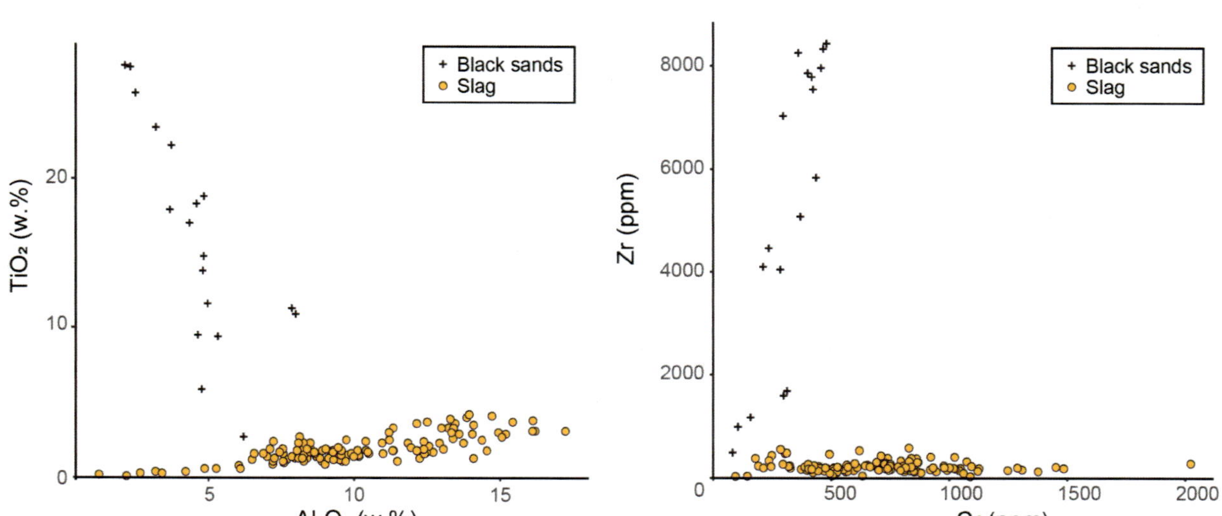

Figure 6.11: Bivariate diagrams $Al_2O_3 - TiO_2$ and $Cr - Zr$ comparing all smelting slag samples and all black sands samples compositions for northeastern Madagascar.

At the same time, a decrease in silica and potassium oxides content can be noticed, corresponding to the elimination of light minerals such as quartz, feldspars ($(K,Na,Ca)AlSi_3O_8$) and muscovite ($KAl_2(AlSi_3O_{10})(OH)_2$). SiO_2 indeed varies between 5% and 74%, depending on the proportion of quartz (SiO_2) in the sample. Alumina content also slightly decreases and varies between 4 and 6 wt.% Al_2O_3. Some of the alumina is probably flushed with clays, which would explain the decrease in strontium (Sr) and barium (Ba).

The comparison between the black sands and the smelting slag clearly shows huge differences. The comparison of the ratios of the elements present in the heavy minerals specific to black sands, such as titanium or zirconium, exclude that the black sands were used as iron ore (Figure 6.11). Moreover, it is difficult to enrich these black sands above 60 wt.% Fe_2O_{3tot} whereas slags iron content is on average 65 wt.% Fe_2O_{3tot}. Indeed, the iron rich minerals (magnetite and ilmenite) cannot be separated from the other heavy minerals and black sands cannot be strongly enriched in iron oxides.

Figure 6.12: (a) Large stone used as an anvil to crush ore in MBR140 (Amboronala).**(b)** Spread of ferruginous concretions south of slag heap MBR142 (Amboronala). **(c)** Details of the ferruginous concretions found south of slag heap MBR142.

Lateritic ores

Much archaeological evidence supports the use of lateritic ore by the ancient metallurgists in the northeast of Madagascar. Spreads of small ferruginous pisolitic concretions, with a diameter ranging between 0.5 to 2cm, were found close to slag heaps on several smelting sites, such as Matavy and Amboronala (Figure 6.12 b-c). These concretions cannot be formed naturally in the Quaternary sand dunes and must have been brought by the metallurgists to the place where they were found. Fragments of similar pisolites were also observed, partially melted, inside several bottom furnace slags. Finally, a quadrangular stone block (35x40x20 cm) was found near the MBR140 clusters. As Amboronala is located along a dune, there is no doubt that this stone was intentionally brought to the workshop by humans. On the two main faces, the block bears very clear traces of percussion and several large chips have been detached on the edges (Figure 6.12 a). All around, in a radius of more than 50 cm, the ground is strewn with small pieces of ferruginous concretions. These small fragments of ore are mixed with orange sand to a thickness of about 5 cm. The stone was clearly used for crushing the ore. This observation is consistent with what is described in the written sources for the later periods (Chapter 1).

Iron oxides and oxi-hydroxides concretions are frequently formed during the process of surface alteration of rocks. In a warm and humid tropical climate, superficial rocks undergo significant hydrolysis of the silicate minerals. The various cations are leached according to their ionic potential (Railsback, 2004; Goldschmidt, 1937). Aluminium and ferritic iron precipitate as oxi-hydroxides (goethite $FeO.OH$, gibbsite $Al(OH)_3$). The result is the formation of a laterite, typically rich in kaolinitic clay ($Al_2O_3.2SiO_2$) and Fe- and Al-oxides (Tardy, 1992). Typically, the less mobile minor and trace elements (Ti, Zr, Mn, V, Cr) are also enriched. In situ laterites are generally well layered, reflecting the percolation of water. A duricrust can form near the surface. Frequently the products of the laterization process, in particular the iron-rich concretions, are reworked by erosion and redeposited.

In our study area, Cretaceous basaltic lava flows, related to the separation between Madagascar and the Indian Plate, crop out at the eastern fringe of the Proterozoic basement, forming the first range of hills inland (Melluso et al., 2005). These could be ideal source rocks for the formation of an iron-rich lateritic cover containing pisolitic concretions. There are few analyses of basalts available in the literature for northeastern Madagascar (Melluso et al., 1997; Storey et al., 1997). No basalt was analysed during this thesis, although it would have been interesting to better understand the geological formation of the ores. Because of the small number of samples and the significant variability of chemical compositions across the study area, it is not possible to confirm that concretions are formed from the alteration of basaltic rocks, but this remains a credible hypothesis. However, the available data on basalts highlights the presence of approximately 2 wt.% TiO_2, which would explain the high titanium content in the ores.

Figure 6.13: Example of a large lateritic block observed in the area close to Bemanevika, not far from the mouth of the Bemarivo River mouth.

No mine or ore extraction area was found or clearly identified during the fieldwork. It could be because pisolites were collected by surface picking. Indeed, concentrations of these pisolites can be observed at the bottom of slopes. During the rainy season, the soil is washed away and the pisolites accumulate. It is then easy to collect them directly at the surface, or possibly by digging holes a few metres deep. At Bemanevika, on a hill called *Antaimby* where large lateritic blocks were observed, shallow pits with a diameter of 2-4 metres were noted, but not investigated in more detail due to lack of time (Figure 6.13). These could be a former ore extraction areas. This would be consistent with 19th century written sources which explain that iron ore was collected in 2 to 4 metre deep pits (Breton, 1898; Hastie, 1817). They also emphasise the importance of settling near a water stream in order to wash these concretions. Surface collection is also described in Kenya for similar ores (Brown, 1995).

The ferruginous concretions sampled show a brown to red shiny surface. Some are magnetic and others are not, without this being linked to a difference in chemical composition. Pisolites were found next to slag heaps only at Matavy (MTY), Amboronala (MBR) and

Ampandrafetana (ISH). For the other sites, we collected pisolites or lateritic blocks directly on outcrops. XRD spectra show peaks of iron oxides, hematite and goethite, as the main components, together with small amounts of kaolinite and quartz (Appendix, Figure B.3). Samples from Bemanevika and Ambodipont Limite also show peaks of gibbsite. The samples have a very high iron oxides grade, especially for the samples collected at smelting sites (average grade: 82.6 wt.% Fe_2O_{3tot})(Figure 6.14). All the samples have remarkably high titanium oxides content. Some of the Bemanevika samples have titanium oxides content up to 3.8 wt.% TiO_2. Magnesium, calcium, potassium and sodium oxides are almost absent, as expected in a laterite. Chromium (Cr) and vanadium (V) are the most abundant trace elements, probably replacing iron in the lattice of the iron-rich minerals. Their concentrations are in the range of 1000 ppm and the ratio Cr:V varies between 1:3 and 1:1. Manganese (Mn: 120ppm-2.3 wt.%MnO) fluctuates strongly from one sample to the other. Copper (Cu: 100-200 ppm) and zirconium (Zr: 300 ppm) can also be noticed.

Sample		MBR 14501	MBR 14502	DMD 501	DMD 502	BMK510	BMK512	MTY501	MTY508a	APL 501	APL 502	APL 503
		Pisolites	Pisolites	Pisolites	Pisolites	Pisolites	Lateritic block	Lateritic block	Pisolites	Pisolites	Pisolites	Lateritic block
SiO_2	(wt.%)	6,68	8,42	16,28	15,62	6,14	3,85	3,98	1,56	5,24	4,50	6,84
TiO_2	(wt.%)	1,30	1,00	1,60	1,60	2,10	2,30	1,20	1,20	1,10	1,20	1,10
Al_2O_3	(wt.%)	7,50	5,63	12,43	12,95	10,48	26,37	12,97	8,66	12,07	8,43	9,84
Fe_2O_3	(wt.%)	83,50	82,80	69,50	68,90	77,90	64,40	78,20	85,70	77,40	80,40	76,90
MgO	(wt.%)	0,04	0,04	0,05	0,05	0,15	0,05	0,05	0,06	0,07	0,06	0,07
MnO	(wt.%)	0,03	0,03	0,09	0,09	0,15	0,28	0,10	0,09	0,13	0,05	0,26
CaO	(wt.%)	0,00	0,00	0,00	0,00	0,20	0,10	0,00	0,00	0,10	0,10	0,10
Na_2O	(wt.%)	0,04	0,05	0,04	0,05	0,10	0,02	0,02	0,34	0,00	0,00	0,00
K2O	(wt.%)	0,02	0,02	0,01	0,01	0,13	0,01	0,01	0,01	0,02	0,02	0,02
P_2O_5	(wt.%)	0,21	0,30	0,10	0,08	0,46	0,36	0,47	0,33	0,36	0,37	0,57
SUM	(wt.%)	99,52	98,48	100,47	99,69	98,14	98,21	97,21	98,34	96,78	95,41	96,08
Ba	(ppm)	20	23	31	34	44	55	37	25	65	27	75
Ce	(ppm)	0	1	13	13	16	0	7	0	0	0	2
Co	(ppm)	8	10	39	38	105	49	45	55	12	11	141
Cr	(ppm)	768	871	1737	1471	549	2264	477	1570	1449	1524	1806
Cu	(ppm)	88	91	125	115	146	282	112	117	213	176	163
La	(ppm)	12	10	6	8	18	14	7	14	8	10	11
Ni	(ppm)	30	33	50	52	89	46	76	67	32	28	95
Rb	(ppm)	0	0	0	0	3	0	0	0	0	0	0
Sr	(ppm)	1	1	1	1	13	5	1	1	2	2	2
V	(ppm)	898	844	1257	1346	1833	1519	988	1809	916	848	1500
Y	(ppm)	5	4	6	5	10	5	3	3	3	2	4
Zr	(ppm)	141	100	211	173	122	229	76	100	138	128	103
SiO_2:Al_2O_3		0,9	1,5	1,3	1,2	0,6	0,1	0,3	0,2	0,4	0,5	0,7
SiO_2:TiO_2		5,1	8,4	10,2	9,8	2,9	1,7	3,3	1,3	4,8	3,8	6,2
V:Cr		1,2	1,0	0,7	0,9	3,3	0,7	2,1	1,2	0,6	0,6	0,8

Figure 6.14: Bulk chemical compositions of some of the collected iron ore samples (see Appendix Figure B.6 for more detailed data).

6.2.4 Relationship between slag and ore

Amboronala (MBR)

Sample		MBR 14001	MBR 14002	MBR 14003	MBR 14004	MBR 14005	MBR 14006	MBR 14007	MBR 14009	MBR 14010	MBR 14011	MBR 14012	MBR 14902	Average Ore
		BFS	BFS	TS	TS	TS	TS	TS	BFS	BFS	BFS	BFS	Substratum	Pisolites
SiO_2	(wt.%)	65,34	26,07	26,52	18,61	23,84	18,13	18,76	26,38	16,47	21,19	24,46	78,61	8,43
TiO_2	(wt.%)	2,00	1,50	1,30	1,60	1,60	1,10	1,40	0,90	1,10	0,10	1,20	0,20	1,16
Al_2O_3	(wt.%)	9,47	10,05	12,24	12,40	10,50	9,59	8,70	9,00	11,48	2,18	9,50	10,75	7,73
Fe_2O_3	(wt.%)	21,60	58,50	55,30	64,00	58,10	67,00	66,10	59,90	67,40	71,20	61,40	2,50	80,92
MgO	(wt.%)	0,15	0,27	0,33	0,34	0,33	0,19	0,34	0,33	0,22	0,88	0,15	0,20	0,07
MnO	(wt.%)	0,04	0,08	0,05	0,13	0,10	0,06	0,14	0,02	0,02	0,01	0,02	0,02	0,04
CaO	(wt.%)	0,40	0,70	0,80	0,70	2,10	1,10	1,50	0,80	0,50	1,30	0,40	0,70	0,06
Na_2O	(wt.%)	0,17	0,35	0,39	0,17	0,58	0,29	0,28	1,69	0,20	0,40	0,26	1,93	0,06
K_2O	(wt.%)	0,21	0,63	0,94	0,20	0,40	0,29	0,35	1,05	0,37	1,23	0,42	4,06	0,04
P_2O_5	(wt.%)	0,12	0,25	0,37	0,33	0,36	0,34	0,39	0,24	0,20	0,19	0,22	0,05	0,24
SUM	(wt.%)	99,76	98,59	98,54	98,72	98,20	98,37	98,18	100,51	98,26	98,79	98,23	99,15	98,96
Ba	(ppm)	61	130	149	129	140	102	147	228	74	189	87	902	29
Ce	(ppm)	67	16	6	3	11	4	4	14	2	6	12	15	4
Co	(ppm)	68	26	28	16	30	21	23	28	26	46	20	37	15
Cr	(ppm)	469	447	740	732	849	1043	552	423	853	94	449	49	803
Cu	(ppm)	39	61	77	88	41	91	85	52	84	297	56	1	110
La	(ppm)	45	14	14	6	12	15	11	11	13	5	15	10	12
Ni	(ppm)	22	9	6	7	4	12	7	5	16	83	9	5	33
Rb	(ppm)	4	13	17	4	8	5	7	25	8	26	9	100	1
Sr	(ppm)	32	71	88	43	135	56	97	95	34	116	29	166	4
V	(ppm)	581	767	1426	1109	1272	1406	1072	798	1498	105	847	67	824
Y	(ppm)	12	10	7	8	6	5	5	6	5	2	6	5	7
Zr	(ppm)	589	224	186	177	199	150	166	108	121	31	194	88	119
SiO_2:Al_2O_3		6,9	2,6	2,2	1,5	2,3	1,9	2,2	2,9	1,4	9,7	2,6	7,3	1,1
SiO_2:TiO_2		32,7	17,4	20,4	11,6	14,9	16,5	13,4	29,3	15,0	211,9	20,4	393,0	7,3
V:Cr		1,2	1,7	1,9	1,5	1,5	1,3	1,9	1,9	1,8	1,1	1,9	1,4	1,0

Figure 6.15: Bulk chemical compositions of smelting slag, sandy substratum and iron ore (average value) collected at the MBR140 workshop (Amboronala). See Appendix B for full chemical analysis.

The site of Amboronala was studied in detail and in particular the workshop MBR140 (Morel and Serneels, 2021). This very small workshop was extensively excavated (Serneels et al., 2019b). The slag from MBR140 is characterised by high iron oxides content, but the amount is quite variable, ranging from 55 to 67 wt.% Fe_2O_{3tot} (Figure 6.15). The ratio between alumina (8.7–12.4 wt.% Al_2O_3) and silica (16.5–26.5 wt.% SiO_2) is highly variable from one sample to another (Figure 6.16). Titanium oxide content is around 1.3% in average. Oxides of calcium (0.4–2.1 wt.% CaO) and potassium (0.2–1.1 wt.% $K2_O$) are also important components. No correlation can be seen between these two elements. The most abundant trace elements are vanadium (V: 750–1500 ppm) and chromium (Cr: 400–1000 ppm). The grades are variable, but the ratios are similar. Manganese shows a strong variability (Mn: 120– 1200 ppm). There are small quantities of barium (Ba), rubidium (Rb) and strontium (Sr).

Because a stockpile of iron concretions was found next to the slag heaps, there is little doubt they were used as ore. The analyses of the pisolites are very homogeneous. The samples have a particularly high iron oxides grade, with an average of 80.92 wt.% Fe_2O_{3tot} (Figure 6.15). The ratio of SiO_2:Al_2O_3 is close to 1:1, which is consistent with the presence of kaolinite. There is a significant amount of titanium oxide, averaging at 1.2 wt.% TiO_2. The most abundant traces elements are chromium (Cr) and vanadium (V). Manganese (Mn: 250 ppm), copper (Cu: 100 ppm) and zirconium (Zr: 100 ppm) are noticeable. In contrast to the sand, barium (Ba), rubidium (Rb) and strontium (Sr) concentrations are very low. Pisolites were also sampled in the surroundings of Amboronala during our surveys, near Ambodimadiro (DMD501 and DMD502, see Figure 6.14). These two samples are not directly associated to an archaeological context. They have iron oxides content around 70 wt.% Fe_2O_{3tot}. These content is low compared to the pieces found at the workshop next to MBR140. On the other hand, the content of the other minor or trace elements is relatively compatible with those of the archaeological ores. They probably come from nearby deposits but the metallurgists knew how to recognise the ores with the best grades and collected them. DMD501 and DMD502 are thus excluded from our dataset for mass balance calculation.

To estimate the contamination by the building materials, a sample of sand was collected next to the slag heaps. During the excavation, no horizontal or sub-surface heterogeneity was detected. We will therefore consider this sample (MBR14901, MBR14902, MBR14903) as a representative indication of the chemical composition of the substratum. It would have been relevant to collect more substratum samples to better understand the variability of the sand at the workshop scale. The X-ray diffraction spectrum shows dominant quartz alongside significant plagioclase and alkali feldspar. The peaks of the muscovite-illite are visible, probably formed as alteration products on the feldspars. The bulk chemical composition shows a high silica content: 78 wt.% SiO_2 (Figure 6.15). Alumina and potassium, sodium and calcium oxides are the other important contributors and reflect the presence of the feldspars. The trace element concentrations are quite low. The most abundant are barium (Ba), rubidium (Rb) and strontium (Sr), typical for feldspars. Titanium (Ti), manganese (Mn), chromium (Cr), vanadium (V) and zirconium (Zr) are low, attesting to the absence of heavy minerals in the sand.

There is no chemical element that is both present in the slag and ore and completely absent from the sandy substratum. Only titanium oxide TiO_2 is present in low concentrations in the sands of Amboronala (0.20 wt.% TiO_2). Alumina is present in similar concentrations in the slag and the substratum. The Al_2O_3:TiO_2 ratio is similar between the slag and the ore (Figure 6.16). For the elements present in high concentrations in the substratum (SiO_2 and Ba), we can observe that the slag concentration is shifted

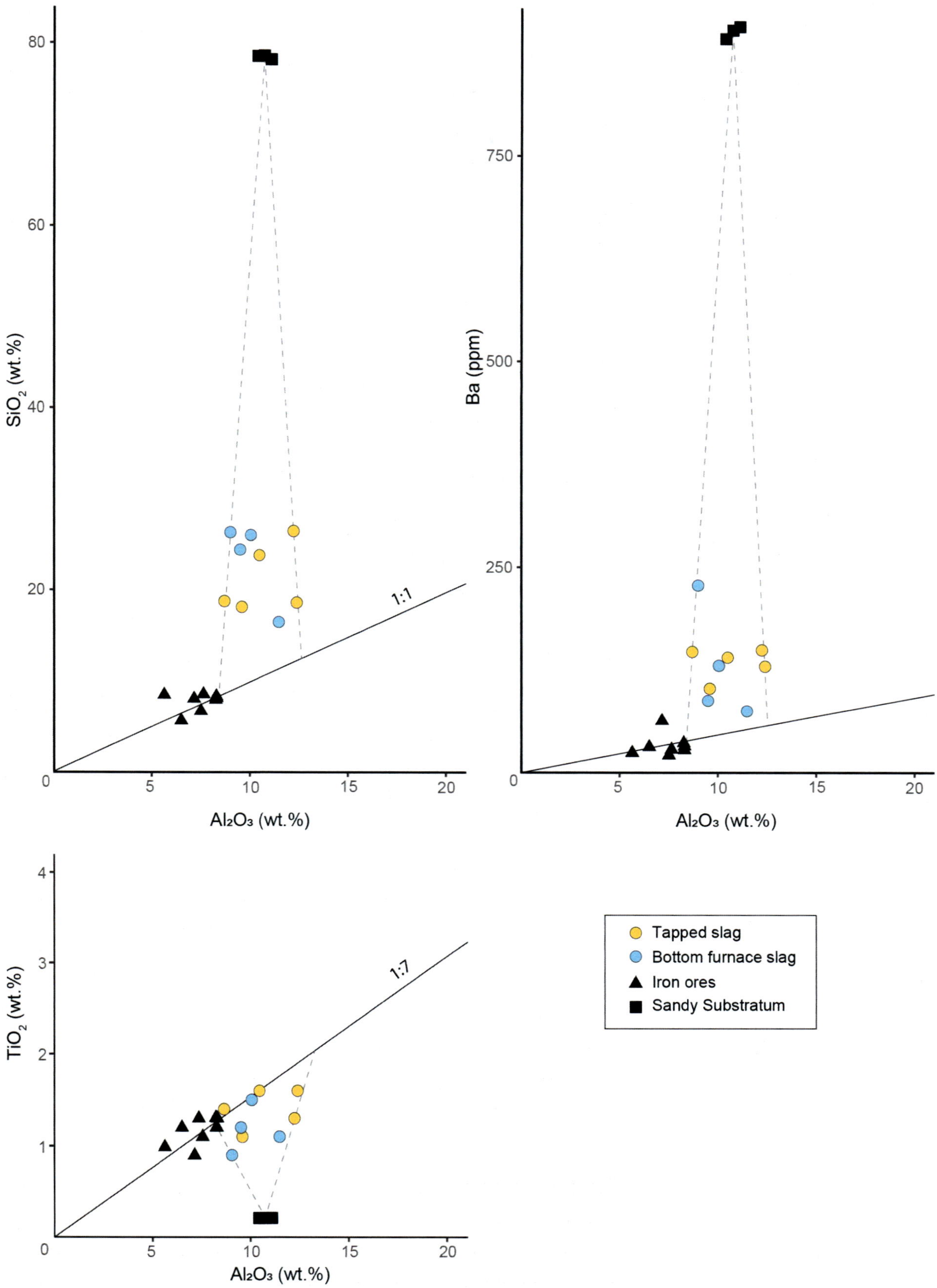

Figure 6.16: Comparison for Amboronala: Bivariate diagrams $Al_2O_3 - SiO_2$, $Al_2O_3 - Ba$ and $Al_2O_3 - TiO_2$ for the bulk compositions of iron ores, sandy substratum and slags from the workshop MBR140. The solid black line represents the mean ratio of the ore samples. The dotted lines are the mixing line joining the point of the sandy substratum, passing through the point of the slag and reaching the line of the ratio of the ore (From Morel and Serneels, 2021).

from the ore ratio toward the substratum concentrations. From one slag to another, contamination is highly variable. The greater the contamination, the more the point is shifted (Figure 6.16). In addition, the alumina content is very variable. The further the chemical composition of a slag is from the ore, while preserving the ratios, the greater the amount of iron oxide which has been reduced. From one slag to another, the amount of iron produced is thus very variable.

This approach is relevant for most of the ratios, except for the $V : Cr$ ratio which is difficult to interpret (Figure 6.17). Chromium and vanadium are elements with a similar behaviour during the smelting process and whose ratios should be constant between the ore and the slag. However, the slag samples from MBR140 appear enriched in vanadium and depleted in chromium. This enrichment cannot be explained by the sandy substratum. It is also unlikely that chromium was reduced under the poor reducing conditions existing in this type of open bowl furnace. Until now, we have not been able to find a satisfactory explanation, except the high variability of iron ores.

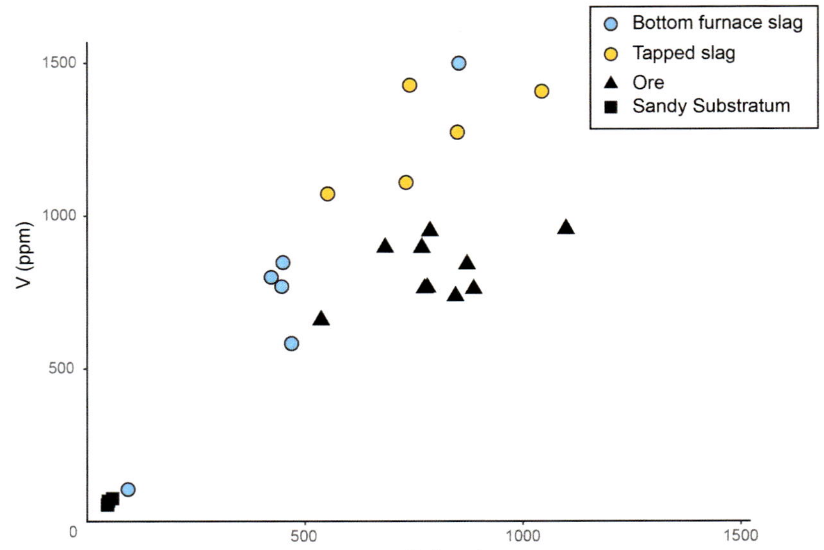

Figure 6.17: Comparison for Amboronala MBR140: Bivariate diagram $V - Cr$ for the bulk compositions of iron ores, sandy substratum and slags from the workshop MBR140.

Four samples were also taken from a wall fragment (MBR11910, MBR14911, MBR14912, MBR149013, see Appendix Figure B.9). The analyses of this wall are compatible with each other but not with the analyses of the sandy substratum. In particular, we can notice that the SiO_2:TiO_2 ratios are completely different between the wall and sandy substratum samples. Moreover, the wall analyses are enriched in chromium and zirconium, but depleted in barium and strontium, compared to the substratum.

The variability of the sandy substratum compositions alone cannot explain such differences between the substratum and wall samples. The compositions of this particular piece of wall are not compatible with the slag. The barium enrichment in the slag, for example, can no longer be explained. On the contrary, if the contaminant material had such high zirconium content, the slag would have much higher content. The heating certainly altered the chemical composition of the wall, even if this explanation is not satisfactory. We therefore decided to exclude the wall samples from the calculations.

The other slag heaps from Amboronala show the same trends as the workshop MBR140, although no ore or substrate was analysed. The spread of alumina grades shows variability in the quality of the smelt from one operation to another while the variability in silica, and other trace elements associated with sandy substratum, shows the heterogeneity of the sandy contamination. However, slags from the same heap have lower chemical variability than the overall site-wide variability. For example, not all slag heaps have the same range of variability in alumina content, which probably reflects the use of slightly different ores (Figure 6.18).

The previous approach can be applied to all studied sites. This approach is particularly useful for understanding the data in cases where we could not collect and analyse substratum samples.

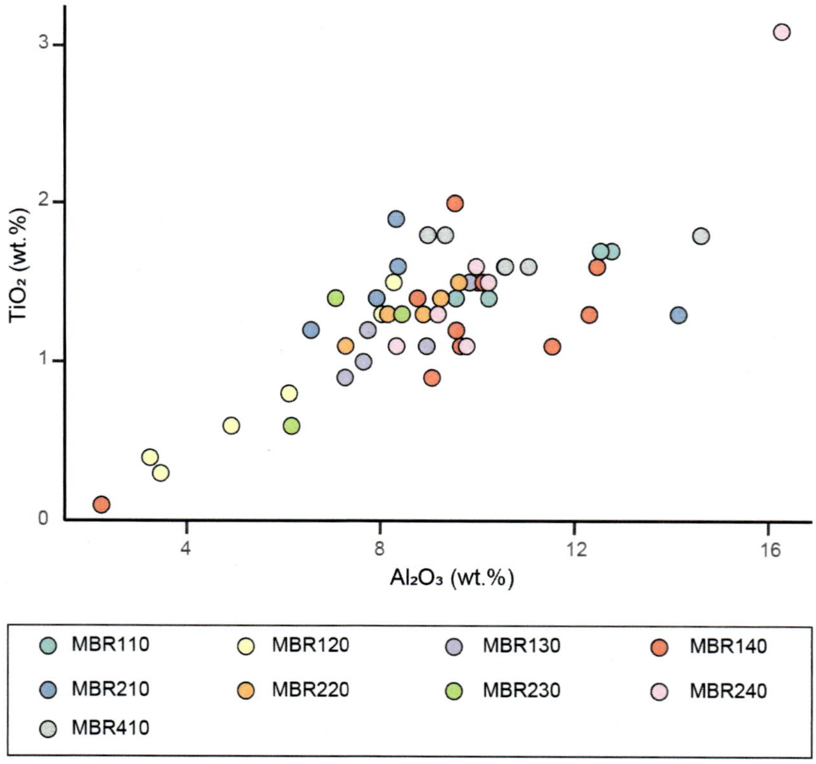

Figure 6.18: Comparison for Amboronala: Bivariate diagram Al_2O_3-TiO_2 for the bulk compositions of smelting slags from all slag heaps from Amboronala (MBR).

Matavy (MTY)

Sample		MTY 11001	MTY 11002	MTY 11004	MTY 11005	MTY 11006	MTY 11011	MTY 11012	MTY 12005	MTY 12006	MTY 12007	MTY 18002	MTY 18004	MTY 18006	MTY 11904	Average Ore
		TS	TS	TS	TS	TS	BFS	BFS	BFS	BFS	BFS	BFS	BFS	BFS	From indurated wall	Pisolites
SiO_2	(wt.%)	15,64	23,81	17,03	20,40	17,09	16,06	21,68	13,12	13,42	9,98	23,81	15,09	9,75	74,19	1,69
TiO_2	(wt.%)	1,80	1,60	3,10	1,90	1,60	1,90	1,80	1,70	1,70	0,40	0,30	1,40	1,80	4,20	1,24
Al_2O_3	(wt.%)	9,56	6,57	17,23	7,55	6,88	12,38	8,04	9,27	10,47	4,21	2,65	9,99	11,35	10,79	8,40
Fe_2O_3	(wt.%)	68,50	63,50	55,10	66,40	70,10	65,20	64,70	71,00	68,90	83,00	67,40	67,50	72,20	10,30	85,54
MgO	(wt.%)	0,14	0,20	0,89	0,22	0,13	0,39	0,17	0,31	0,30	0,47	0,98	0,48	0,35	0,14	0,05
MnO	(wt.%)	0,11	0,12	0,21	0,14	0,11	0,13	0,12	0,12	0,14	0,04	0,03	0,14	0,22	0,07	0,09
CaO	(wt.%)	0,20	0,60	1,60	3,00	0,30	0,40	0,40	0,40	0,40	0,60	1,20	0,80	0,30	0,10	0,01
Na_2O	(wt.%)	0,06	0,17	0,32	0,18	0,04	0,14	0,05	0,13	0,29	0,18	0,35	0,18	0,09	0,10	0,11
K_2O	(wt.%)	0,04	0,13	0,46	0,15	0,07	0,28	0,09	0,23	0,34	0,47	1,43	0,49	0,18	0,03	0,01
P_2O_5	(wt.%)	0,35	0,37	0,44	0,50	0,53	0,38	0,37	0,44	0,43	0,26	0,23	0,48	0,33	0,08	0,33
SUM	(wt.%)	96,77	97,27	97,00	100,62	97,06	97,73	97,73	97,09	96,79	99,88	98,55	96,91	97,00	100,12	97,86
Ba	(ppm)	33	52	61	43	41	42	39	34	50	11	51	39	70	71	25
Ce	(ppm)	19	28	12	22	14	13	22	6	-1	-5	5	4	-1	133	0
Co	(ppm)	118	68	60	37	31	111	49	77	57	92	95	39	61	39	51
Cr	(ppm)	1019	213	2023	246	405	1485	840	991	1247	1089	505	1121	1308	76	1529
Cu	(ppm)	174	193	57	121	108	118	130	118	120	312	334	117	182	70	132
La	(ppm)	23	23	15	23	15	11	18	15	14	12	10	11	12	50	10
Ni	(ppm)	28	26	9	8	7	14	12	13	11	57	80	12	14	37	64
Rb	(ppm)	0	1	3	0	1	2	1	2	5	4	22	6	2	0	0
Sr	(ppm)	16	50	106	45	23	35	28	32	35	66	88	41	24	2	1
V	(ppm)	1606	702	3121	982	1188	2142	1465	1975	2142	996	444	1791	2265	336	1792
Y	(ppm)	6	6	9	7	7	7	7	7	7	1	2	6	8	12	3
Zr	(ppm)	193	194	282	221	179	187	219	155	150	32	48	137	161	525	100
SiO_2:Al_2O_3		1,6	3,6	1,0	2,7	2,5	1,3	2,7	1,4	1,3	2,4	9,0	1,5	0,9	6,9	0,2
SiO_2:TiO_2		8,7	14,9	5,5	10,7	10,7	8,5	12,0	7,7	7,9	24,9	79,4	10,8	5,4	17,7	1,4
V:Cr		1,6	3,3	1,5	4,0	2,9	1,4	1,7	2,0	1,7	0,9	0,9	1,6	1,7	4,4	1,2

Figure 6.19: Bulk chemical compositions of smelting slag, sandy substratum and iron ore (average value) collected on three slag heaps at Matavy (MTY11, MTY12 and MTY18). See Appendix B for full chemical analysis.

Slag samples from Matavy have similar chemical compositions to the ones from Amboronala (Figure 6.19). They show high but heterogeneous iron content, ranging from 55 to 83 wt.% Fe_2O_{3tot}. Silica (10.0-23.8 wt.% SiO_2) and alumina content (2.6-17.2 wt.% Al_2O_3) is also highly variable. Except for a few samples, the average titanium oxide content is about 1.7 wt.% TiO_2. Calcium and potassium oxides contents are low, except for two samples where leucite ($KAlSi_2O_6$) can be observed. The principal trace elements are chromium (Cr: 200-2000 ppm) and vanadium (V: 700-3000 ppm) with extremely variable grades.

At Matavy as well, pisolitic concretions were collected in the sand next to a slag heap (MTY11). The bulk analysis show very high iron grades (85 wt.% Fe_2O_{3tot}). The silica content is low (1.7 wt.% SiO_2) whereas alumina shows an average content around 8.4wt.% Al_2O_3. The $SiO_2 : Al_2O_3$ ratio is close to 1:5. Red iron-rich rocks were also collected in the surroundings of the smelting site and analysed. Some of these samples (MTY503, MTY504) have iron content far too low to be exploited as iron ore and were immediately excluded from the data set (<31 wt.% Fe_2O_3). Two other samples (MTY512, MTY513) have iron content up to 80 wt.% Fe_2O_3 but their $TiO_2 : Al_2O_3$ ratios are not compatible with the slag.

The large variability in the $SiO_2 : Al_2O_3$ ratio indicates that the slag is significantly but variably contaminated by the sandy wall.

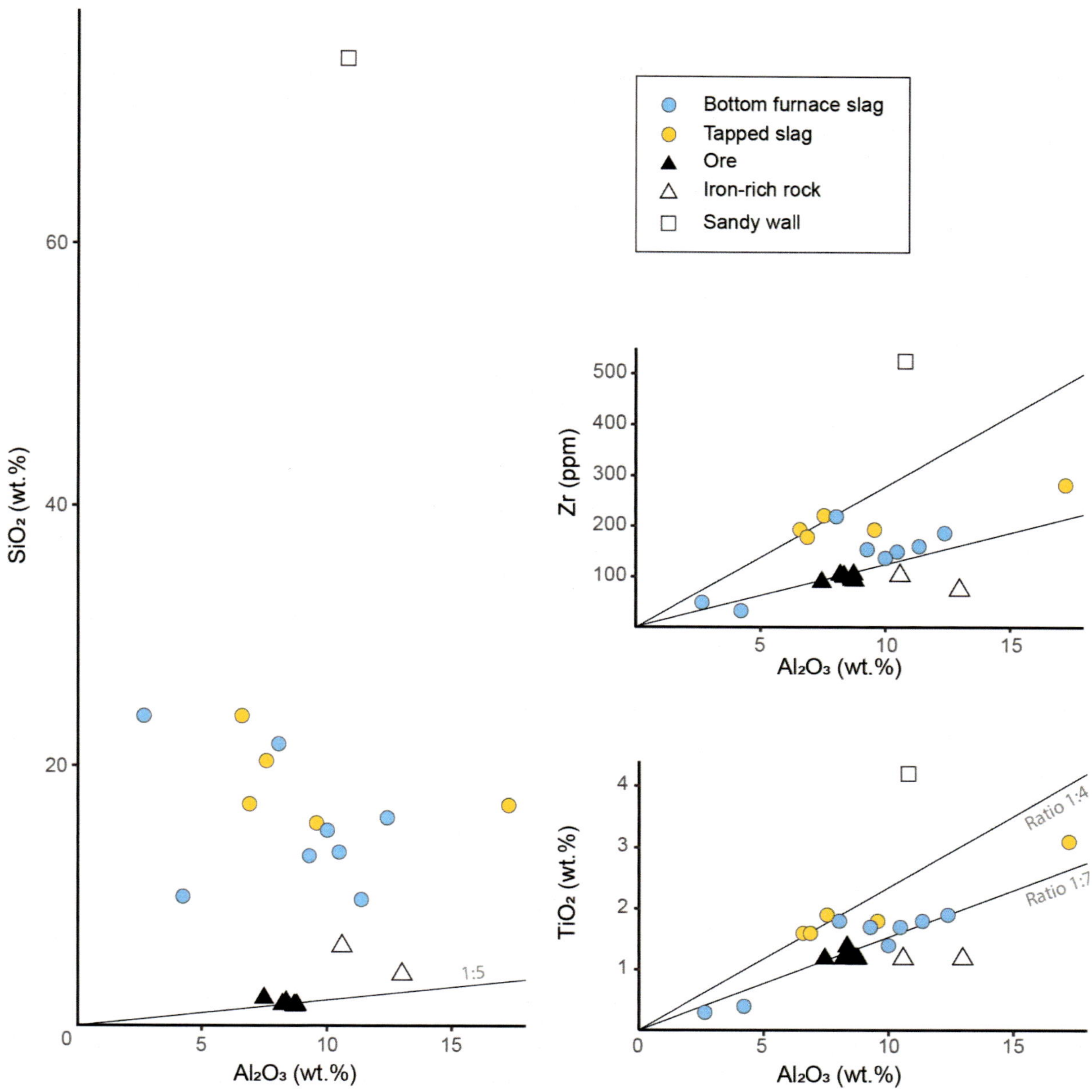

Figure 6.20: Comparison for Matavy: Bivariate diagrams $Al_2O_3 - SiO_2$, $Al_2O_3 - Zr$ and $Al_2O_3 - TiO_2$ for the bulk compositions of iron ores, sandy substratum and slags from Matavy. The solid black line represents the mean ratio of the ore samples.

No sand sample was collected during the excavation. A sample (MTY11901) was collected on a sandy wall fragment. However, the analyses were not as expected. The Matavy sand observed during the excavations was particularly white and therefore probably devoid of any heavy minerals. MTY11901 is rich in titanium oxide and iron oxides, which is, on the contrary, characteristic of heavy minerals. It also contains trace elements present in heavy minerals such as cerium (Ce), manganese (Mn), vanadium (V) and zirconium (Zr). These analyses are also not compatible with the analyses of the slag, which should have higher and variable contents of TiO_2 and Zr. This sample should therefore be excluded from our dataset, for the same reasons as the Amboronala wall samples.

The ratio TiO_2 : Al_2O_3 is constant for most of the slags and compatible with the pisolites (Figure 6.20 - ratio 1:7). The variability of the TiO_2 or Zr contents is probably not related to substratum contamination but rather to the use of ores from two different deposits. Four samples are aligned along a second TiO_2 : Al_2O_3 ratio (Figure 6.20 - ratio 1:4), which corresponds to another ore. These samples cannot result from the analysed pisolites. These four slags must have been produced from an ore with higher iron grades and with a different TiO_2 : Al_2O_3 ratio.

Finally, two samples have extremely low Al_2O_3 content. In one case, the smelting probably failed as the sample (MTY12007) has iron content similar to ore (83 wt.% Fe_2O_3) and therefore has hardly been smelted. In the second case (MTY18002), the sample is very heavily contaminated with silica.

Bemanevika

No excavations were undertaken in Bemanevika but only test pits. Metallurgical waste was studied, but sampling was limited. The analysed slag pieces have high iron content (61.4 wt.% Fe_2O_{3tot}), but lower than in Amboronala and Matavy (Figure B.7). They are characterised by remarkably high titanium oxide content (3.2 wt.% TiO_2). The slag from Bemanevika has higher alumina content than the ones from Amboronala and Matavy. Microscopic studies indeed show a large number of spinels up to 200 μm in diameter in the slag from Bemanevika. Silica content is variable (12.3-22.2 wt.% SiO_2).

Iron oxides-rich rocks were collected on a hill called *Antaimby* (*"the place of slag"*), located few hundred metres away from the slag heaps (Figure 4.27). The toponymy is linked with a ancient metallurgical activity but no slag heap was found. This hill could be an ancient mining area. Pisolites were collected, some were carefully washed to remove all the clay from the soil (BMK510) and others were not cleaned (BMK501 and BMK502). The unwashed pisolites have extremely high alumina content (average 42.5 wt.% Al_2O_3) and far too low iron oxides (average 47.8 wt.% Fe_2O_{3tot}). They are unusable as iron ore. On the other hand, the washed pisolites are characterised by high iron grades (77.9 wt.% Fe_2O_{3tot}) and especially high titanium oxides content (2.1 wt.% TiO_2). The ratio of silica (6.1 wt.% SiO_2) to alumina (10.5 wt.% Al_2O_3) is close to 1:2. The material is a Fe-rich bauxite. The main trace elements are chromium, vanadium and zirconium. Barium content is very low. Finally, fragments were collected from large scoriaceous boulders measuring more than one meter in diameter (BMK503, BMK511, BMK512). These samples have iron oxide content that is too low to be used as ore (59.4 wt.% Fe_2O_{3tot}).

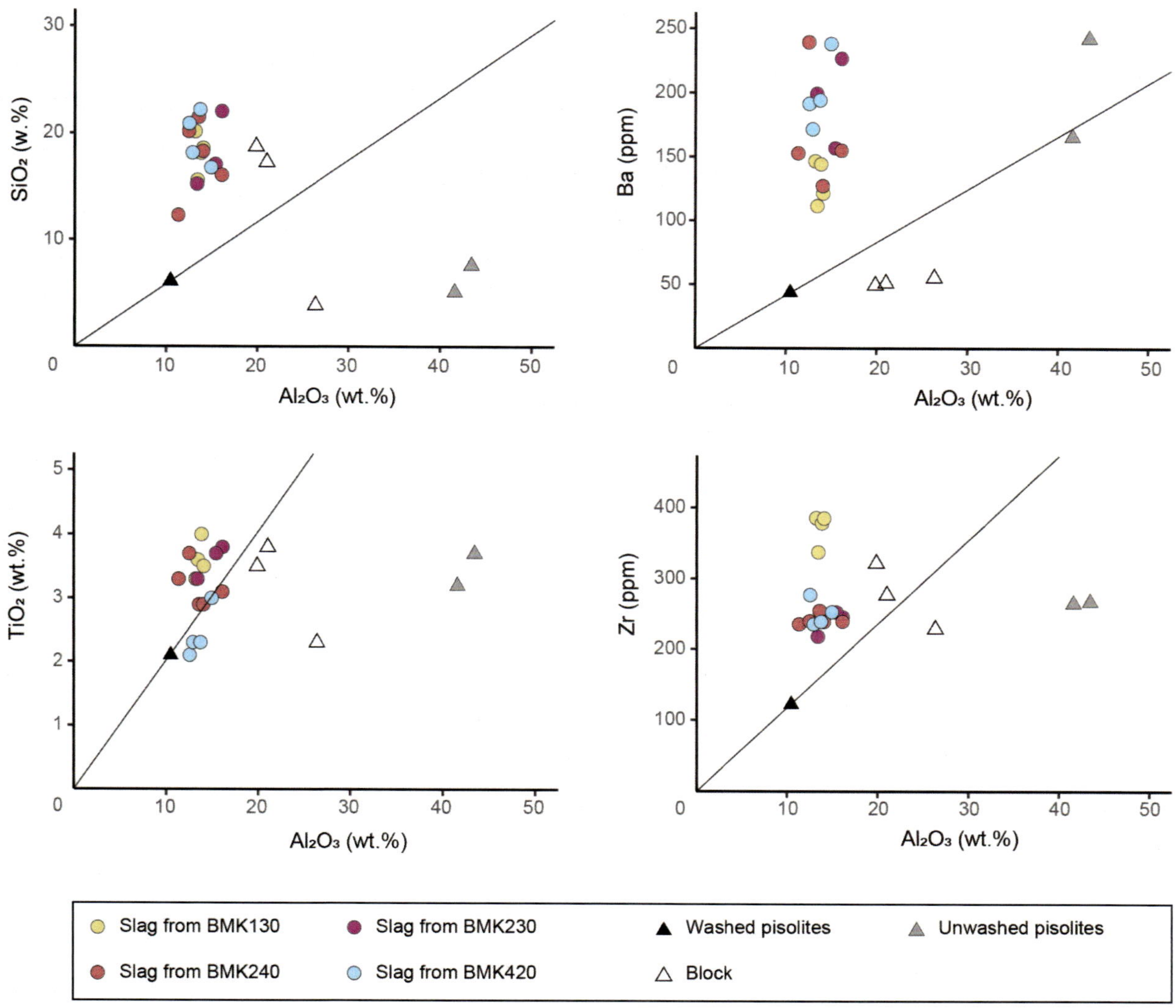

Figure 6.21: Comparison for Bemanevika: Bivariate diagrams $Al_2O_3 - SiO_2$, $Al_2O_3 - Ba$, $Al_2O_3 - TiO_2$ and $Al_2O_3 - Zr$ for the bulk compositions of iron ores, sandy substratum and slags from Bemanevika. The solid black line represents the mean ratio of the washed pisolites sample.

No substratum samples were taken and no walls were found during the surveys. However, the sandy substratum in this area is greyish and most likely contains dark minerals such as magnetite or ilmenite. Concentrations of black sands were also observed in the rivers.

Slag samples from Bemanevika are enriched in silica, barium and zirconium (Figure 6.21). The silica contamination is heterogeneous from one slag to another and no groups can be established according to the slag heaps or the metallurgical sectors. On the other hand, the slags are not enriched with the same trace elements, depending on the sector from which they were taken. This probably reflects the variability of the substratum. Slags from the BMK200 sector (slag heaps BMK230 and BMK240) tend to be enriched in barium more than in zirconium for example.

A single ore sample is by far not enough to draw strong conclusions. The $TiO_2 : Al_2O_3$ ratio could be consistent with slags from BMK420 and some of the slag from BMK240. The substrate probably contains titaniferous minerals, so it is difficult for now to establish a clear relationship between slag and ore. It is likely that several ore deposits were exploited.

Benavony

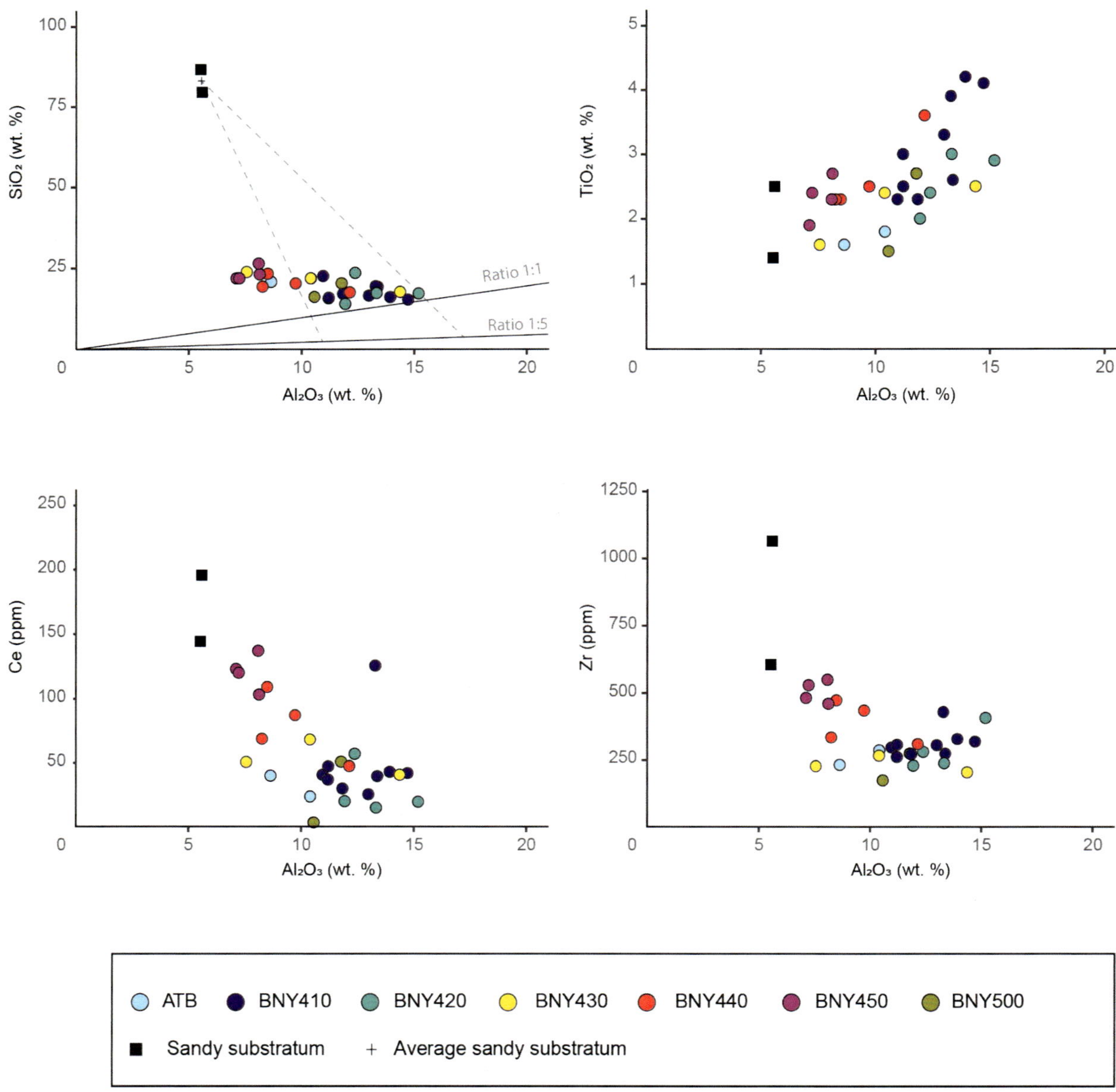

Figure 6.22: Comparison for Benavony: Bivariate diagrams $Al_2O_3 - SiO_2$, $Al_2O_3 - Ce$, $Al_2O_3 - TiO_2$ and $Al_2O_3 - Zr$ for the bulk compositions of sandy substratum and slags from Benavony. The solid black lines represent the ratio variation of pisolites samples from other studied smelting sites.

Two slag heaps from Benavony were excavated (BNY410 and BNY420) but samples were collected for each of the five slag heaps near the ancient settlement (Figure 4.7). Slag pieces from Benavony also have high iron grade, averaging at 62 wt.% Fe_2O_3). The ratio between silica (14.0-26.4 wt.% SiO_2) and alumina (7.1-15.2 wt.% Al_2O_3) is variable from one slag to another. Titanium oxides content is significantly high (2.7 wt.% TiO_2). Calcium and potassium are remarkably high in comparison with slags from the previously described sites. The main trace elements are chromium (200-1100 ppm) and vanadium (650-1900 ppm), but their content is very variable and the ratio V:Cr is not stable. Slag pieces from Benavony are characterised by high zirconium and REE content. There are differences in chemical composition that distinguish some heaps from others, in particular BNY440 and BNY450, which tend to have lower alumina content than the others, but higher Zr and REE contents.

No ore samples were found during the excavations and the chemical composition of the ore used at Benavony is not known. A sand sample was collected at the same depth as the furnaces (BNY41901, BNY41902, BNY41903). Macroscopically, the sand is greyish and therefore contains heavy minerals. The presence of magnetite grains was observed in the field with a magnet.

Ores found in archaeological contexts on other sites show $Al_2O_3 : SiO_2$ ratios oscillating between 1:1 and 1:5 and have alumina content around 8 wt.% Al_2O_3. This range of ore could be consistent with the majority of the Benavony slags. However, for the slags of BNY440, BNY450 and one of BNY430, an ore with higher iron content must have been used. Slags from the same slag heap have, in most cases, constant ratios $Al_2O_3 : TiO_2$ but slightly different from the other slag heaps. It would appear that ores collected from different outcrops, of different chemical composition, were used. Given that Benavony is the largest production site found in northeastern Madagascar, production must have been slightly higher and it is not surprising that the workers had to seek ore from several locations.

The contamination of the sandy substratum can be observed through the high variability of silica content (Figure 6.22). The chemical compositions of the slag are also variably enriched in zirconium or REE, which are trace elements characteristic of the heavy minerals (monazite) of the substratum and present at low levels in lateritic ores.

General conclusions

The study of ores is essential in understanding a metallurgical practice. It is however very difficult to find the ore deposits exploited by smelters. In our case, our fieldwork was not sufficiently oriented towards the study of ores. Despite this, several conclusions can be drawn from the study of the ores and slag for the four sites detailed above:

- Only pisolites were found on archaeological sites so it seems that the metallurgists favoured the use of pisolites relatively to blocks of consolidated lateritic duricrust. Lateritic crusts are hard rocks and therefore probably more difficult to collect. Moreover, the lateritic crust samples analysed systematically show lower iron content than the pisolites. It is therefore possible that these blocks are lower quality ores, although a larger sampling would be needed to discuss this point.

- Once collected, the pisolites were washed. The reduction workshops are systematically installed near a watercourse. This is consistent with the historical sources which describe that the ore is carefully washed to separate it "from the muddy parts" (Vacher, 1899:544).

- The ferruginous concretions found on the smelting sites have remarkably high iron oxides content. The study of the slags shows that several sources of ores could have been used on the same smelting workshop. However, all the iron-rich rocks collected during our surveys outside of the smelting sites tend to have lower iron content than those used by the metallurgists. They were therefore able to identify good ores during their collection.

- Each slag heap has chemical compositions which are less variable than the overall variability of the site, and of the whole district. It would appear that in most cases, for each slag heap, the ore used came from the same deposit. In the following part of this chapter we will explain that one slag heap is not the result of an intensive production campaign but rather of a succession of smelting operations spaced out over time. A slag heap, even of such a small size, was probably built up over several decades. Yet the slags from the same slag heap were produced with the same ore. This could mean that the same human group, or family, returned to produce iron in the same place and collected the ore from the same deposit.

- The sandy substratum constitute an important part of the final chemical composition of the slag. Specific minerals such as zircon or monazite can thus give a fairly specific chemical signature to the slag. However, this contamination is very heterogeneous. No link between the amount of contamination and the amount of reduced iron could be deduced from our data. Heavy contamination does not imply either good or bad reduction, nor high and low production of iron (see mass balance calculations below).

Once the link between ores, substrates and slag has been established, metal production can be studied in detail.

6.2.5 General Approach of the Variability Inside a Single Technology

The data from northeastern Madagascar led us to a more general approach to the variability in slag composition (Morel and Serneels, 2021). Smelting slag is the result of repetitive smelting operations following a given technique. A technique is characterized by the raw materials used as well as the way the smelt is carried out (Rehren et al., 2007). The composition of the slag reflects both aspects. Both must be investigated at the level of a single sample but also within an assemblage. The ore is the first contributor to the slag. There is a huge range of natural ores. Some of them are homogeneous with a very constant composition; others are highly variable. The benefication process modifies the composition of the ore. The building materials of the furnace interact frequently during the process and provide a contribution to the formation of the slag. Other forms of contamination are possible, including the voluntary addition of a flux, for example (Chirikure, 2014, Miller et al., 2001). Together, the ore and the contaminating materials provide the chemical stock to the slag. If the two components present a chemical contrast, it is possible to identify their contribution and decipher the slag composition. An important point related to the management of the furnace is the efficiency of the reaction of reduction. If the conditions are strong (high temperature, long duration, high availability of reducing agent), a greater proportion of the metallic oxide will be reduced to the metallic state. In the bloomery process, the conditions are not very strong, and a variable proportion of the iron oxide is not reduced and enters the slag. When good archaeological and chemical data are available, the mass balance calculation is an interesting tool to understand the variabilities. One very critical point is the correct characterization of the ore. It is also possible to approach the question using simple graphical representation. Several bivariate diagrams using two non-reducible elements present at a measurable grade in the different materials, but with contrasted ratios, can be used. Frequently, the diagram $SiO_2 - Al_2O_3$ is very useful. Figure 6.23 illustrates schematically the behavior of a set of data, ore, slag and contaminant regarding the two main types of variability encountered in smelting systems.

If there is no contamination, the ratio between the two unreducible elements in the ore will be the same in the slag (Figure 6.23 a,b). As it is expected that the ore shows a certain variability, one can observe a similar variability in the slag (Figure 6.23 a—shaded zone). The distance between the point representative of the ore and the position of a slag will be proportional to the efficiency of the reduction (Figure 6.23 —red arrows). If all the slag samples cluster on a single point, the reduction is very homogeneous throughout the smelting activity (Figure 6.23 a). The process is in this case very stable and very well mastered. If the slag points are aligned along the line of the ore ratio, this is the evidence of a variability in the efficiency of the reduction from one smelt to the other (Figure 6.23 b). The effect of the contamination will be to shift the points away from the line of the ore ratio (Figure 6.23 c–f—blue arrows). The shift will take place on a mixing line joining the point of the contaminant material, passing through the point of the slag and reaching the line of the ratio of the ore. Using the lever rule, it is possible to estimate the proportion between the two components in the mix. One can read the diagram to evaluate the variability of the efficiency of the reduction and the proportion of the contamination. A very well mastered process will show a fixed reduction rate and a fixed proportion of contaminant (Figure 6.23 c). A poorly mastered process is characterized by variable reduction and contamination (Figure 6.23 f). One expects a high rate of reduction for an efficient process. The role of the contamination is trickier to evaluate. In many cases, the contaminating material will contribute to flux the slag and lower the temperature to be reached to obtain the separation between the slag and the bloom.

The Figure 6.23 shows these theoretical trends on bivariate plots, but the same trends can be observed on ternary plots as well.

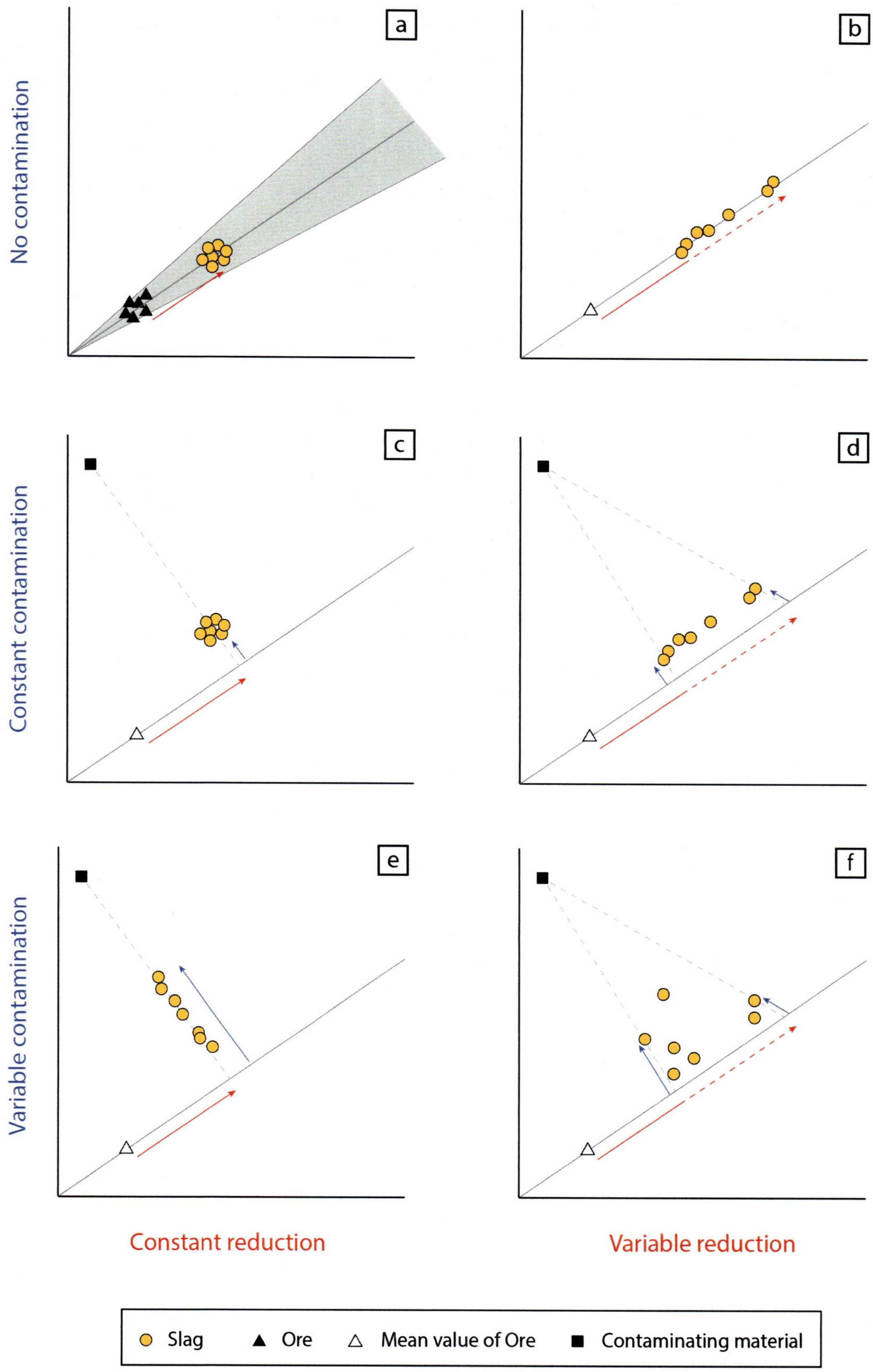

Figure 6.23: Schematic representations of the variability of the contamination and of the efficiency of the reduction reaction on bivariate diagrams for unreducible elements in a smelting system (ore-contaminant-slag). (a) Bivariate diagram for no contamination and a constant reduction, (b) Bivariate diagram for no contamination and a variable reduction, (c) Bivariate diagram for a constant contamination and a constant reduction, (d) Bivariate diagram for a constant contamination and a variable reduction, (e) Bivariate diagram for a variable contamination and a constant reduction, (f)Bivariate diagram for a variable contamination and a variable reduction (From Morel and Serneels, 2021).

6.3 Estimation of the metal production

With archaeological and compositional data available from north-eastern Madagascar, it is possible to perform mass balance calculations (Leroy et al., 2015; Charlton et al., 2010; Kronz, 2003; Serneels, 1993; Eschenlohr and Serneels, 1991). This quantitative approach leads to the estimation of the amount of iron produced. This section is taken from an article already published (Morel and Serneels, 2021).

6.3.1 Mass Balance Calculations

The mass balance approach is based on the principle of the conservation of mass. The mass of the materials introduced into the furnace must be equal to the mass of the materials coming out of it. In a simplified approach, only four components will be considered to make the calculation:

$$\text{Ore} + \text{Contamination} = \text{Slag} + \text{Iron} \tag{6.6}$$

Air and fuel are also introduced in the furnace and participate in the reaction to produce the combustion gases. The combustion gases will evacuate the volatile substances liberated by the different materials at high temperature. Water from hydrated minerals like clays is lost. There is also always a significant quantity of oxygen lost from the reduction of the iron oxides of the ore. For this reason, if the calculation is made including oxygen, then the two sides of the equation will not be of equal mass. To avoid this bias, it is recommended to recalculate the compositions as elements, without oxygen. The analyses for each sample must then be normalised to 100%, on an anhydrous basis. The rest of the calculation can then be done in g/100g of sample. The aim of the calculation is to estimate the quantities of the different components in the system, relatively to one of them.

$$(\text{M}\times\text{Ore}) + (\text{W}\times\text{Contamination}) = (\text{S}\times\text{Slag}) + (\text{F}\times\text{Iron}) \tag{6.7}$$

The factors M, W, S and F are the relative quantities of the respective components. As the fieldwork data often allow to estimate the total mass of slag found at a workshop, it makes sense to calculate the factor M, W and F relatively to one mass unit of slag (S = 1) to evaluate the global production of iron. The same calculation could be made for M, W or F equal to 1. When M is equal to 1, the factor F reflects the yield directly (see Chapter 2). Theoretically, for one given system, the 4 factors should be fixed for the system as a whole and for each element separately. Using 3 equations, for

3 different elements including iron, it is possible to find a unique solution for the 3 unknown factors. It is recommended to use the chemical composition data obtained for the more abundant elements, because they will show less relative variability than minor components. In most cases, the calculations must be based on Fe, Si and Al. The measured or averaged grade in wt.% of a material can be presented as g per 100 g of this material.

Si (M × g of Si in 100 g of Ore) + (W × g of Si in 100 g of Contamination) = (1 × g of Si in 100 g of Slag) (6.8)

Al (M × g of Al in 100 g of Ore) + (W × g of Al in 100 g of Contamination) = (1 × g of Al in 100 g of Slag) (6.9)

Fe (M × g of Fe in 100 g of Ore) + (W × g of Fe in 100 g of Contamination) = (1 × g of Fe in 100 g of Slag) + (F × g of Fe in 100 g of Iron) (6.10)

Silicon and aluminum are assumed to be absent from the metallic phase of the bloom, because they are not reduced to the metallic state in the conditions reached during the bloomery process, and their oxides cannot be dissolved in the solid metallic phase. For this reason, the iron component is omitted in equations (6.8) and (6.9), and the F factor is calculated following the equation (6.10).

The potential contamination must be investigated as it can be a complex parameter. There is always a contamination from the ashes of the fuel (Ben-Yosef and Yagel, 2019; Jackson et al., 2005; Serneels, 2002; Crew, 2000). The most common fuel is charcoal, and it contains about 1-5 wt.% of ashes, mainly calcium and potassium oxides (CaO, K_2O). This contamination always affects Ca and K, even though Mg, Na and P can be impacted. In special cases, other elements can be affected, including trace elements. In general, not all the ashes are incorporated in the slag. The ore fed into the furnace is the result of a more or less complex process of benefication (Pleiner, 2000). It is frequent that some materials from the gangue or from the host rock are still present in the final concentrate. The building materials of the furnace walls undergo partial melting during the process. Such partially melted building materials are found during excavation, and their composition can be measured experimentally (Veldhuijzen, 2007; Eschenlohr and Serneels, 1991). Finally, the deliberate addition of another substance can be considered. It is expected that this addition aims to lower the melting point of the slag, and in this case, the substance can be termed a flux. In the archaeological record, there is only limited material evidence supporting the proper use of a flux for the production of iron by the bloomery process (Pleiner, 2000). Instead, this practice is well documented for the production of cast iron in blast furnaces during the late Middle Ages in Europe. Finally, blending two types of ore with contrasted compositions can be considered as way of fluxing.

Theoretically, the factors M, W and F calculated from 3 elemental equations should apply to all other elements. In practice, it is never the case, for several reasons. First, during the smelting process at high temperature, several elements will fractionate between the combustion gas, the solid metallic product and the melted slag (Desaulty et al., 2008; Coustures et al., 2003; Serneels, 1993; Tylecote et al., 1977). It is not possible to measure the composition of the gas and the composition of the metallic product can hardly be evaluated from archaeological remains. Hence, the measured values for the fractionating elements will not fit the calculated results. Several substances can be partially or totally volatilized, such as elemental sodium (Na), potassium (K) and zinc (Zn) or oxides of sulfur (SO_3) and arsenic (As_2O_3) (Craddock, 1995; Misra et al., 1993; Craddock, 1985). Several other elements can be reduced to the metallic state in the conditions of the bloomery process (Ellingham, 1944). Copper (Cu), nickel (Ni) and cobalt (Co) will enter easily in the metallic phase. Other precious (Ag, Au, Pt, Os) and base metals (Pb, Sn) should behave similarly but are very rarely present at a significant level in iron ores (Zitzmann, 1977). Other elements can be partly affected because they are difficult to reduce in the conditions of the bloomery process. Phosphorus (P) is reduced under conditions very close to the ones of iron (Vega et al., 2002). Chromium (Cr), manganese (Mn) and vanadium (V) demand stronger conditions, but these are not impossible to reach during the bloomery process (Truffaut, 2014; Alipour et al., 2021). At very high conditions, silica (SiO_2) begins to react, and a few percent silicon (Si) are a common feature in cast irons.

On the other hand, the knowledge about the archaeological materials involved in the process is not always satisfactory. In general, the slag and the furnace building materials can be well studied, but the ore and the metallic product are more difficult to characterize. Finally, it is more or less impossible to chemically define the ashes. In the end, only a small number of elements can eventually provide valuable information for the calculation of the mass balance. On the other hand, the behaviour of the fractionating elements can provide useful evidence to understand the efficiency of the smelting process. The understanding of the contamination is also essential for the characterization of the smelting process. The ideal components for the calculation would not be affected by any of the problems presented above, which are contamination and fractionating. In practice, no element is ideal, but many can be used to produce satisfactory calculations for a given system. The suitable elements will be the ones with a significant contrast in the ore and the other materials involved in the system. For lateritic iron ores, the low Si:Al ratio allows to use aluminium. In marine oolithic ores, high calcium is expected, sometimes with magnesium. Other elements are typical for specific types of iron ores and rare in most other natural materials, like manganese (Mn), chromium (Cr) and titanium (Ti).

6.3.2 Mass Balance Calculations for the smelting process at Amboronala MBR140

	Sample	Mean Value Ore Pisolithes	MBR 14901 Sandy Substratum	MBR 14010 Bottom Slag	MBR 14006 Tapped Slag	MBR 14007 Tapped Slag	MBR 14004 Tapped Slag	MBR 14012 Bottom Slag	MBR 14009 Bottom Slag	MBR 14002 Bottom Slag	MBR 14005 Tapped Slag	MBR 14003 Tapped Slag
	Si (wt.%)	4,98	36,70	8,44	9,27	9,60	9,45	12,44	13,07	13,17	12,10	13,37
	Al (wt.%)	4,04	5,90	6,66	5,56	5,04	7,13	5,47	5,05	5,75	6,03	6,99
	Fe (wt.%)	57,89	43132,00	51,68	51,30	50,62	48,62	46,74	44,42	44,22	44,13	41,72
1	Calculated Amount of Material for 100 g of Slag											
1.1.	Ore *M* (g)	-	-	154	118	101	162	100	85	105	118	140
1.2.	Wall *W* (g)	-	-	8	13	16	9	24	27	25	21	22
1.3.	Slag *S* (g)	-	-	100	100	100	100	100	100	100	100	100
1.4.	Iron *F* (g)	-	-	37	17	8	45	12	5	17	25	40
1.5.	Measured Ti (%)	0,71	0,12	0,72	0,72	0,92	1,04	0,78	0,57	0,97	1,04	0,84
1.6.	Calculated Ti (%)	-	-	1,10	0,85	0,74	1,17	0,74	0,64	0,78	0,87	1,03
2	Calculated Amount of Materials for 100 g of Ore											
2.1.	Ore *M* (g)	-	-	100	100	100	100	100	100	100	100	100
2.2.	Wall *W* (g)	-	-	5	11	16	6	24	32	24	18	16
2.3.	Slag *S* (g)	-	-	65	85	99	62	100	118	95	85	71
2.4.	Iron *F* (g)	-	-	24	15	8	28	12	6	16	21	28
2.5.	Yield (%)	-	-	41	26	14	48	21	10	28	36	48

Figure 6.24: Table of results of the mass balance calculations for the smelting process at MBR140. Slag cannot be averaged (from Morel and Serneels, 2021).

The Amboronala MBR140 workshop is the only case in our study where both sandy substratum and iron ores were sampled. This is therefore the only case where mass balances can be reasonably calculated. The composition of the iron ore samples are very homogeneous whereas the analyses of the slag pieces show a very strong variability. The number of samples (only 9) is much too small to statistically demonstrate the representativity of the mean values. In this case, a single calculation based on the mean values is not significant. In contrast, the calculation can be performed for each piece, and the results reflect the variability from one smelt to another, both in terms of the efficiency of the reduction reaction and of the proportion of contamination by the sand (Figure 6.24). At Amboronala MBR140, the variability within a single operation cannot be investigated because it is impossible to identify the different fragments of a single smelting operation. In other case studies, this internal variability has been shown to be limited (Humphris et al., 2009; Crew, 2000; Eschenlohr and Serneels, 1991).

Silicon and aluminium are present both in the ore and in the sand,

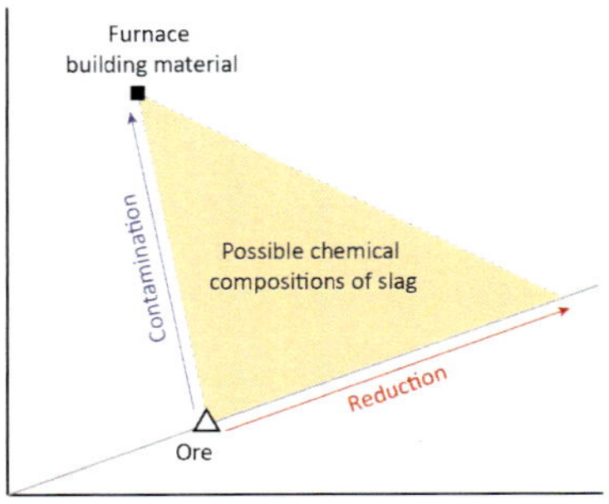

Figure 6.25: Diagram summarising the range of possible chemical compositions (yellow area) for a slag for a given ratio between two chemical elements. If the chemical compositions of the slag are not plotted in this area, it means that the ores or furnace construction materials (or fluxes) have not been correctly identified.

at measurable grades and with highly contrasted ratios (Figure 6.16). These allow the calculation of M, W and F using Si, Al and Fe equations (Figure 6.24, lines 1.1, 1.2 and 1.4). One must keep in mind that in such a system of equations, with any defined ore, sand and slag, there is always one single mathematical solution, but the significance of the calculated numbers is given only if the materials are identified correctly. Here are some examples where the results of the calculation are not sensible:

- In some cases where the materials have been incorrectly defined, the result of the calculation may be a negative metal production. This is of course inconsistent and impossible.
- Some combined materials, ore and wall, cannot chemically have given some slag chemical compositions. Since the slag is a material resulting from the mixture of ore and wall, the elemental ratios of the slag must be somewhere between the ratios of the ore and the wall (Figure 6.25).The calculations will have a solution in this case, but which will be meaningless.

To demonstrate that the ore and the sand are really the ones involved in the formation of the slag, at least one other element should give results in agreement with the calculated M and W. The value of this element must be recalculated from M and W and compared to what has been measured analytically. At Amboronala, titanium can be used to confirm the correct identification of the materials (Figure 6.16). The grades of Ti are high enough in all materials, and the ratios $Si{:}Ti$ and $Al{:}Ti$ are different in the ores and the sand. For 8 pieces of slag out of 9, the measured and calculated grades for Ti are quite similar, within 10% relative uncertainty (Figure 6.24, lines 1.5 and 1.6). Only for the sample MBR14010, is there no good agreement. This particular case can be explained by the use of raw materials with a small difference in composition, possibly pisolites collected in another area. This sample is excluded from the panel for the rest of the discussion.

Variability of the Iron Production at MBR140

The amount of iron produced per unit of ore varies significantly from one sample to the other: the calculated values range from 6 g to 28 g for 100 g of ore (Figure 6.24, line 2.4). With the lower values, below 10 g, the product was really small. The smelters were close to failure, especially since the metal remained dispersed, and no bloom is produced. On the contrary, the highest values, near 30 g, were a nice success. This variation in the amount of iron produced reflects the variability of the efficiency of the reaction of reduction. Three main physico-chemical parameters, linked one to another, are responsible for the efficiency of the reaction: temperature, time and the partial pressure of carbon monoxide. These parameters are related to the technical gestures of the smelters and the way they drive their furnace. At Amboronala, the smelters used small and simple bowl furnaces. Compared to other bloomery furnaces of the same period, the architecture is not very efficient for saving heat and concentrating the reducing gases (Leroy et al., 2020; Juleff, 2009; Chirikure and Rehren, 2004). The use of bellows can be inferred from the shape and size of the tuyere (Figure 5.8). Apparently, the smelters were not able to control the reduction process systematically. Fortunately, because the ore is high grade, it is not too difficult to produce a minimum amount of iron even under poor conditions.

There are not many arguments to estimate the quantity of ore processed during one single smelt. Considering the size of the furnace, maybe a batch of 10 L, about 20 kg, is a reasonable estimate. The worst smelt (MBR14009) would have given only 1.2 kg of metal, which would have been a porous block of the size of a small fist. The best bloom (MBR14003) reached more than 5 kg, an uncompacted block of about 1 L.

Variability of the contamination at MBR140

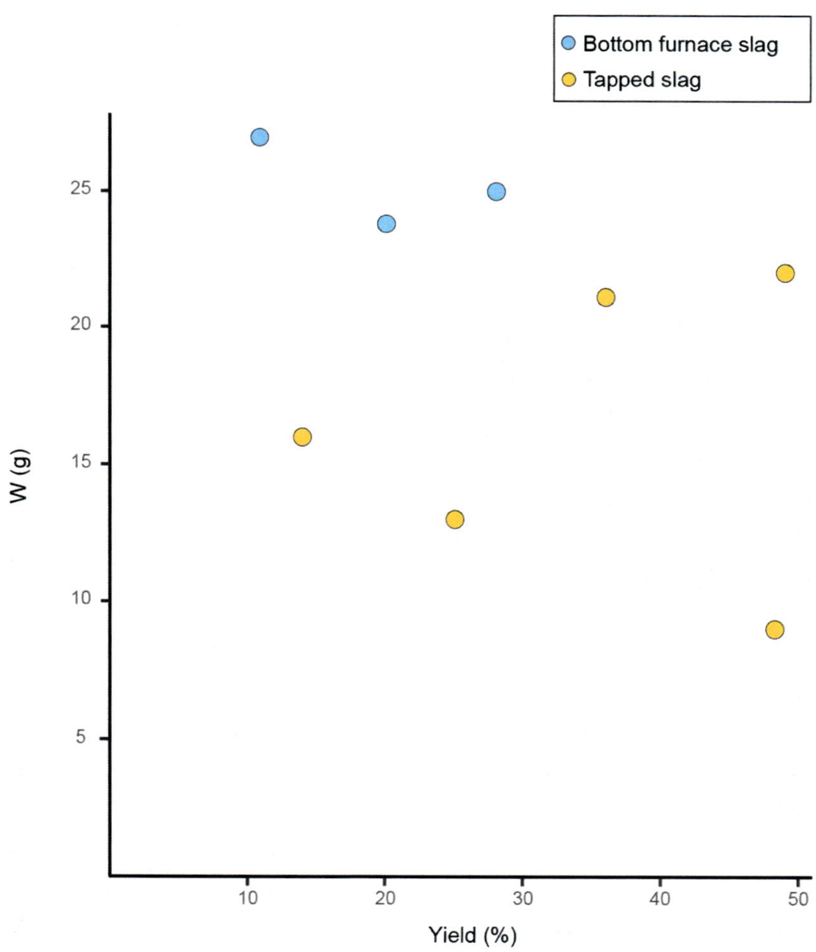

Figure 6.26: Bivariate diagrams of the mass of sandy substratum involved in the formation of 100 g of slag (W) as a function of the yield.

At Amboronala, the sand always contributes to the formation of

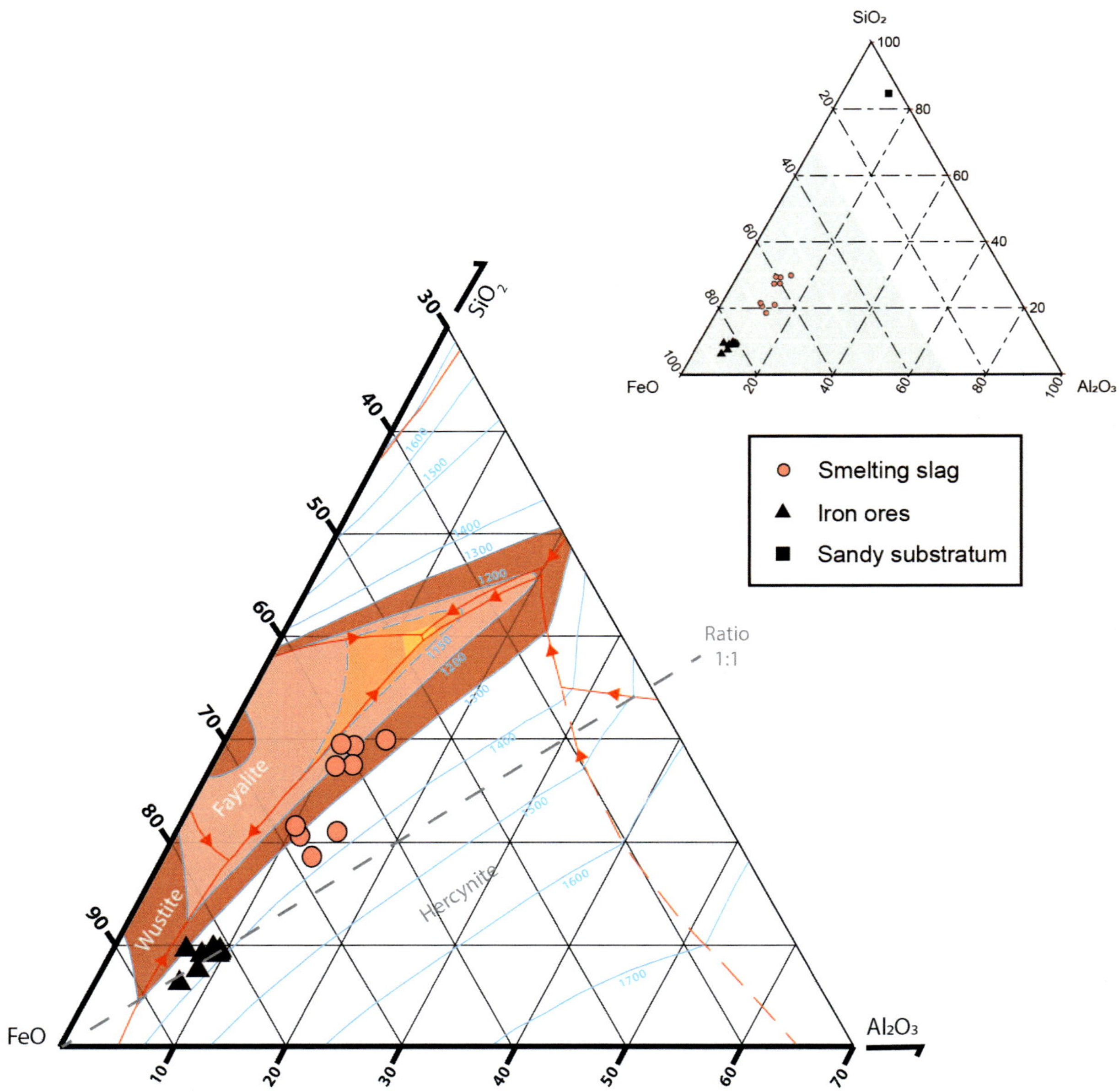

Figure 6.27: $FeO_n - SiO_2 - Al_2O_3$ ternary phase diagram for the bulk compositions of iron ores, sandy substratum and slags from MBR140 (Kowalski et al., 2000).

the slag. The contamination is highly variable: between 6 g and 32 g per 100 g of ore (Table 3, line 2.2). The amount of slag always raises with increasing contamination (Table 3, line 2.3). The bottom furnace slags show a slightly more contamination than the tapped slag, which is consistent with the macroscopic observations. As the pisolitic ore containing kaolinite is rich in aluminium with a low Si:Al ratio, the effect of the contamination with a silica-rich sand is a lower melting temperature of the slag. On the ternary phase diagram for the system FeO_n-SiO_2-Al_2O_3, it can be seen that a slag with a SiO_2:Al_2O_3 ratio of 1:1 will melt above 1400°C (Figure 6.27). This thermal range is rather high and difficult to reach in a small and open bowl furnace. In all cases, the addition of silica will decrease the melting point of the slag. At Amboronala, most of the slag compositions plot above the 1200 °C isotherm, and a few pieces are close to 1400°C. Apparently, it was difficult for the smelters to melt the slag, and this is reflected by the irregular shape of many of the fragments. However, the contamination by the sand is not correlated to the yield (Figure 6.26). A similar yield is calculated for MBR14003 and MBR14004, but the amount of sand incorporated is about three times higher for MBR14003 than for MBR14004. On the other hand, the level of contamination is similar for MBR14003 and MBR14007, but the production of iron is 28 g and 8 g, respectively. The efficiency of the reduction reaction is not directly linked to the proportion of contamination.

6.3.3 Total iron production of metallurgical district of northeastern Madagascar

Available data for other slag heaps from northeastern Madagascar do not allow to perform accurate mass balance calculations. These calculations would be the result of too many approximations to be meaningful. More work is needed to evaluate local production variations.

About 150 slag heaps were located during our fieldworks, which corresponds to approximatively 450 tons of slag. It is likely that we did not find all the existing slag heaps. The amount of slag produced should reasonably be between 450 and 900 tonnes of slag. The average amount of iron produced per 100g of slag in the MBR140 workshop is 23g and the median is 17g. If we take the extreme values, we can estimate the total production in northeastern Madagascar over the four centuries of activity at between 75 and 200 tonnes of iron. It corresponds to between 191 and 500 kilograms of iron produced each year. It is a very low production, far from the intensive productions described in other parts of Africa (Robion-Brunner, 2008).

		Minimum estimated mass of slag	Maximum estimated mass of slag
	For 100g of slag	450 T	900 T
Median production	17 g	76 T	153 T
Average production	23 g	103 T	207 T

Figure 6.28: Estimation of total iron production in the whole metallurgical district over the whole chronological period. The data obtained on MBR140 (Amboronala) is projected over the whole study area.

The current state of research does not allow an accurate estimate of the *Rasikajy* coastal population between the 11th and 15th centuries. The production of chlorite schist was at its peak and trade with the Indian Ocean was active; it was thus a favourable economic period. Surveys have located 23 settlement sites, some of which are probably villages and others larger towns, such as Vohémar or Benavony. There is no doubt that some of the settlements have not yet been discovered. Probably about 20,000 people were living along the coast in the studied area.

At the arrival of the first Europeans in the early 16th century, the Malagasy populations used little iron according to written sources. Iron was needed for spearheads, knives and objects such as iron hooks. Spearheads can be lost during hunt or at war, but knives are generally objects kept for a long time and sharpened regularly. In contrast, the production of chlorite schist tripods requires iron tools that wear out quickly.

Hence, the local iron production was not sufficient to satisfy the iron needs of the population estimated at 20,000 inhabitants. If our estimates are correct, production would correspond to barely 10 to 50g of iron per person and per year. They did not produce enough to export their iron, neither to other Malagasy populations, nor around the Indian Ocean. It was a local production, intended for local needs. The missing iron was most certainly imported. The Persian world, India and China were indeed very large iron producers at that time and would have been able to provide some of the iron needed. Even the African hinterland of the East coast could have been able to provide iron.

6.3.4 A poorly controlled technology?

The metallurgical wastes from northeastern Madagascar show an unexpected pattern. On one hand, the smelters more or less always use the same raw materials in the frame of a single technique based on simple bowl furnaces. They were able to identify excellent ores with exceptional iron grades among other rocks with lower iron content. The technique is moreover shared at the scale of the whole district. By contrast, the chemical composition of the slag is highly variable due to the combined effects of a strong variability in the efficiency of the reduction and the changing proportion of the contamination by the sandy wall.

The temperature rise was also poorly controlled. Indeed, the addition of silica in the system lower the melting temperature of the slag. In our case, the furnace is such a small structure that the ore was probably loaded once or twice. When the melting point is reached, the slag flows and the reduction stops. Above 1000°C, reducing conditions are optimal and will not increase anymore with the rising of the temperature (Figure 2.3). The duration of the operation is thus the decisive factor to reduce a greater quantity of iron oxide to metal. The melting temperatures of the studied slag pieces are between 1150°C and 1450°C, following the plots in the ternary diagram (Figure 6.29). If the temperature rise was controlled and constant, a slag with a higher melting point should be the result of a better reduction process because more time is needed to reach higher temperatures. However in our case, there are some slag pieces from the same slag heap which have different melting point but resulting from operations that reduced the same amount of oxides. The two smelts therefore lasted approximately the same amount of time but with a varying temperature rise. It is very likely that the smelters did not control the temperature rise.

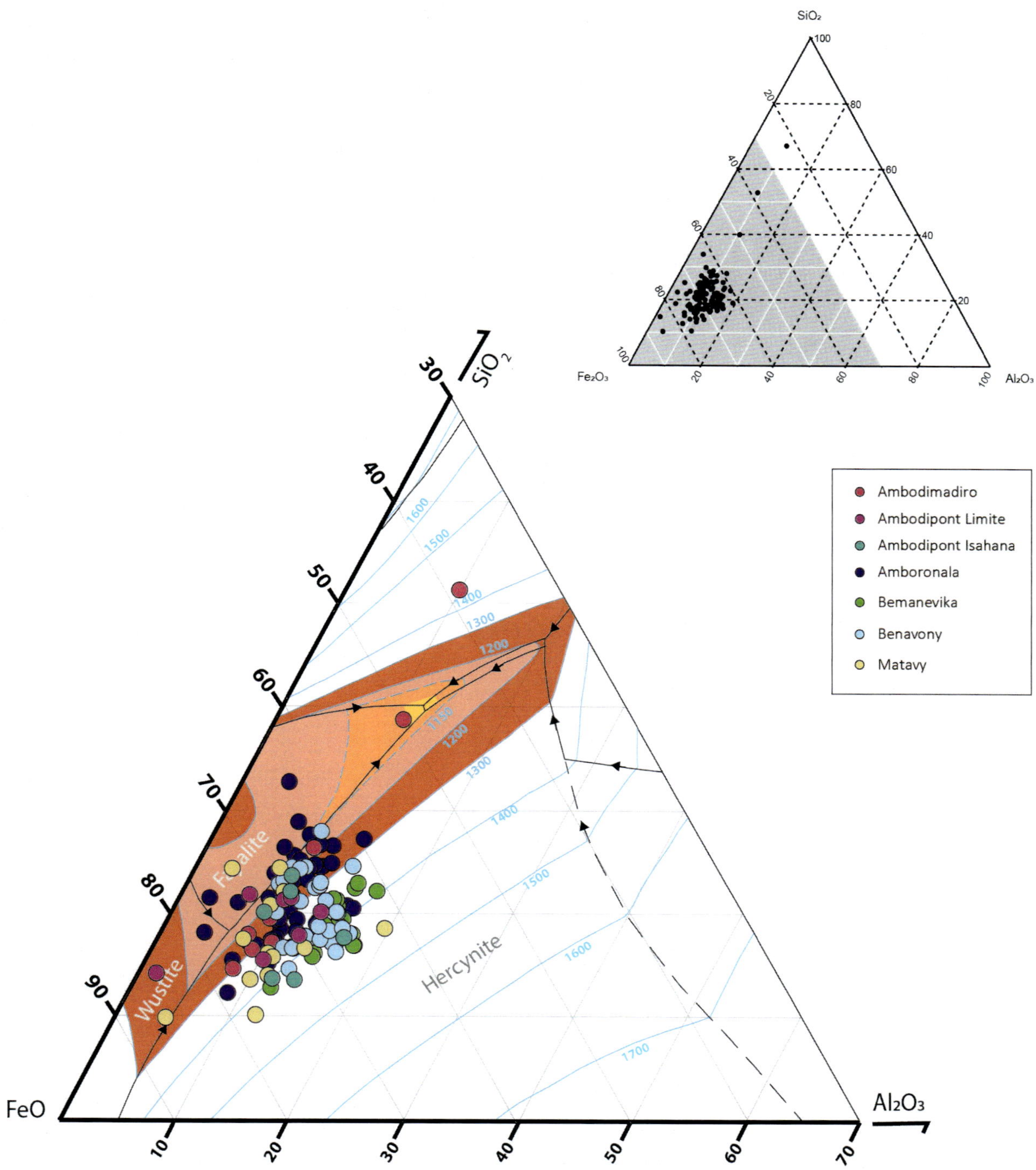

Figure 6.29: $FeO_n - SiO_2 - Al_2O_3$ ternary phase diagram for the bulk compositions of all slag from northeastern Madagascar (Kowalski et al., 2000).

From this variability, one can conclude that the smelters did not reproduce systematically the same operation. The gestures were not replicated properly. Each smelt was different, from quasi-failure up to very successful operations. The people at work were apparently poorly mastering the metallurgical process. This is rather surprising and raises questions about the identity of the smelters and the organisation of the activity.

For an experienced smelter, a professional, it is expected that after a few trials with the same type of furnace and the same raw materials, he will master the process and be able to reproduce it within a narrow range of variability. Inexperienced people working intensively will probably need a few tens of trials to reach a constant level of practice.

For example, the number of operations carried out at MBR140 must have been above 250. This estimate is based on the presence of 4500 kg of slag including at least 250 bottom furnace blocks. The amount of work can be roughly estimated. Each smelt probably lasted less than a day. The procurement of the raw materials, ore and charcoal, may require double that time but can be shared between several people. A total of 250 smelts could thus be the output of a single year of intensive work for a small team of three to five people. Regarding the local climatic conditions, it is more probable that the smelting activity is seasonal, and the two heaps of slag could reflect two successive seasons of activity, but such an intensive activity is unlikely because it would have led to the homogenization of the technique, which is not observed. The site is dated between 1000 and 1200 CE, based on six radiocarbon samples (Figure 4.5 and Figure A.2). Unfortunately, these dates set a time frame but cannot help to evaluate the duration and the frequency of the activity.

It is then also possible to assume less intensive work, longer than annual intervals between the campaigns and the involvement of poorly experienced people. If the period of activity lasts for 100 years or more, then the number of smelts per year falls to two or three. The site can also be the result of several short campaigns, of a few tens of smelts, repeated at interval of several years. In this case, the process will not homogenize. The basic knowledge of smelting is shared by the successive workers, but none of them can be considered as highly experienced. No one is a professional. As the ores are quite rich, even this low level of expertise is sufficient to provide the small amount of iron requested for the local community. This situation can be generalized along the coast in the area where good iron ores are available. In this area, everybody can make the iron he needs for himself. Elsewhere, the supply is provided by the maritime trade network. There is no economic pressure for an increasing and more efficient production.

This socio-economic approach can be challenged by a more technical approach. The bowl furnace as it is reconstructed seems to be a very simple structure. It must indeed be difficult to control the contamination of the crumbling walls made of loose sand. The very simple shape does not allow the strict control of the reduction and temperature conditions. Maybe, such a crude technology cannot be efficiently mastered. But in this case, it is difficult to understand why such a poor technique was maintained for several centuries.

Mineralogical Study of Iron Slag | 7

Slags are complex materials that do not really have any parallel among natural rocks. Iron slag is formed under strong reducing conditions and cools extremely rapidly, especially in the case of tapped slag. The most similar natural rocks would be volcanic rocks, which result from the rapid cooling of a lava.

The chemistry of the slag has a direct impact on its mineralogical composition. The temperature conditions and the cooling rate also play a role. Studying the mineralogy of the slag, i.e. both the minerals present and their chemical composition, is thus essential to understand the behaviour of the chemical elements during the cooling of the slag. It allows to better understand the cooling rate, as well as the reduction and temperature conditions in a bloomery furnace.

The mineralogy of the slag is therefore directly related to the way a furnace operates. However, iron slag mineralogy is understudied in the literature. There is a lack of in-depth and systematic studies that would allow comparisons and better modelling of what happens in a furnace during smelting.

In this chapter, we will present a preliminary study of the mineralogy of the slag from northeastern Madagascar.

7.1 Methodological approach to study slag mineralogy

A combined approach using X-ray diffraction (XRD), optical microscope observations, and scanning electron microscope observations coupled with chemical point analyses (SEM-EDS) was chosen to study slag mineralogy. To better understand mineralogical data, a constant back and forth was conducted with the bulk chemical analyses measured by XRF.

A total of 41 samples were observed under optical microscopy and 32 samples under electron microscopy. With a few exceptions, both optical and electronic observations were undertaken on the same set of samples. The samples were selected to best represent the range of chemical compositions for the different excavated sites (Amboronala - MBR, Ambodimadiro - DMD, Benavony - BNY, Bemanevika - BMK, Matavy - MTY) and for the two slag morphologies (TS and BFS).

The analytical methodology applied to measure the slag mineralogical composition using XRD has already been detailed in the previous chapter (Chapter 6).

7.1.1 Optical microscopy (OM)

Observations under the optical and electron microscopes are made on polished sections of about 5x2 cm square and 1 cm thick. As far as is possible, these sections are sampled to observe both a cooling surface (bottom or top) and the core of the slag, in order to be able to observe the cooling gradients. The sections are then polished to obtain a mirror-like surface.

They are then observed under a metallographic microscope under reflected light. Since most of the minerals in the slag are opaque, reflected light gives more valuable information than observations under transmitted light on thin sections.

The identification of minerals is based on the shape of the crystals, and therefore their habitus, as well as their reflectance, i.e. their ability to reflect all or part of the wavelengths composing white light. The visual identification of minerals is then confirmed with SEM observations and analyses.

7.1.2 Scanning Electron Microscopy (SEM)

Observations on the thick sections were conducted using a Scanning Electron Microscope (SEM - Thermo Fischer FEI XL 30 Sirion FEG) coupled to an Energy Dispersive Spectroscopy detector (EDS - Silicon Drift Detector (SDD) X-MAX). The working distance adopted for all analyses is 8.5 mm, with an energy of 15 keV and an analysis time of 50 seconds. The AZtec software (Oxford Instrument, 2019, version 4.2) was used to process the data.

SEM-EDS analyses allow to perform spot analysis of a phase and thus to know the chemical composition of a particular mineral. However, EDS analyses are not quantitative but only semi-quantitative. To make our analyses as reproducible as possible and above all comparable with other data, I applied the analytical protocol developed by Ildiko Katona-Serneels at the University of Fribourg, following the recommendations of Oxford Instrument. Before each set of 50 analyses, a calibration was performed with a cobalt standard. Furthermore, in order to maintain control over the quality of the analyses, no normalisation of the data to 100% was automatically performed by the software. All analyses whose total sum was not between 98 and 102% were excluded from the dataset.

These precautions are not sufficient to consider our EDS analyses as quantitative, it would have been necessary to perform microprobe analyses. Nevertheless, other samples from other study areas were analysed following this protocol by Ildiko Katona-Serneels, and then analysed with a microprobe. It would seem that when the protocol is rigorously followed the differences between EDS and microprobe analysis are minimal.

7.2 Global description of minerals present in slag

Smelting slags from northeastern Madagascar have a variable but specific chemical fingerprint. As previously detailed (Chapter 6), the analysed samples have high Fe_2O_{3tot} content, high TiO_2 content, but rarely high CaO or K_2O content, and an absence of MnO or P_2O_5. Concerning trace elements, the two most remarkable elements are Cr and V.

The slag chemical composition determines its mineralogical composition and the order of crystallization of the minerals. In the studied samples, the same minerals were recurrently observed, in varying proportions and sizes (Appendix - Figure B.4). The main minerals are spinel (magnetite-like spinels and aluminous spinels), wustite, fayalite and, in some particular cases, leucite.

The XRD data show that magnetite and wustite are not the only iron oxides present in slag samples. Goethite and lepidocrocite, which are iron hydroxides and therefore related to post depositional iron alteration, were also detected. Most of the samples have significant traces of corrosion, notably along fractures or large pores, or where metallic iron fragments are entrapped in the slag.

Other minerals such as quartz and rutile were detected. These minerals are inherited from the contamination of the sandy substratum (see Chapter 6 and Appendix - Figure B.2).

In this section we will describe the main minerals present in the slag, and then, in the next section we will discuss the texture and the order of crystallization of these minerals.

7.2.1 Spinels

The vast majority of the studied samples have spinels in their mineralogical assemblage (Appendix - Figure B.4). This mineral is very often described in archaeological or experimental slag mineral assemblages (Paynter et al., 2015; Killick and Miller, 2014; Ige and Rehren, 2003).

The spinel group is a large family of minerals with a complex chemistry and several poles of chemical compositions (Figure 7.1). The group can be divided into three series according to the chemical composition of the minerals: the aluminous series, of which hercynite ($FeAl_2O_4$) is part of; the ferrous series, of which magnetite (Fe_3O_4) or the ulvöspinel is part (Fe_2TiO_4); and the chromium series, of which chromite is part ($FeCr_2O_4$).

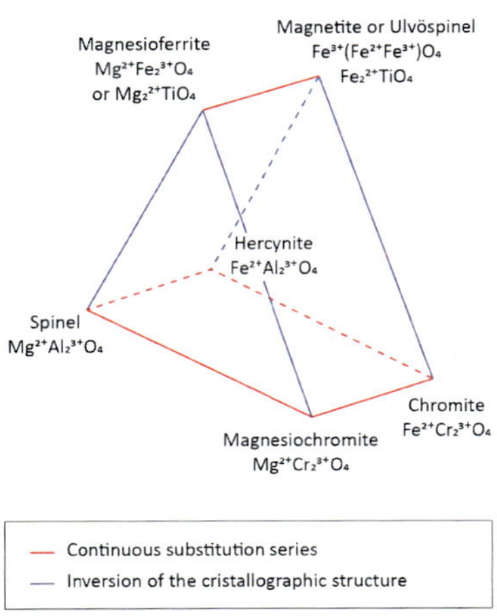

Figure 7.1: Simplified diagram representing the nomenclature and the evolution of the chemical compositions of the different series of the spinel group (Deer et al., 1992). The bases of the triangles are defined by normal spinels and the vertice by inverse spinels.

The general formula of spinels is $A^{2+}B^{3+}B^{3+}O_4$. The cations in the A position occupy tetrahedral sites in the crystal lattice while the cations in the B position occupy octahedral sites. There are spinels with a so-called inverse structure ($A^{3+}B^{2+}B^{3+}O_4$) where a trivalent cation occupies a tetrahedral position. This is the case for magnetite ($Fe^{3+}(Fe^{2+}Fe^{3+})O_4$). This inversion affects the crystallographic structure and therefore the substitutions from one element to another.

There is a continuous series between the spinels *sensu stricto* ($MgAl_2O_4$) and the hercynites ($FeAl_2O_4$), which means that Fe^{2+} and Mg^{2+} atoms can substitute each other without concentration limits. The same applies to Cr^{3+} and Al^{3+} (Deer et al., 1992, Figure 7.1). Vanadium can also replace a 3+ cation in small quantities. Titanium Ti^{4+} can also substitute for Fe^{3+} such as:

$$2Fe^{3+} \rightleftharpoons Ti^{4+} + Fe^{2+} \tag{7.1}$$

Important Al^{3+} to Fe^{3+} substitution may occur in hercynite, though there does not appear to be a complete continuous and natural series to magnetite. The crystal structure of spinel must change from a so-called normal structure to an inverse structure and not all intermediate compositions are observed.

Three types of spinels of different chemical compositions can be distinguished in the studied samples.

Sp1: titaneous-rich ferro-hercynite

A first spinel composition, which will be referred to as **sp1** in the following, is identified in most of the samples. These crystals are polygonal and show a greyish hue under reflected light and BSE microscopy (Figure 7.3). The edges of these grains are generally straight and sharp. The largest of these grains can reach 100 μm in some samples, but in most cases they are between 20 and 60 μm. It is common to see a dozen sp1 grains clumped together, even if in some cases the grains are scattered. Except for textures 7 and 9 (see texture details below), sp1 spinels are the first phase to crystallise when present.

These sp1 spinels have an iron and titanium enriched hercynite composition, though not enriched enough to be magnetite or ulvöspinel. The chemical composition of sp1 in a single sample is variable. But the variability within a sample remains smaller than the overall variability between all samples. The alumina content varies between about 35 and 50 ox.% Al_2O_3 in all the samples analysed and the iron oxides between 40 and 55 ox.% FeO. Titanium is also an important element as the content varies between 1 and 6 ox.% TiO_2. Chromium and vanadium are present significant quantities in sp1 spinels. Chromium can reach locally 4 ox.% Cr_2O_3 and vanadium up to 3 ox.% V_2O_5. In smaller quantities,

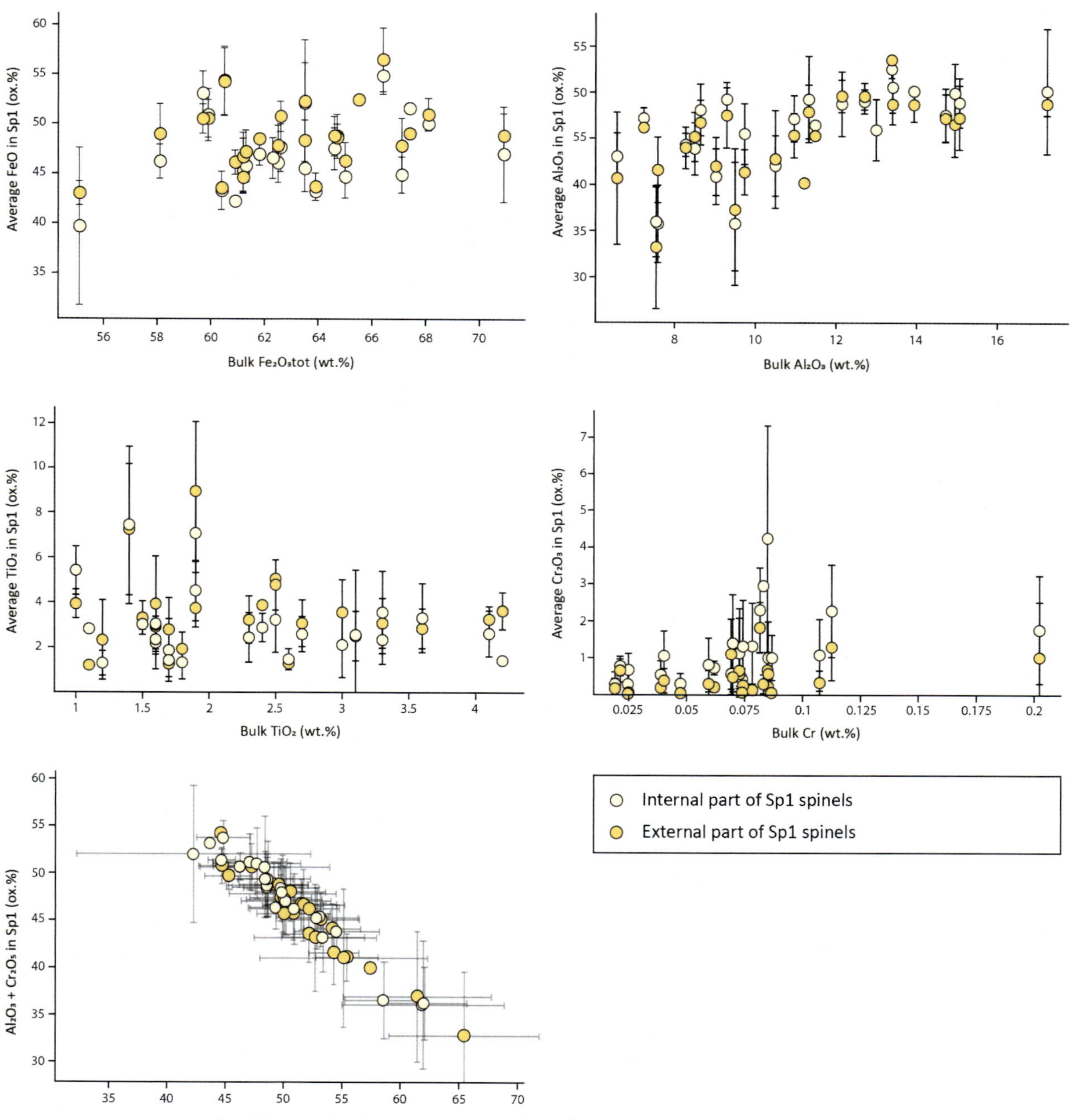

Figure 7.2: Bivariate diagrams illustrating variations of the chemical compositions of Sp1 spinels in the slag samples from northeastern Madagascar. Bulk chemical composition data of a sample (measured by XRF) are compared to EDS spot analysis on spinels from the same sample. Each point therefore represents the average of the EDS point analyses for one sample, both for the internal parts and for the edges of the spinel grains.

MgO (1 ox.% MgO) and SiO_2 (0.5 ox.% SiO_2) are also detected. As noted above, Cr^{3+}, or other 3+ cations, can substitute Al^{3+}. Ti^{4+}, and other 4+ cations, also substitute two 3+ cations (Al^{3+} or Fe^{3+}) by associating with a Fe^{2+} (equation 7.1).

There is no clear correlation between the bulk chemical composition measured by XRF and the average chemical composition of sp1 spinel in a sample. A slight trend is observed for alumina: samples whose bulk composition is enriched in Al_2O_3 tend to have sp1 spinels whose compositions are enriched in Al_2O_3 (Figure 7.2). This trend is not observed for FeO, TiO_2 or Cr_2O_3 (Figure 7.2).

On the other hand, when the grains of a sample are enriched in TiO_2 and FeO, they are depleted in Al_2O_3 and Cr_2O_3 (Figure 7.2). The composition within a single sample progressively reaches a composition closer to that of magnetite or ulvospinel.

The chemical composition of a sp1 grain is however not homogeneous. First, the edges of the grains tend to be enriched in TiO_2 and FeO whereas the core of sp1 grains tends to be enriched in Cr_2O_3. Moreover, concentric circular zonations are observed on sp1 crystals in most samples (Figure 7.3 - see Cr and Ti). These zonations are related to the variation of TiO_2 and Cr_2O_3 contents. Indeed, during crystallization, as the grains grow, the chemical composition of the liquid evolves. By consuming aluminum to crystallize, the liquid is locally enriched in chromium and *vice versa*. This portion of the newly crystallized mineral will therefore be enriched in chromium. The same phenomenon takes place with titanium.

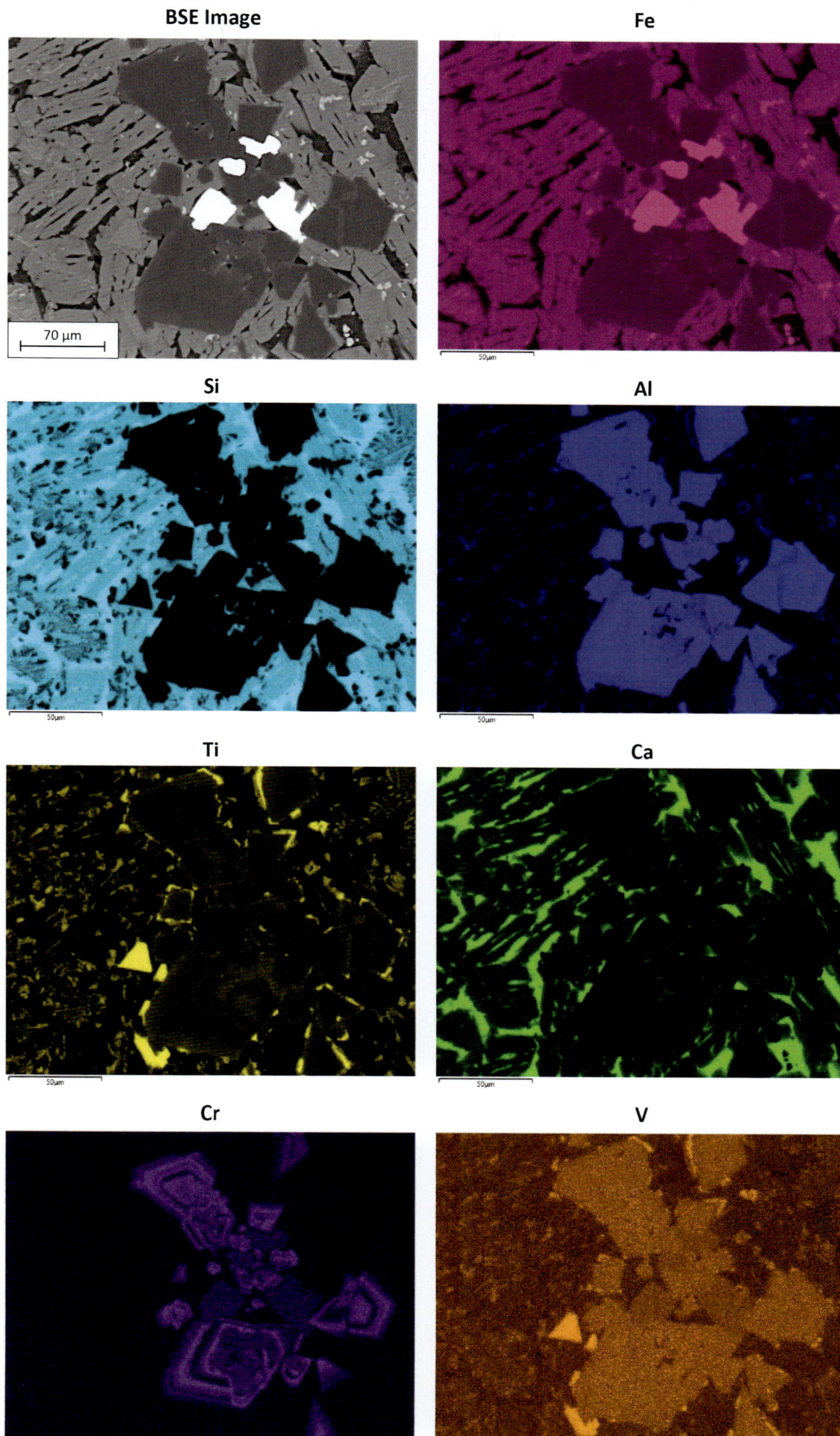

Figure 7.3: SEM-EDS Elementary mapping of the sample BNY41001 (Benavony). The sp1 spinels do not show homogeneous compositions. The edges of the grains are enriched in titanium. Chromium zonations are also clearly visible on this figure. A very slight iron zonation can also be observed.

Sp2: Aluminium-rich magnetite/ulvöspinel

The second composition of spinels in our samples are spinels with high iron and titanium contents. These spinels will be called **sp2**. Sp2 spinels are in most cases found adjacent to sp1 spinels, growing along the edges or corners of their grains. Sometimes they form thin bands along the sp1 grains, sometimes the grains are more developed and form polygonal crystals 5 to 25 μm wide. In the case of fast-cooling textures, the sp2 crystallises as small crystals, independently of the sp1 grains. They appear white in reflected light and in BSE microscopy.

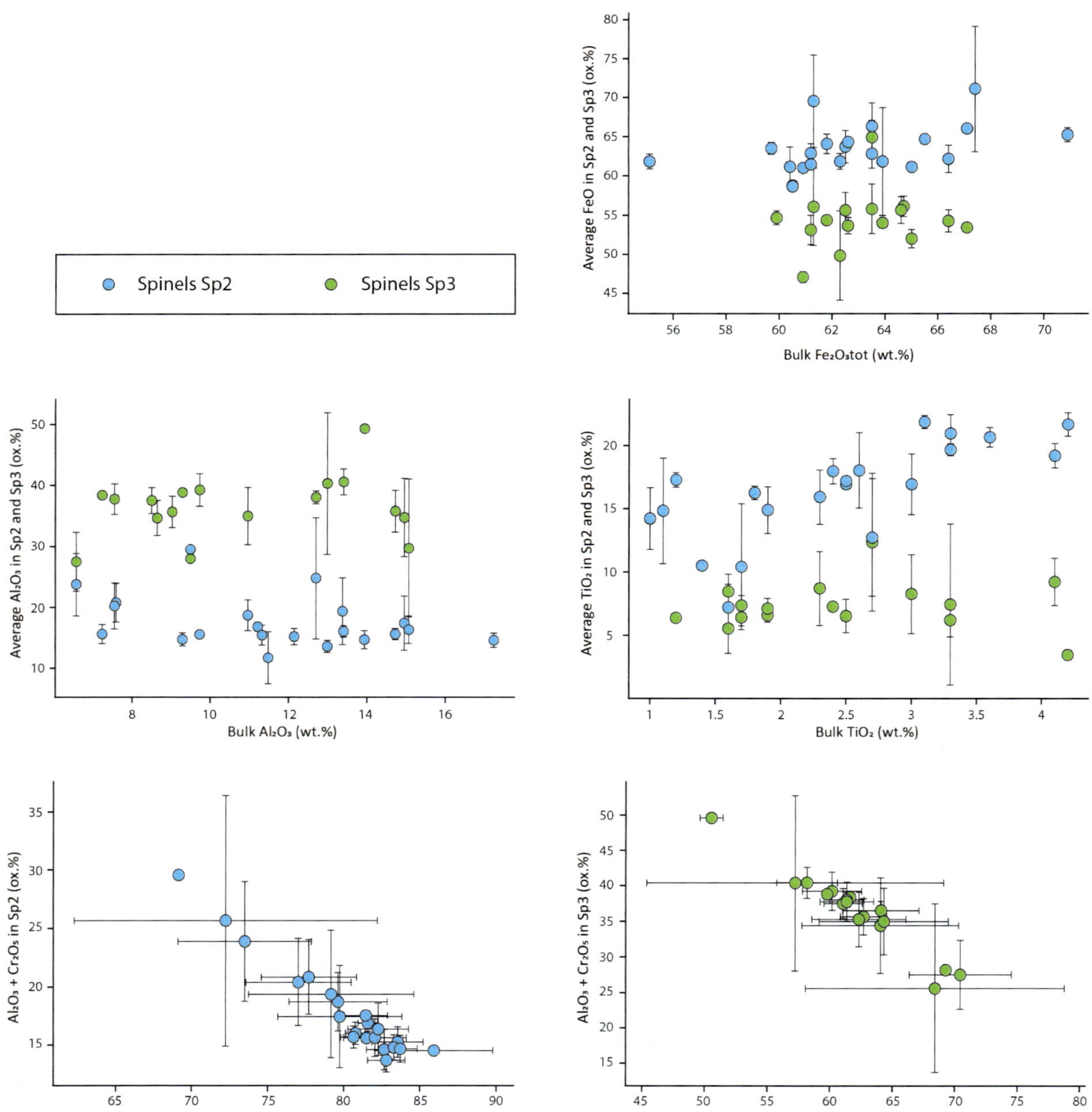

Figure 7.4: Bivariate diagrams illustrating variations of the chemical compositions of Sp2 and Sp3 spinels in the slag samples from northeastern Madagascar. Bulk chemical composition data of a sample (measured by XRF) are compared to EDS spot analysis on spinels from the same sample. Each point therefore represents the average of the EDS point analyses for one sample.

The sp2 crystals show a variability in their chemical composition, both within the same sample and between different samples. But there is a clear difference with sp1 spinels. The iron oxides content varies between 60 and 70 ox.% FeO and the alumina content between 10 and 30 ox.% Al_2O_3. The TiO_2 content is remarkably high and range from 10 to 23 ox.% TiO_2. Chromium is absent from these minerals and vanadium content is between 0.5 and 3 ox.% V_2O_5.

It can be clearly seen that unlike sp1 spinels, there is a clear correlation between the bulk TiO_2 composition of a sample and the TiO_2 content of the sp2 grains (Figure 7.4). The richer the sample is in titanium, the richer the sp2 is. Moreover, as for sp1 spinels, when Al_2O_3 content increases, $FeO + TiO_2$ contents decrease.

These compositions do not clearly correspond to any pole of the spinel group (Figure 7.1). The high iron and titanium contents tend to interpret these crystals as spinels with an inverse crystallographic structure, between magnetite and ulvöspinel but rich in aluminium. Moreover, there is a marked gap in composition between the sp1 and sp2 spinels (Figure 7.5). This is not a gradual evolution of the chemical composition in the same grain but distinct grains of different compositions. One explanation could be the transition from a normal structure to an inverse structure.

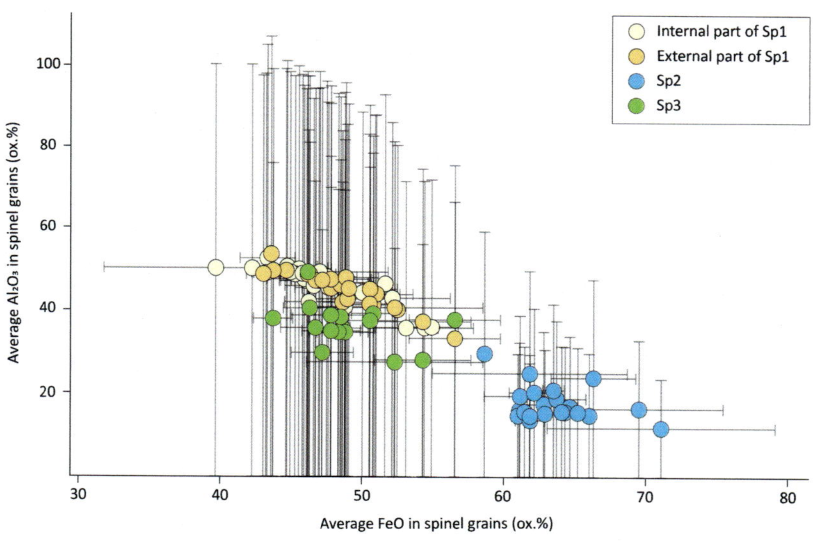

Figure 7.5: Bivariate diagrams FeO-Al_2O_3. There is a clear difference of the FeO:Al_2O_3 ratio between sp1 and sp2 spinels. It is possible that this significant difference is due to the transition from a so-called normal crystallographic structure for sp1 to a so-called inverse structure for sp2. Each point represents the average of the EDS point analyses for one sample.

Sp3: Spinels with an intermediate composition

The last type of spinel appears as small and thin margins, maximum 5 μm thick, found within sp2 grains. These **sp3** spinels have a grey tint and are located at the edge of the sp2 spinels. The separation between these two spinels is saw-toothed and could be an uncompleted re-equilibration structure. The chemical composition of sp2 would have evolved into sp3 towards the end of the crystallization of the grain. Then, near solidification, diffusion took place at the interface of the two spinels compositions to form two more stable phases.

Since Sp3 seems to have a epitaxial growth on sp2 grains, it is probable that sp3 also has an inverse crystallographic structure, even if the contents are close to those of sp1. The FeO content is between 50 and 60 ox.% FeO, the alumina content between 25 and 40 ox.% Al_2O_3 and the titanium content between 5 and 10 ox.% TiO_2. Almost no vanadium and chromium are detected.

The compositions of the sp3 spinels are also variable from sample to sample. However, the bulk compositions of the samples do not seem to impact the sp3 compositions.

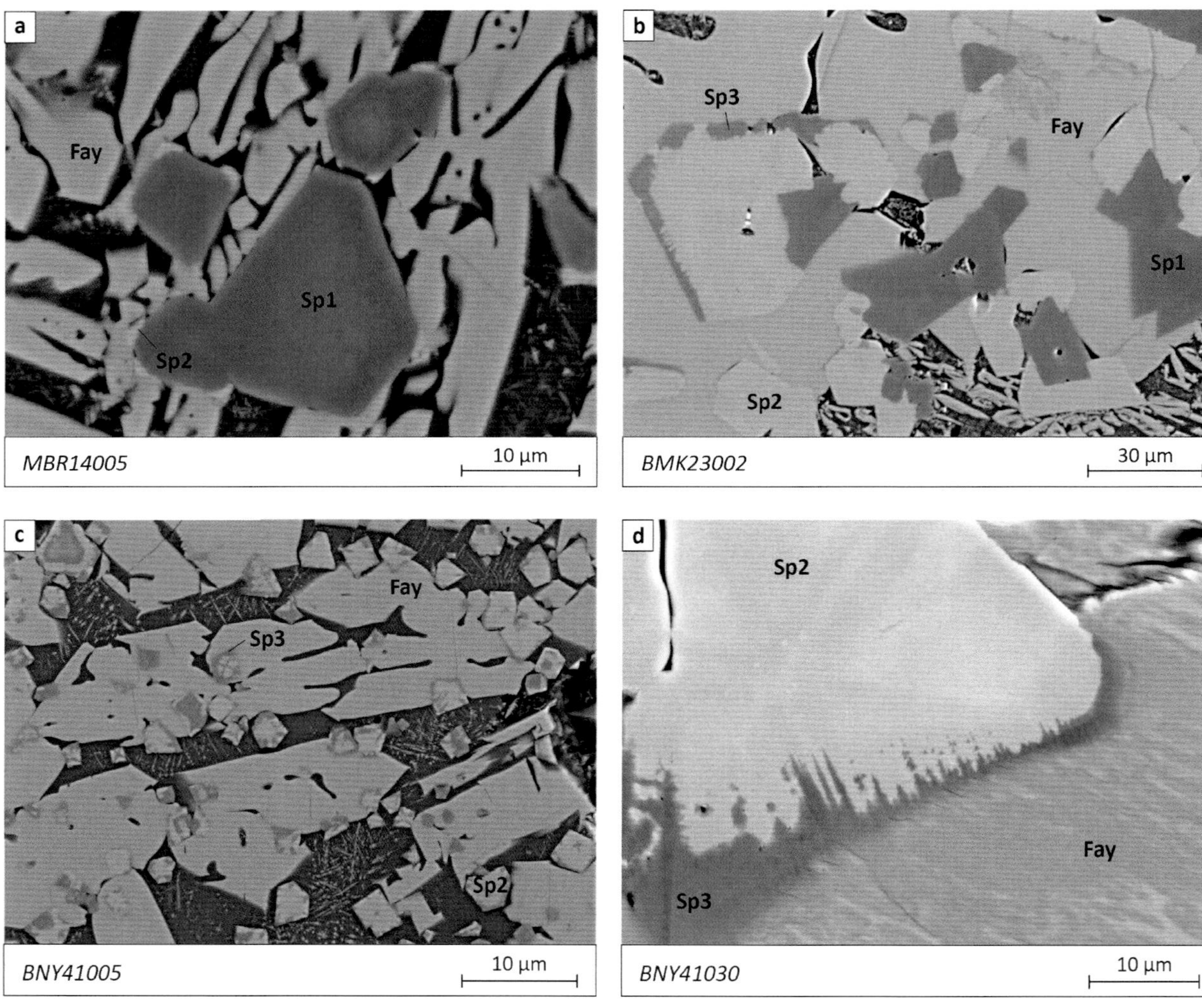

Figure 7.6: BSE Images of spinels with a sp2 or sp3 chemical compositions (Fay: Fayalite, Sp: Spinel). **a.** Example thin sp2 crystals attached to sp1 grains. **b.** Example of large sp2 crystals with a sp3 spinels chemical composition on their edges. **c.** Example of tiny mixed sp2 and sp3 crystals, typical for high cooling rate textures. **d.** Detailed image of the interface between sp2 and sp3.

Symplectite spinels

Finally, we observe some special structures which crystallize among the last phases. These are small grains made of two phases, one grey and one white, intermingled. These are small spinel crystals that crystallise late and are generally observed in the interstices left by the grains already formed. They are for example recurrently observed entrapped inside fayalite grains (Figure 7.7 - a and b).

It seems that these are symplectite phases which crystallize at the eutectic. The temperatures are too low at this stage of cooling for the chemical elements to be mobile in the liquid. Minerals therefore crystallise according to the chemical elements available in the immediate environment. The two phases thus crystallise into interpenetrating dendrites at the same time, giving these particular interpenetrated structures (Figure 7.7 - b, c and d).

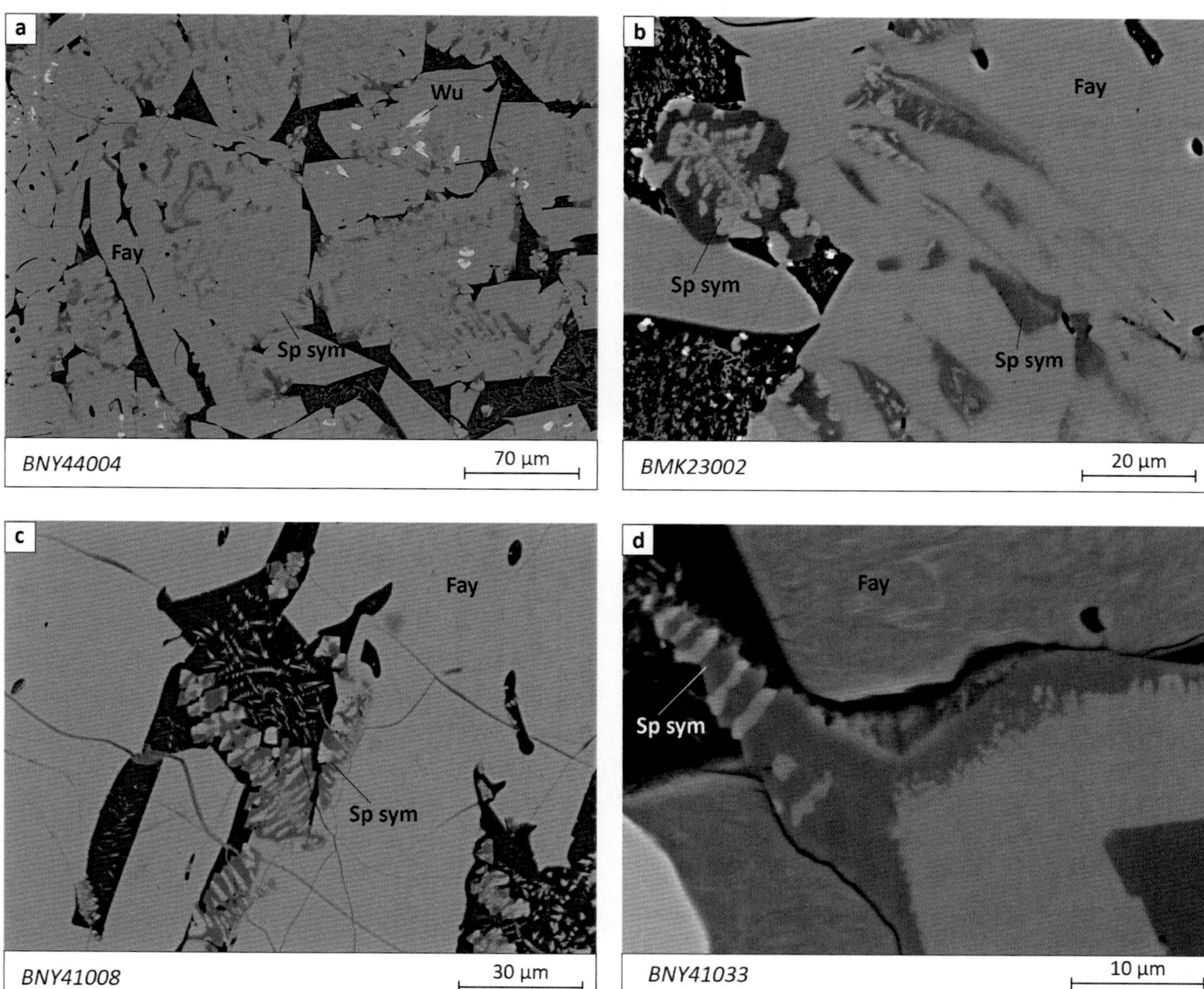

Figure 7.7: BSE Images of symplectite spinels (sp sym) crystallizing at the eutectic (Fay: Fayalite, Sp sym: Symplectite spinel, Wu: Wustite). **a.** Example of spinels entrapped in the interstices of a growing fayalite. **b**, **c** and **d.** Detailed images of the two intermingling spinel phases growing at the same time.

It is not really possible with SEM-EDS technique to analyse these phases as they are too small. However, it seems that the white phase correspond to the sp2 composition and the grey phase to the sp3 composition. In the remainder of this section, we will call this mixed structure a symplectite spinel (sp sym).

7.2.2 Wustite

Wustite (FeO) is the main iron oxide found in the samples studied. This mineral is not systematically present in all the samples, it was observed by optical microscopy in 25 samples out of 41. This mineral appears white in both reflected light microscopy and SEM backscattered electrons. It has a dendritic shape with varying degrees of thickness or rippling (Figure 7.10). The faster the cooling, the thinner and shorter the dendrite. In cases where the wustite is one of the last phases to crystallise, it takes the form of fine vermicelli or small dots (Figure 7.10 - c). In some cases, wustite is oxidised to hematite (Fe_2O_3). This oxidation process generally takes place when the furnace is still hot and the atmosphere is no longer reducing enough. The shape of the wustite dendrites are then still visible, but their surface is no longer white but greyish (Figure 7.10 - d).

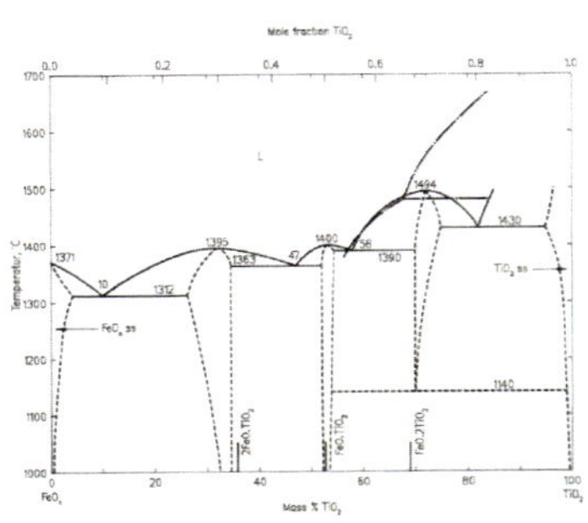

Figure 7.8: FeO-TiO_2 phase diagram: On the left hand side of the diagram it can be seen that the wustite at high temperature (1300°C) is stable with 5% TiO_2. As the temperature decreases, the stability range of the wustite decreases and the crystal can incorporate less and less TiO_2. Titanium-bearing exsolutions are then formed (Kowalski et al., 2000).

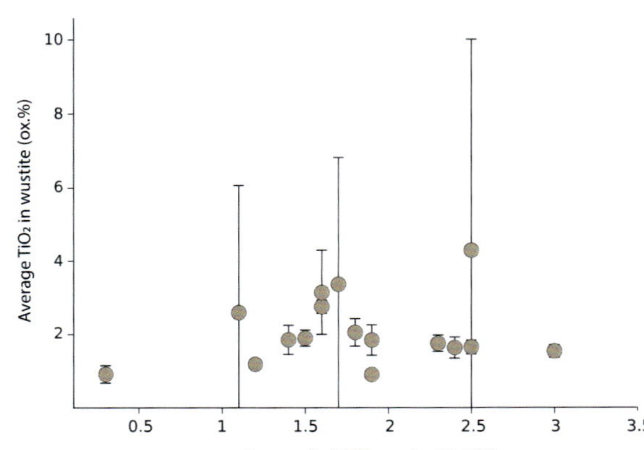

Figure 7.9: Bivariate diagram comparing the bulk TiO_2 of a sample (measured by XRF) and TiO_2 content in wutiste grains within the same sample (EDS spot analysis).

Wustite can incorporate up to 5% of titanium oxide as a substitute of iron at high temperatures, i.e. temperatures above 1300°C (Figure 7.8). EDS spot analyses on wustite grains show that the TiO_2 content varies between 2 and 4% (Figure 7.9). No correlation could be observed between the average TiO_2 content of a sample and the average TiO_2 content of the wustite grains within the same sample. When the slag cools, the wusite crystals are no longer stable with such high titanium content. Titanium-bearing exsolutions are formed. In the solid state, the FeO-TiO_2 phase diagram illustrates this phenomenon. However, it should be kept in mind that it models a simplified system since the slag is not only composed of FeO and TiO_2, but also of Al_2O_3, SiO_2, CaO, etc., which can affect the stability ranges of the minerals.

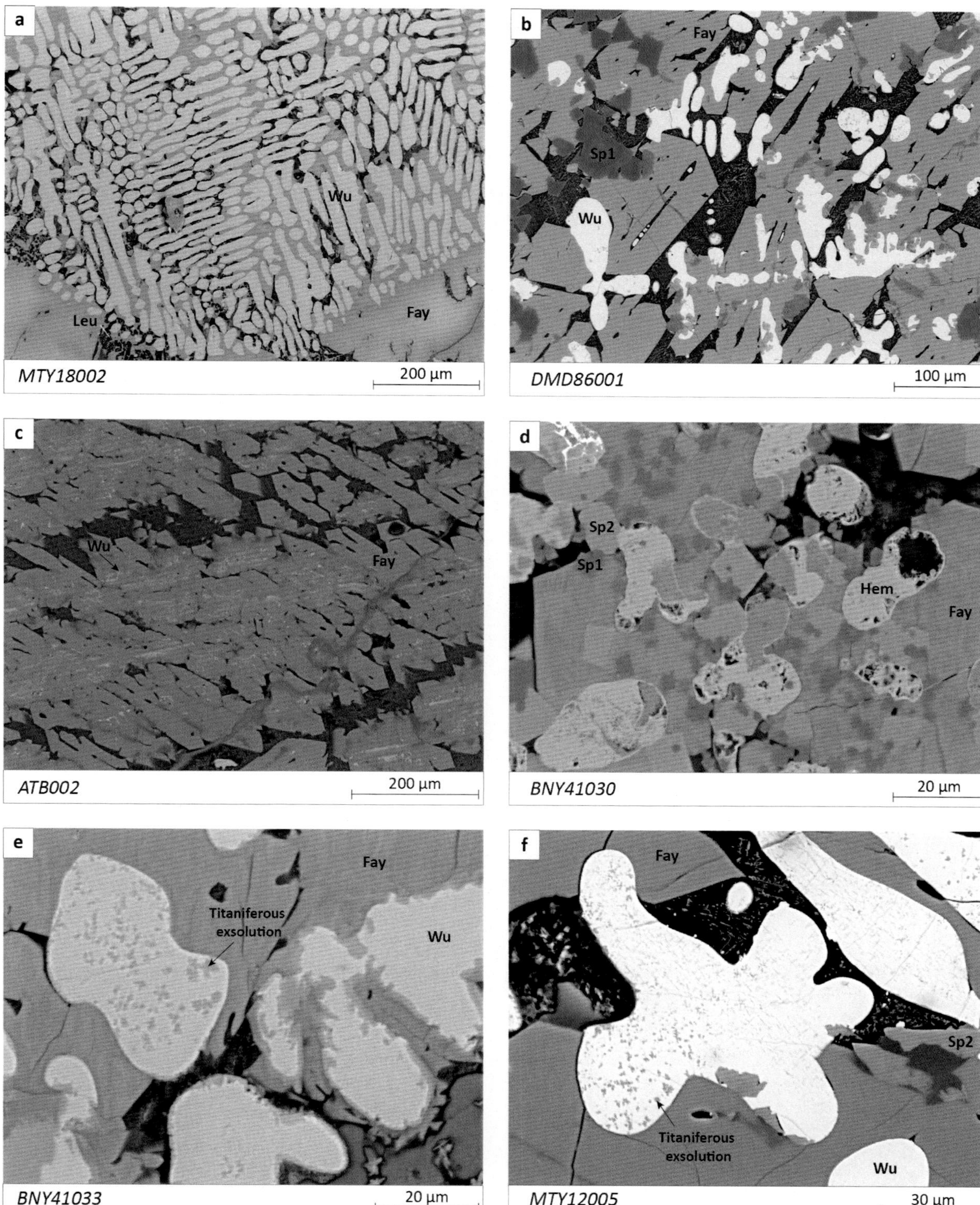

Figure 7.10: BSE images of wustite (Fay: Fayalite, Hem: Hematite, Sp: Spinel, Wu: Wustite). **a.** Example of compact and large wustite dendrites. **b.** Example of long and scattered wustite grains. **c.** Wustite in the shape of long and thin vermicelli. It could be a symplectite crystallization with fayalite. **d.** Example of wustite dendrites which were altered in hematite (Fe_2O_3). **e** and **d.** Titaniferous exsolutions in wustite.

The titaniferous phases are included in the wustite dentrite, sometimes with polygonal shapes, sometimes in the form of very small vermicelli (Figure 7.10 - e and f). These phases rarely exceed 5 μm and are therefore too small to be properly analysed with the SEM-EDS technique. According the diagram, the exsolutions should be made of $FeTiO_4$ (Figure 7.8). On the EDS elementary maps, these phases appear to contain a significant proportion of TiO_2 and also some iron (Figure 7.11). These titaniferous exsolutions are a good indicator of high temperatures (>1300°C) in the furnace, without being an accurate thermometer. The presence of these phases confirms that the liquid has reached high temperatures. However, their absence can mean that the liquid did not reach these temperatures, but also that the slag cooled too quickly for the exsolutions to occur. It may also mean that the sample has low TiO_2 content. Finally, the fact that the wustite grains still contain up to 4 ox.% TiO_2 shows that the cooling rates are too high for all the titanium to be exsolved.

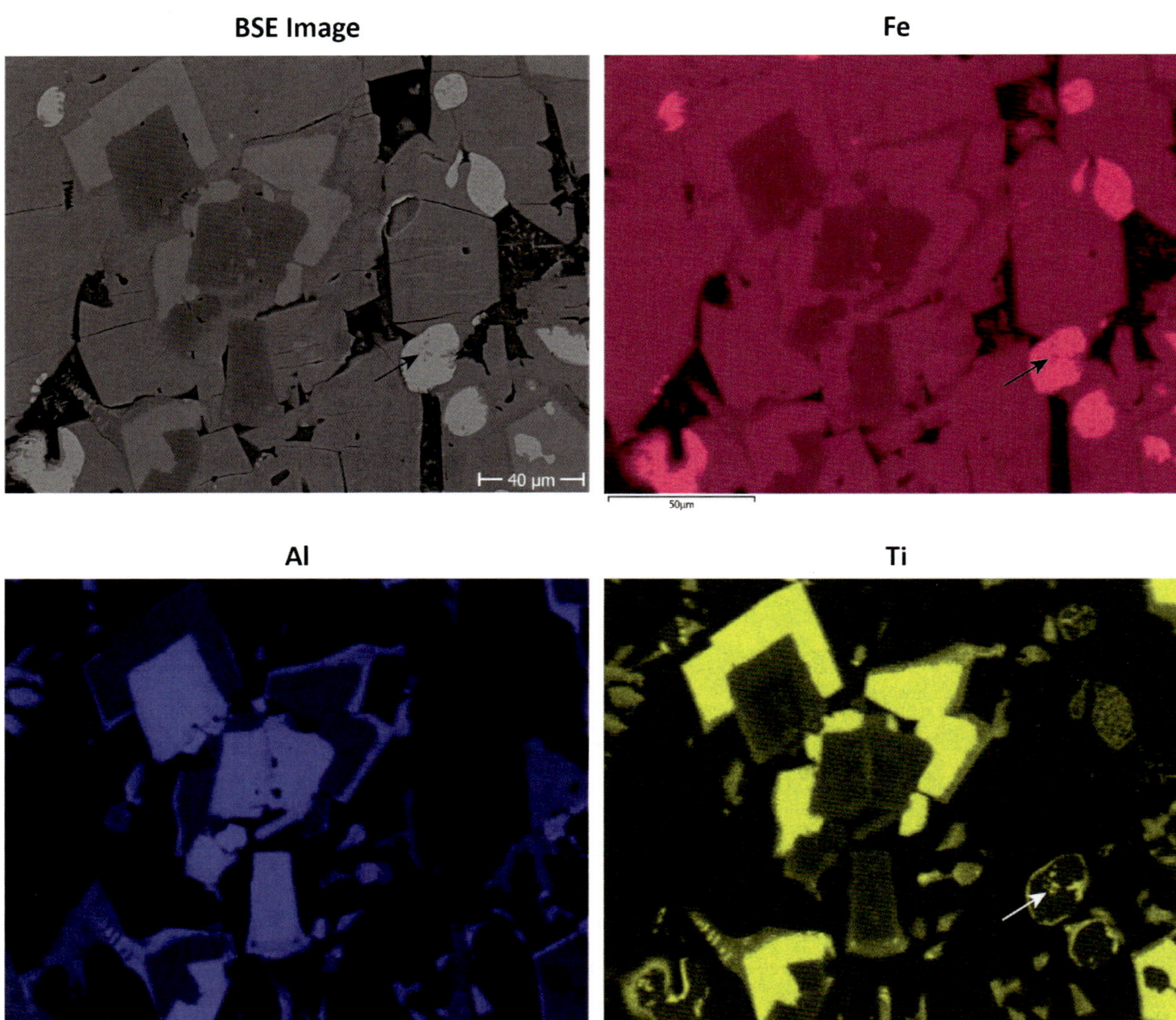

Figure 7.11: SEM-EDS Elementary mapping of the sample BNY41033 (Benavony). It is observed that the exsolutions in wustite grains are rich in titanium and that they also contain iron (see arrows).

7.2.3 Fayalite

Fayalite (Fe_2SiO_4) is systematically found in the slags from north-eastern Madagascar. This mineral belongs to the family of olivines. Fayalite appears with a light grey hue under both optical and electron microscopy. It can crystallise under several shapes depending on the cooling rate. Grains that had time to crystallise show a stubby prismatic shape (Figure 7.13 - a and b), while those that crystallised quickly have a so-called skeletal rod shape (Figure 7.13 - c, d, e and f).

Iron in fayalite in found under its Fe^{2+} oxidation state and can therefore be substituted by other cations with the same charge and a similar size. There is a continuous solid solution between the Fe and Mg pole (Mg_2SiO_4 - forsterite) or between the Fe and Mn pole (Mn_2SiO_4 - tephroite). On the other hand, calcium can only be integrated in small quantities in fayalite. There is however an olivine with calcium, which is called kirschsteinite ($CaFeSiO_4$), but this mineral does not show any appreciable variation from the ideal composition (Deer et al., 1992).

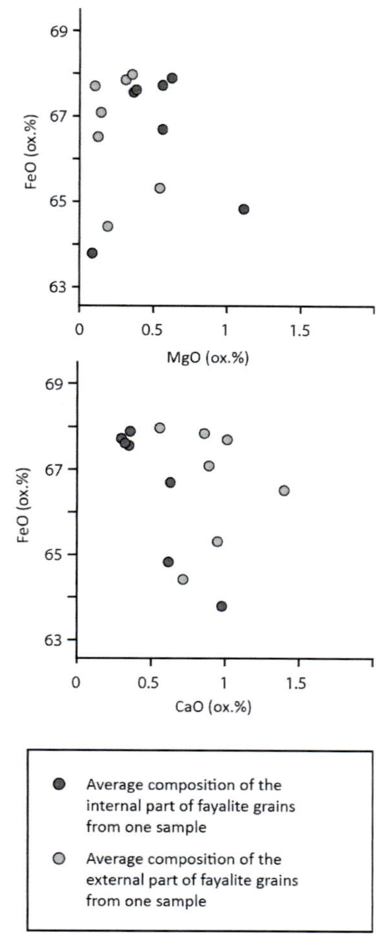

Figure 7.12: Diagram showing the evolution of magnesium and calcium contents in fayalite grains (SEM-EDS measurements). Each point represents the average value for one sample of the chemical composition of the internal or external parts of the fayalite grains present in this sample.

In the studied samples, the chemical compositions of the fayalites are homogeneous and show little variability, whatever the chemical composition of the slag. On average, the fayalites contain 29-32 ox.% SiO_2, 65-69 ox.% FeO and some minor elements in substitution, which are titanium (0.2-0.3 ox.% TiO_2), aluminium (0.2-1.0 ox.% Al_2O_3), calcium (0.2-4.0 ox.% CaO), magnesium (0.1-2.0 ox.% MgO), manganese (0.1-0.7 ox.% MnO) and phosphorus (0.1-0.4 ox.% P_2O_5). Calcium and magnesium are the chemical elements that show the greatest variability not only from one sample to another, but even from one fayalite grain to another within the same sample. No correlation was observed between the bulk calcium and magnesium contents measured by XRF and the calcium and magnesium contents of the fayalites. The other chemical elements show very little variability within the same sample.

Moreover, the edges of the fayalites tend to be enriched in calcium, but never as much as in kirschsteinite, and to be depleted in magnesium (Figure 7.15 and Figure 7.12). Proportionally, iron content decreases, and generally also titanium and aluminium contents. Silicon and manganese remain stable. Finally, we note that fayalites grains with skeletal textures, of which second generation fayalites, and therefore associated with high cooling rate, tend to have higher calcium content, even in the core of the grain (see BMK42001). On the contrary, phosphorus tends to be incorporated in first generation fayalite rather than in second generation crystals. Phosphorus is integrated as a substitution for silicon and is typical of high temperature phases. No forsterite was found and only one sample observed by SEM contains kirschsteinite (MBR13006,

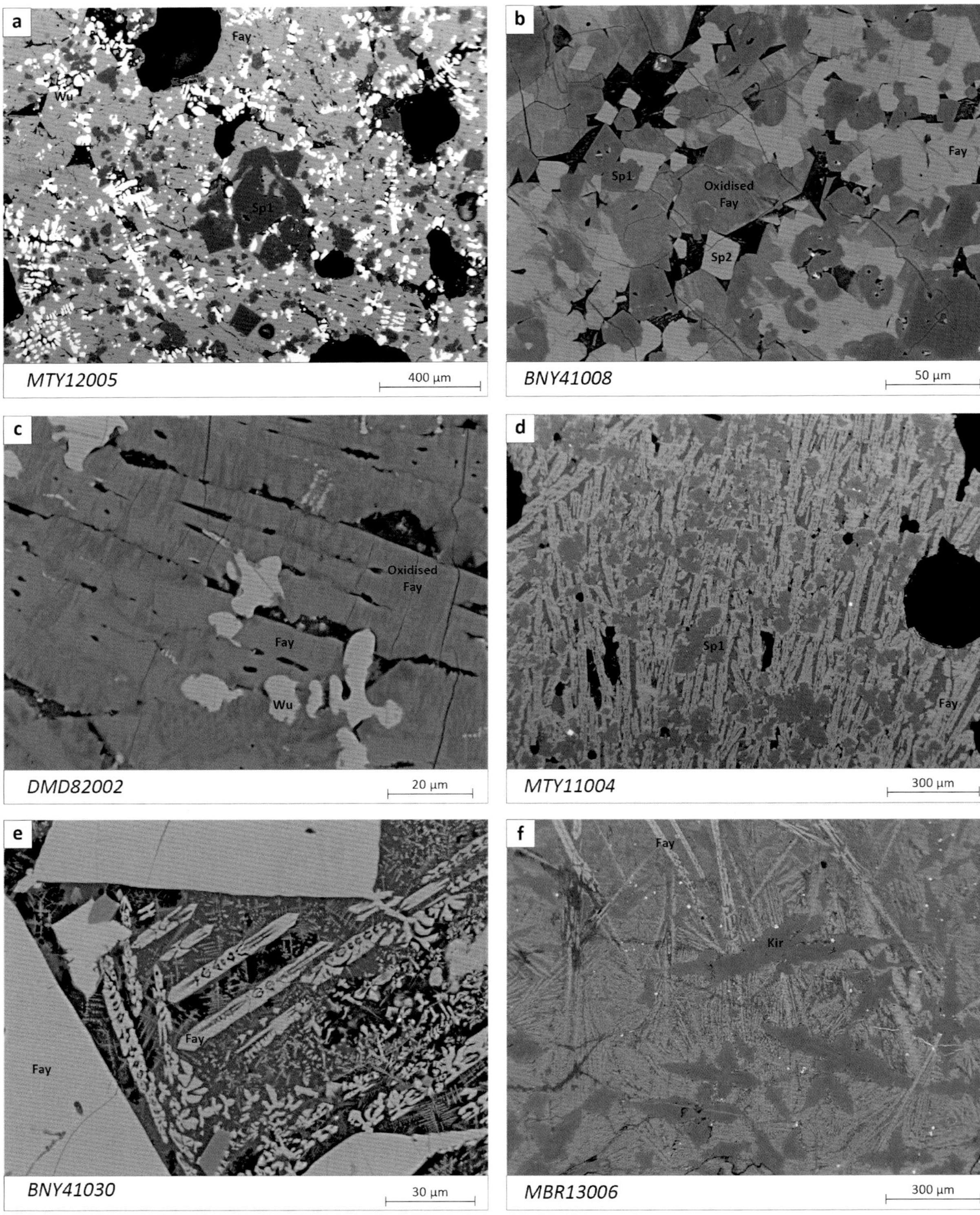

Figure 7.13: Backscattered electron image (BSE) of the different crystallization forms in which fayalite can be observed (Fay: Fayalite, Kir: Kirschsteinite, Sp: Spinel, Wu: Wustite). **A.** Sample with stubby large fayalite crystals. **B.** Prismatic crystal of fayalite. The darker grey area are altered part of the fayalite grain. **C.** Fayalite crystals with a skeletal shape. There are also dark grey bands which correspond to the oxidised areas of the fayalite. **D.**. Fayalite crystals with a skeletal shape. **E.** Second generation fayalite grains and crystallization with a skeletal shape. **F.** Sample with grains of Kirschsteinite and long and thin skeletal fayalitic rods.

Figure 7.13 - f) which has indeed high CaO content (1.6 wt.% CaO). This sample is thus an exception as it contains fayalite with 5-6 ox.% MgO and 4-5 ox.% CaO, and therefore has lower iron oxides (55-56 ox.% FeO).

Fayalite can contain iron under the Fe^{3+} oxidation state but only in small quantities. This phase is mainly an alteration product, as fayalite is sensitive to water circulation (Deer et al., 1992). These oxidation structures could be observed on most of the studied samples. Indeed, these structures are visible under a microscope as they make the fayalite slightly darker (Figure 7.13 - b and c). When comparing the EDS analysis of these dark zones with the analysis performed on the same fayalite grain but on a 'normal' zone, we can see that there is no variation in minor element compositions. There is only a slight variability in iron and oxygen contents, reflecting the presence of Fe^{3+} (Figure 7.14). Some of these structures are found in the vicinity of pores or fractures, so there is a high probability that they are related to post-depositional alteration. But since we know that the studied slag samples contain a lot of Fe^{3+} (Chapter 6, Figure 6.3), we can wonder if the fayalites did not integrate Fe^{3+} as soon as they crystallised.

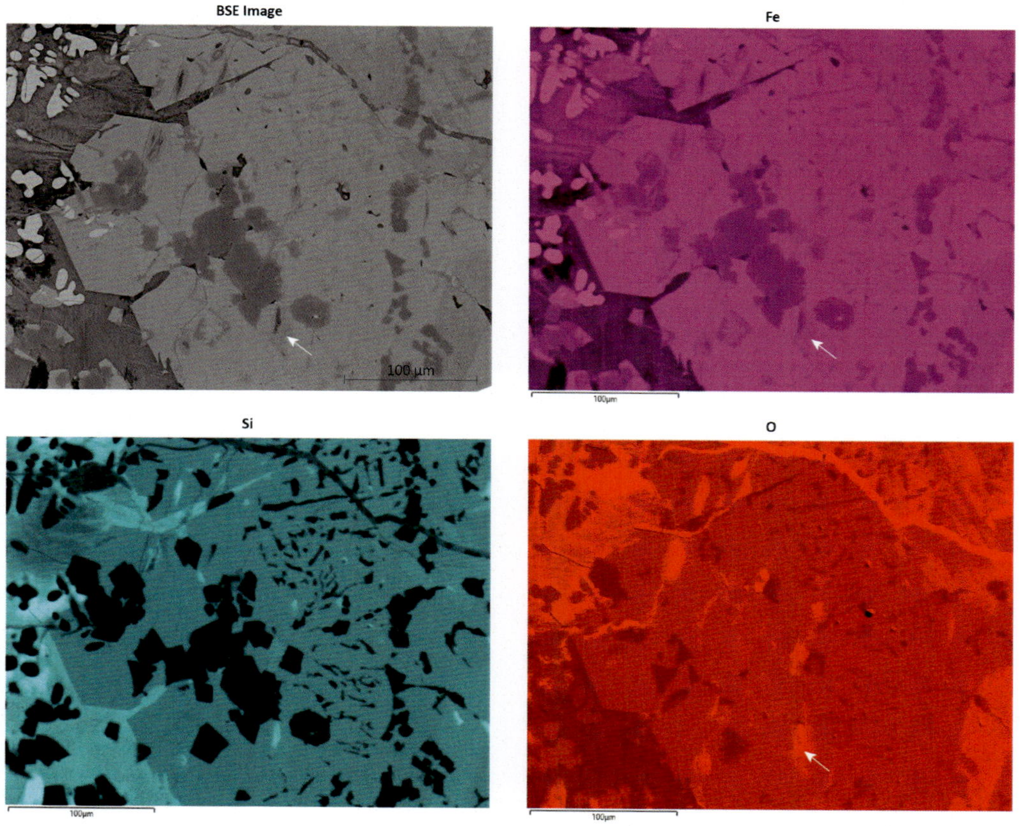

Figure 7.14: SEM-EDS Elementary mapping on the sample ATB002 (Antaimby, Sambava). Dark grey structures can be seen on the fayalite grains (e.g. the area marked by an arrow). No variation of minor elements content, such as calcium or magnesium, can be observed. There is however a slight variation in oxygen and iron content. This variation means that some of the iron in the fayalite is under the Fe^{3+} oxidation state.

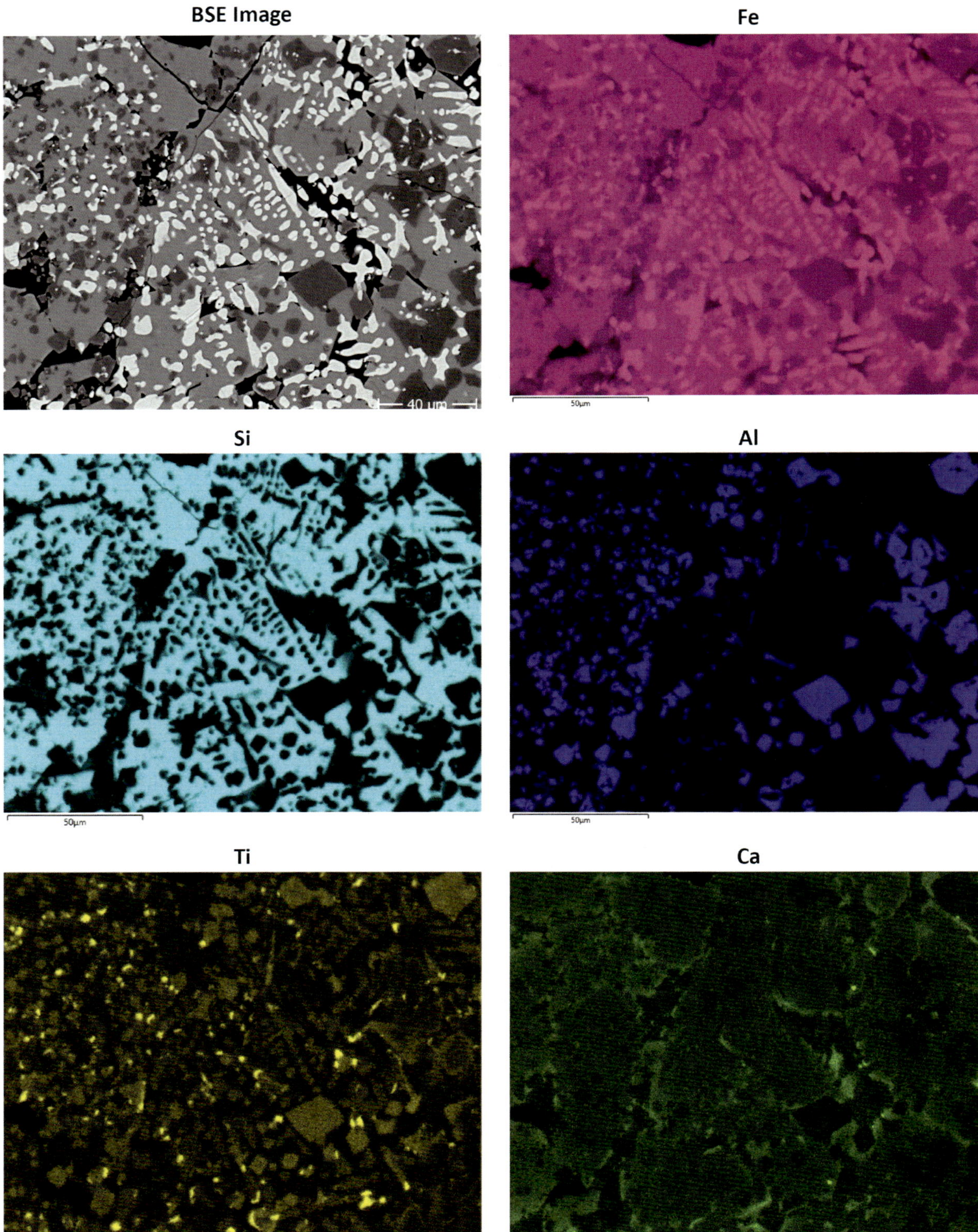

Figure 7.15: SEM-EDS Elementary mapping on the sample DMD92002 (Ambodimadiro). It can be seen that the fayalite grains are enriched in calcium on the edge of the crystals.

7.2.4 Leucite

Leucite ($KAlSi_2O_6$) is a feldspathoïd observed in some samples as one of the last phases to crystallize. With a few exceptions, this phase is detected in samples with the highest K_2O content (0.6-1.5 wt.% K_2O).

Leucite appears as a black phase both under optical and electron microscopes. It is observed in the interstices left between minerals that have crystallised before, and notably between the fayalite rods. Most of the time, it crystallises in symplectite with another mineral, either fayalite (Figure 7.16- **b** and **c**), or wustite (Figure 7.16- **d**). Visually the two minerals are intermingled because they crystallize at the same time. The crystals occupy the remaining space and cannot have their own shape.

In the few cases where the K_2O content is high, leucite has its own shape and relatively large crystals can grow (Figure 7.16- **a**).

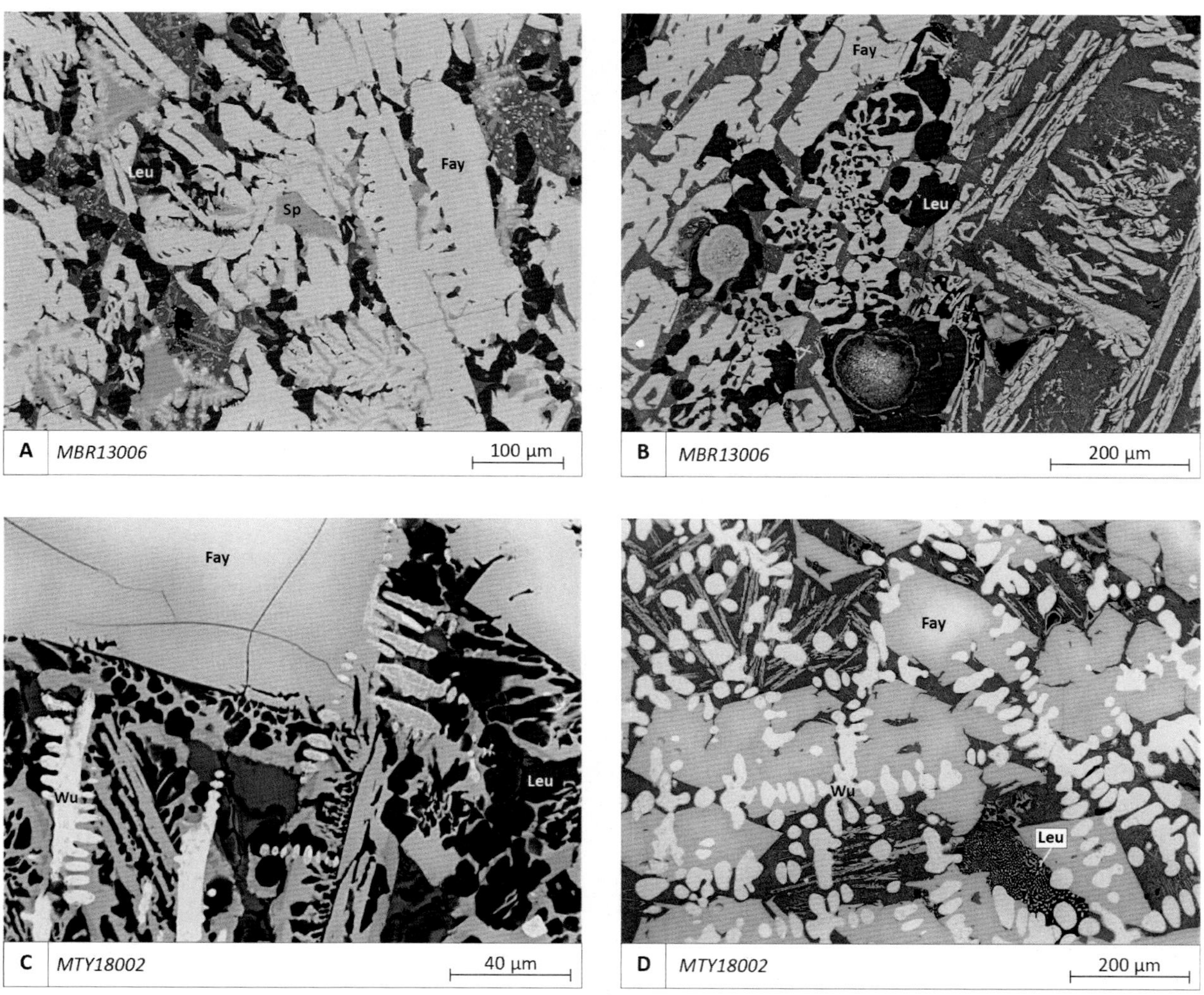

Figure 7.16: Backscattered electron image (BSE) of the different crystallization forms in which leucite can be observed (Fay: Fayalite, Leu: Leucite, Sp: Spinel, Wu: Wustite). **A.** Sample where leucite crystals with a proper shape are observed. **B. C.** and **D.** Sample with a leucite-wustite symplectite area at the eutectic.

7.2.5 Glassy phases

The last phase to solidify is always a non-crystallised glassy phase. This phase includes all the chemical elements that have not been incorporated into the previously crystallised minerals, in particular what remains of silicon, aluminium, iron or calcium, but also in smaller quantities sodium (Na), phosphorus (P), potassium (K) and in some cases a little sulphur (S).

Small crystalline phases, such as small fayalite, or small beads rich in sulphur, are often found entrapped in the glass (Figure 7.17). It is not uncommon to detect small metallic iron particles as well.

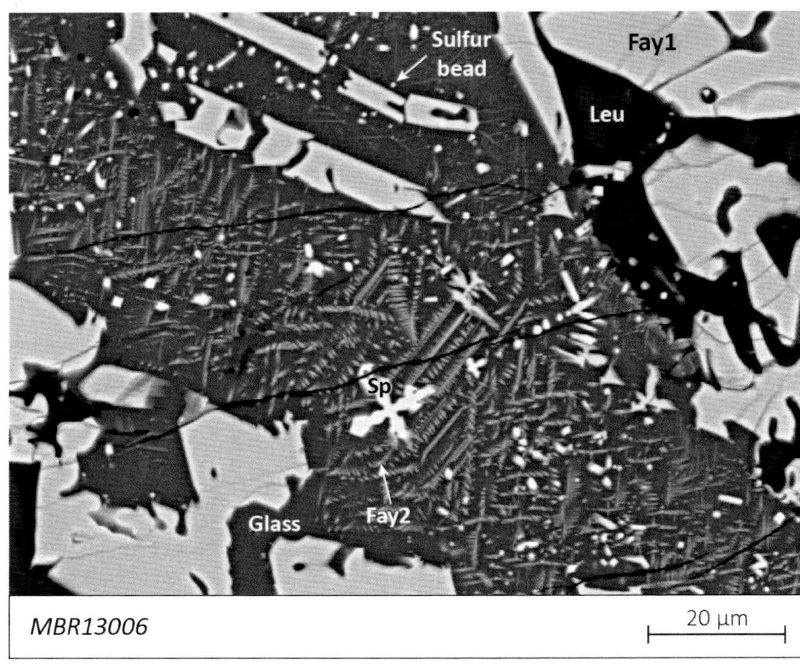

Figure 7.17: Backscattered electron image (BSE) of the last phases to solidify in the sample MBR13006 (Fay: Fayalite, Leu: Leucite, Sp: Spinel, Glass: glassy phases).

7.3 Textures

Not all the minerals described above are necessarily present in every sample. The crystallization order and size also vary, depending on the chemical composition of the liquid and the cooling rate.

Smelting slag can be likened to a liquid magma that cools and solidifies with minerals that gradually crystallise. Slag, however, cools particularly quickly, within minutes or hours, depending on the technical processes (internal slag, external tapping slag, etc.). In general, the faster a liquid cools, the smaller the minerals will be as they will not have had time to crystallise and grow.

Furthermore, the chemical composition of the cooling liquid decides which mineral will crystallise first. The order of crystallization in slag can be modelled using a FeO-SiO_2-Al_2O_3 phase diagrams, although this is a simplified system that does not take into account some elements such as titanium (Figure 7.19 to Figure 7.33). When the first mineral crystallises, the liquid is depleted in the chemical elements composing the mineral. The chemical composition of the liquid therefore progressively changes (see red arrow on Figure 7.19 to Figure 7.33). After a while, the chemical composition of the liquid touches the cotectic lines (black lines on Figure 7.19 to Figure 7.33) and both minerals on either side of the line crystallise at the same time. The chemical composition of the liquid evolves along the cotectic line until it reaches the eutectic where a third mineral crystallises at the same time. In practice, in the case of smelting slags, at some point, the temperature becomes too low to crystallise any mineral. A glassy phase then forms, containing all the chemical elements remaining in the final liquid[1].

1: This approach has been developed for several years by V. Serneels (see Serneels et al., 2019a).

The arrangement of the minerals in relation to each other, their size, the order of crystallization of each mineral family and their proportion to each other were documented. This led to the definition of 13 different textures which are described in this section (Figure 7.18). These textures provide insight into the variability of slag formation processes.

Sample	Morphology	Texture 1	Texture 2	Texture 3	Texture 4	Texture 5	Texture 6	Texture 7	Texture 8	Texture 9	Texture 10	Texture 11	Texture 12	Texture 13
DMD82002	TS			xx			x							
DMD86001	BFS					xx								
DMD92002	TS					xx								
DMD92003	TS						xx							
ATB002	BFS	xx				x				x				
BNY41002	TS	xx	xx	x										
BNY41005	TS			xx						x				
BNY41006	BFS		xx		x	x		x						
BNY41008	BFS		xx			x								
BNY41030	BFS		xx					x						
BNY41033	BFS					xx				x			x	
BNY44001	BFS		xx											
BNY44004	BFS	x	x			x								
BNY45005	BFS					xx				x				
BMK13002	BFS		xx											
BMK23001	BFS	xx	x											
BMK23002	BFS		xx											
BMK42001	BFS					xx								
BMK24002	BFS					xx								
BMK24005	BFS	xx												
BMK24006	BFS		xx											
BMK42002	BFS	xx												
BMK42004	BFS	xx			x			x			x			
MTY11002	TS			xx										
MTY11004	TS	xx												
MTY11005	TS			xx		x								
MTY11006	TS			x						xx				
MTY11012	BFS	x				xx								
MTY12005	BFS					xx		x						
MTY12007	BFS						x		xx					
MTY18002	BFS								x			xx		
MTY18006	BFS					xx								
MBR11001	TS	x		x										
MBR11005	TS	xx		xx		x								
MBR13006	BFS			x										x
MBR13010	TS					xx								
MBR14004	TS	x				xx								
MBR14005	TS			xx		x								
MBR14007	TS					xx								
MBR14010	BFS					xx							x	
MBR14011	BFS								x					

xx Main texture of the sample   x Texture present in the sample but in a small proportion

Figure 7.18: Table of textures present in the smelting slag of northeastern Madagascar. These textures have been defined from reflected light microscopy observations.

Several textures can coexist in the same sample. In the majority of cases there is a dominant texture coexisting with another texture present in smaller proportions. Sometimes a single texture is observed over the whole sample. In rarer cases, the sample is so heterogeneous that it is not possible to define a dominant texture but several textures present in significant proportions (MBR13006 for example). The overall chemical composition of the sample is therefore a mixture of the chemical compositions of all the textures present.

Each texture is illustrated by a figure with a typical sample as an example, a representation of the crystallization process on the FeO-SiO_2-Al_2O_3 phase diagram and a schematic representation of the crystallization order (Figure 7.19 to Figure 7.33). The representation of the order of crystallization illustrates the order of appearance of the minerals according to the spatial relationships of the crystals with each other. It does not represent the time taken for each group of minerals to crystallise. A selection of EDS analyses for the main minerals of each texture is also associated.

Texture 1 (Figure 7.19):

This texture is characterised by large sp1 spinels crystallising first, followed by very thin sp2 spinels at the edge or corners of the sp1. Sometimes very small sp3 spinels can be seen at the edge of the sp2. Fayalite grains (fay1) then start to crystallise, sometimes with a prismatic shape when the cooling rate is slow, sometimes with the shape of skeletal rods when cooling is faster. Spinels (sp int) keep crystallising and are observed in the interstices of the fay1 structure. A second generation of small fayalites (fay2) then crystallise and finally a glassy phase forms.

Texture 2 (Figure 7.20):

This texture is very similar to the previous one. The difference is that the sp2 spinels are bigger and euhedral, but always sticking to the edges of sp1 spinels, sometimes completely encompassing them. The sp3 spinels also form a thicker border on the sp2. The fayalite crystallises after the spinels and the glassy phase is then solidified.

Texture 3 (Figure 7.21):

Texture 3 is mainly observed in TS or next to the cooling surfaces of BFS. It is therefore a typical texture for high cooling rates.

The spinels crystallise first, with a low-titanium core (sp1 composition) which rapidly evolves into a high-titanium content spinel (sp2 composition) at the edges. The transition is not progressive but rather sharp. The two spinel compositions however cohabit in the same grain. The edges of this spinel have a jagged aspect.

The further away from the cooling surface, the larger the spinels become and the larger low-titanium core of the grain becomes.

Skeletal fayalite rods (fay1) and possibly second generation fayalites (fay2) crystallise. The glassy phases then solidify.

Texture 4 (Figure 7.22):

In this texture, a large number of very small sp1 spinels crystallise first. Then larger polygonal sp2 spinel grains form, without sp3 on their edges. These sp2 spinels are larger than the sp1 spinels and do not form on any of their edges. First (fay1) and second (fay2) generation fayalites then form, followed by the glassy phases.

Texture 5 (Figure 7.24):

Texture 5 is characterised by the presence of large sp1 spinel crystals. Then wustite dendrites and sp2 spinels start to crystallize more or less at the same time. The wustite certainly starts to crystallise slightly before the sp2 in most cases, but it is not so clear. The sp2 spinels are observed on the edges or corners of the sp1. Sometimes no sp2 is observed. Fayalite grains then crystallise (fay1 then fay2), followed by the solidification of the glassy phases.

Texture 6 (Figure 7.25):

The texture 6 is the equivalent of texture 3, but with wustites. It is a typical texture for fast cooling rate and is found next to cooling surfaces in TS.

A spinel with an sp1 composition crystallises first, then, still in the same grain, the composition abruptly evolves towards an sp2 composition. The edges of the spinels grains show a jagged aspect. Fine wustite dendrites then begin to crystallise, followed by fayalite rods. The spinel continues to crystallise and in the interstices of the fayalite are found sp sym spinels. The glassy phases solidify last.

No samples of this texture were observed in SEM-EDS, only in optical microscopy.

Texture 7 (Figure 7.26):

This texture is similar to texture 4, except that it contains wustite. The fayalite crystals are large and stubby and probably begin to crystallize first. Then, large wustite dendrites begin to form and are progressively engulfed by the fayalite which continues to grow. Smaller sp1 spinels then crystallize agglutinated to each other, at the margin of the fayalite grains. They are followed by larger sp2 crystals, with sp3 rims. Second generation fayalite rods form, followed by the glassy phases.

It is interesting to note that the wustites associated with this texture 7 are always strongly altered, probably to hematite. As this is certainly a texture characteristic of slow cooling, the material remained hot for a long time once solidified. During this time, the wustites were transformed into hematite.

The chemical composition of the texture measured over a large area in EDS is not compatible with the order of crystallization deduced from microscopic observations. The reason could be because of the strong alteration of the wustites, or because these textures generally contain 5-6 ox.% TiO_2, which is not taken into account in the phase diagram.

Texture 8 (Figure 7.27):

Texture 8 has large wustite dendrites which crystallise first and are close together. Fayalites form next, followed by the glassy phases. No spinel is observed in this texture.

Texture 9 (Figure 7.28):

Thick wustite dendrites crystallize first in texture 9. They are followed by sp2 spinels whose edges show sp3 spinel compositions. No sp1 spinel is observed. Fayalite crystallises next and spinels (sp int) are recurrently found trapped in the interstices of fayalite grains. The glassy phases solidify last.

Texture 10 (Figure 7.30):

Texture 10 is relatively rare. Long falite rods crystallise first. Then in the interstices crystallises small sym sp spinels. Then second generation fayalites are observed, which form at the same time as the glassy phase.

Texture 11 (Figure 7.31):

This texture is characterised by the crystallization of very large fayalite grains. The fayalite grains keep growing while wustite dendrites begin to form in parallel. Afterwards, second generation large fayalites appear, followed by a third generation of small fayalites. Finally, the glassy phases are formed.

texture 12 (Figure 7.32):

This texture has only been observed in slag with a BFS morphology. Texture 12 shows long and thick fayalite rods. They no longer have a skeletal shape but have saw-tooth edges. Then fine vermicelli of wustites form in the central interstices of the fayalite grains. They follow the elongation of the fayalite crystals. The wustite quickly stops growing. This makes room for spinels (sp int) which intercalate in the outer interstices of the fayalite grains. Very small other phases are then seen in the glass, probably second generation fayalites. Finally, the glassy phases solidify.

Texture 13 (Figure 7.33):

Only one sample was observed with this texture, MBR13006, which show high bulk CaO content. Long fayalite rods crystallize, followed by kirschsteinite crystals. A new generation of fayalite develops, characterised by thin and long rods organised in sheaves, starting from the edges of the kirschsteinite grains. Finally, the glassy phase is formed.

There are also numerous white dots in this sample which are metallic iron particles embedded in the slag.

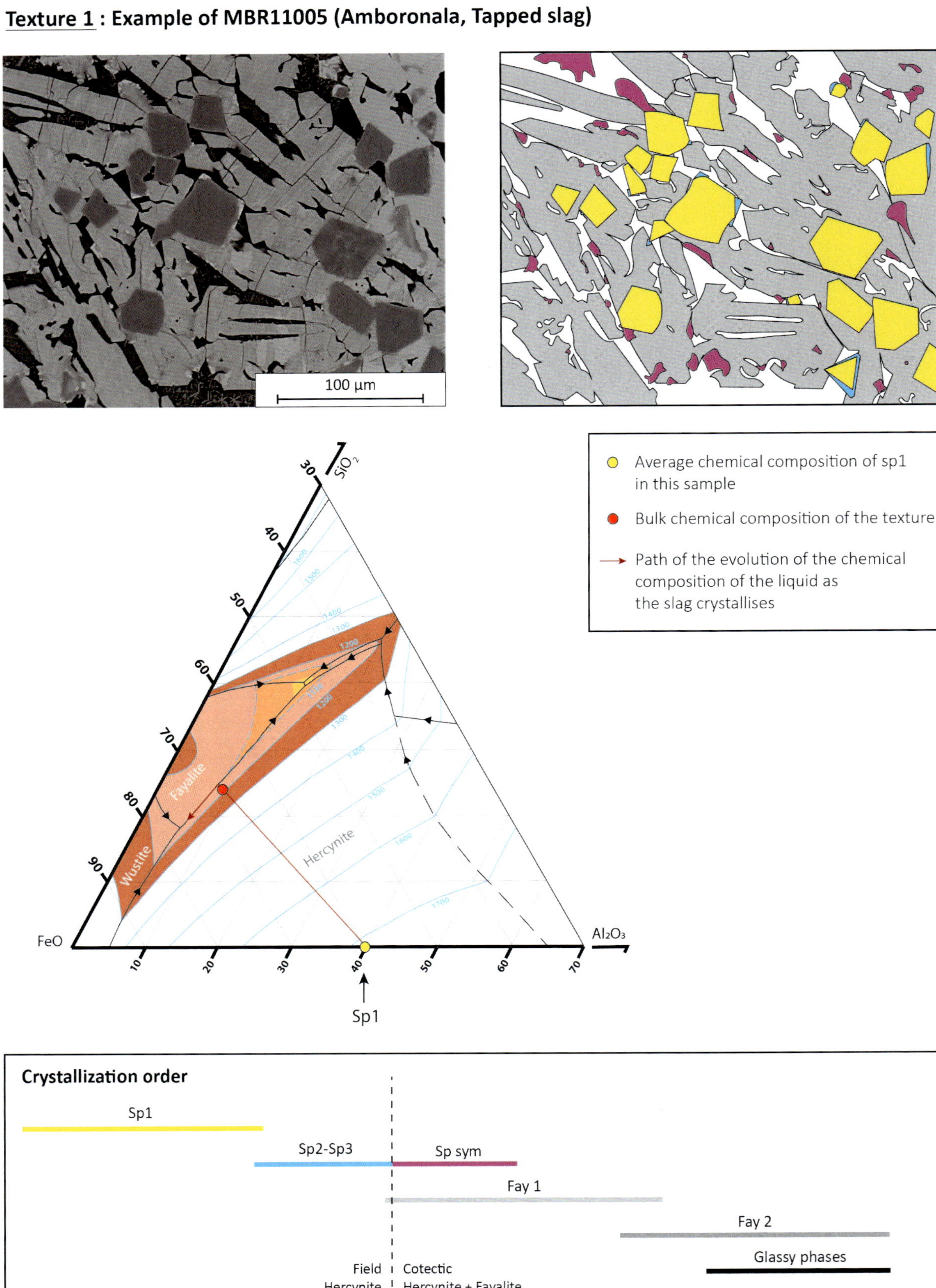

Figure 7.19: Summary sheet of **texture 1** (Example of MBR11005, Amboronala). **Top**: A typical BSE image of texture 1, associated with a sketch of the texture. **Middle**: Crystallization path on the FeO_n-SiO_2-Al_2O_3 phase diagram (Kowalski et al., 2000). The chemical composition of the texture was measured by EDS on a large area of the MBR11005 sample. **Bottom**: Reconstruction of the crystallization order of the minerals.

Texture 2 : Example of BMK23002 (Bemanevika, Bottom Furnace Slag)

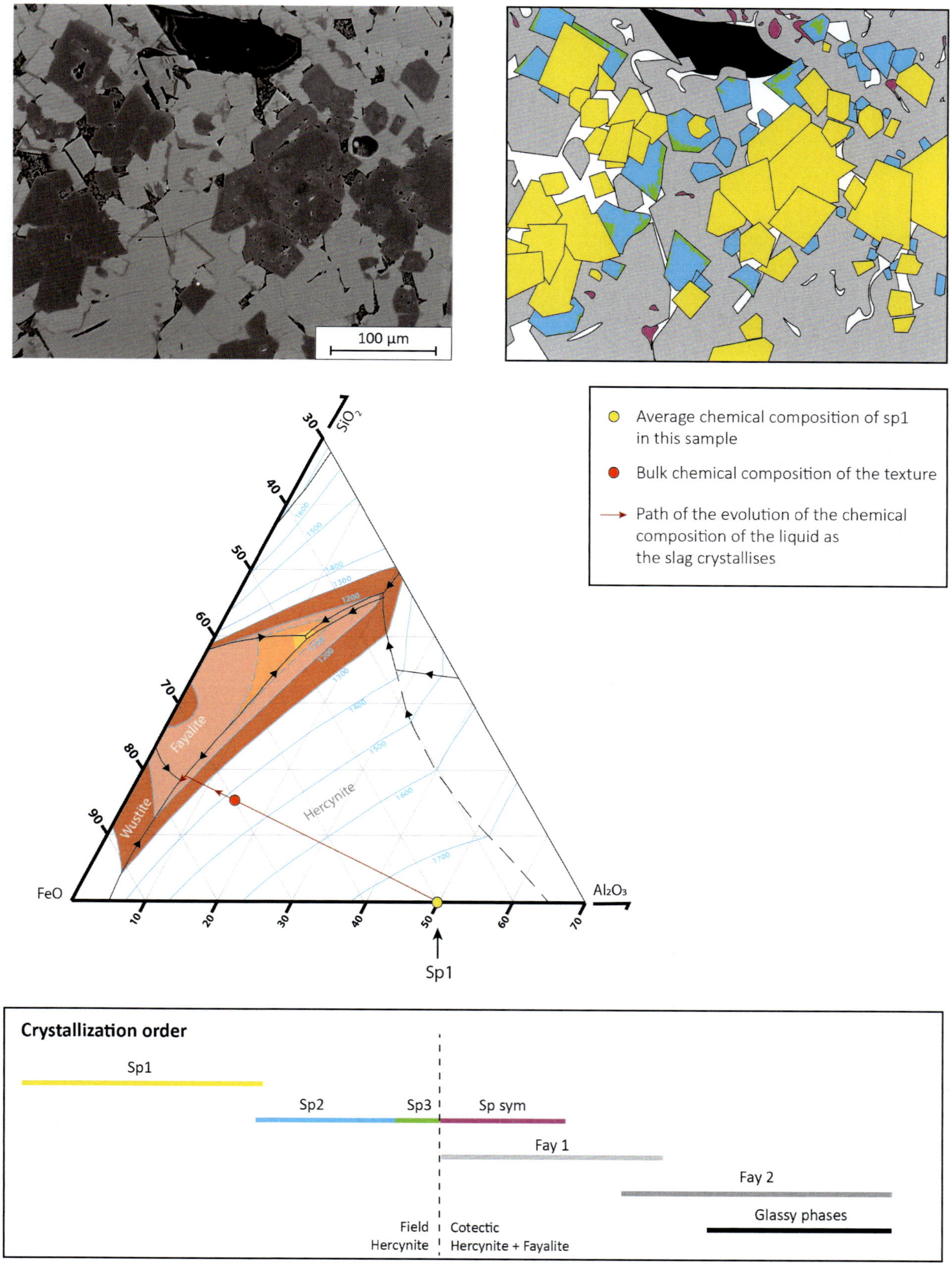

Figure 7.20: Summary sheet of **texture 2** (Example of BMK23002, Bemanevika). **Top**: A typical BSE image of texture 2, associated with a sketch of the texture. **Middle**: Crystallization path on the FeO_n-SiO_2-Al_2O_3 phase diagram (Kowalski et al., 2000). The chemical composition of the texture is the bulk XRF analysis of BMK23002. **Bottom**: Reconstruction of the crystallization order of the minerals.

Texture 3 : Example of DMD82002 (Ambodimadiro , Tapped slag)

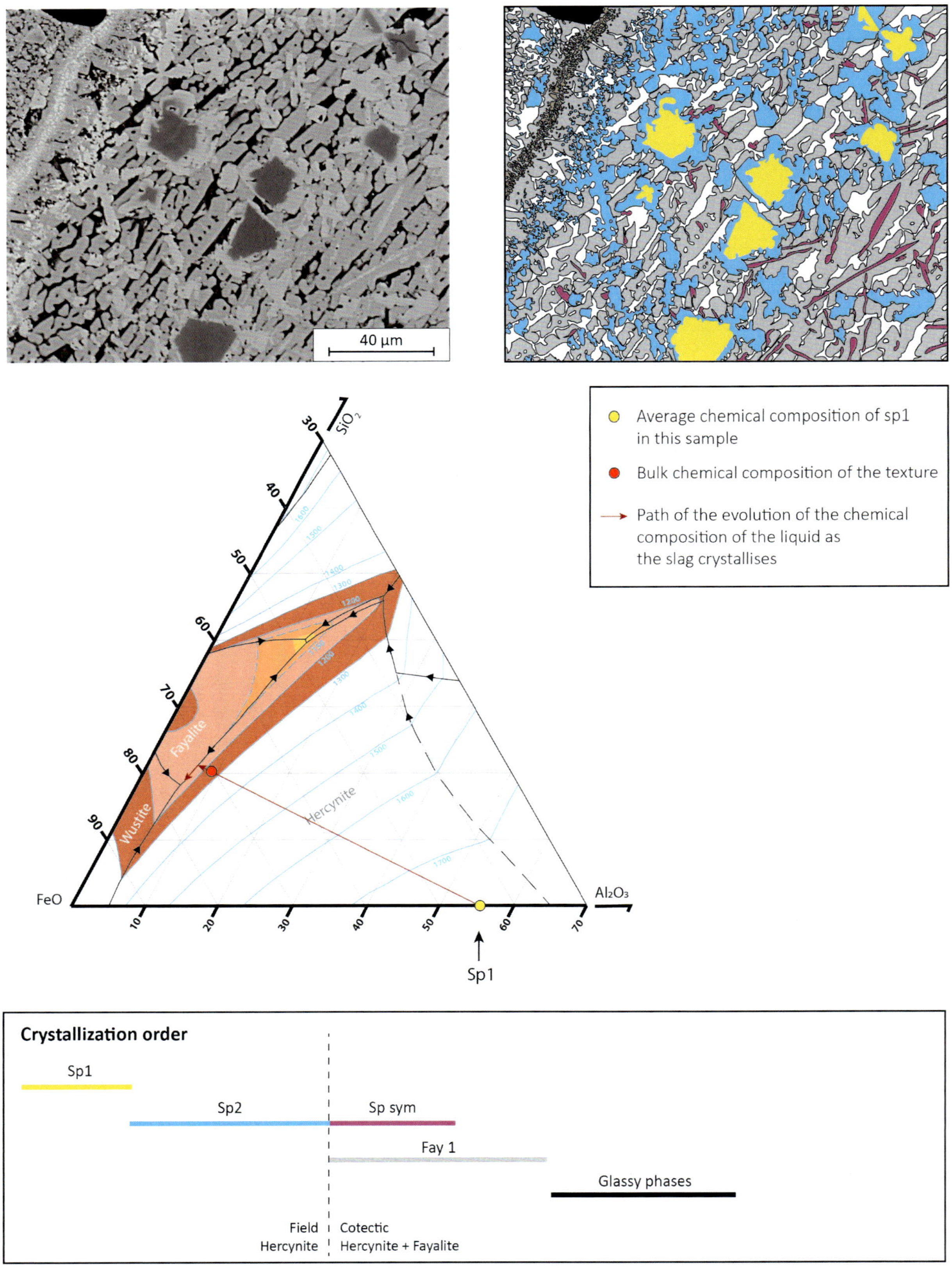

Figure 7.21: Summary sheet of **texture 3** (Example of DMD82002, Ambodimadiro). **Top**: A typical BSE image of texture 3, associated with a sketch of the texture. **Middle**: Crystallization path on the FeO_n-SiO_2-Al_2O_3 phase diagram (Kowalski et al., 2000). The chemical composition of the texture was measured by EDS on a large area of the DMD82002 sample. **Bottom**: Reconstruction of the crystallization order of the minerals.

Texture 4 : Example of BNY41006 (Benavony, Bottom Furnace Slag)

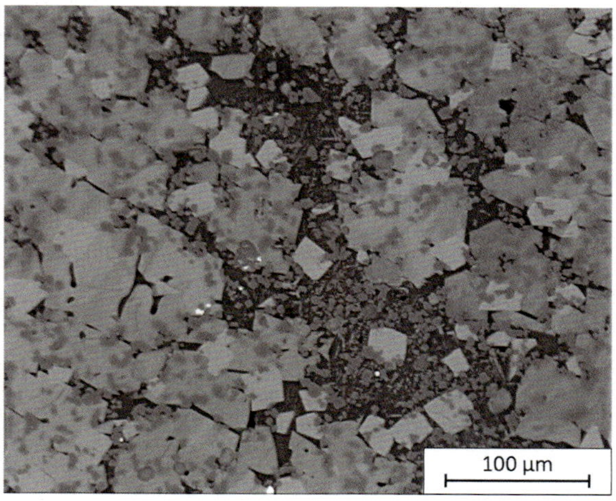

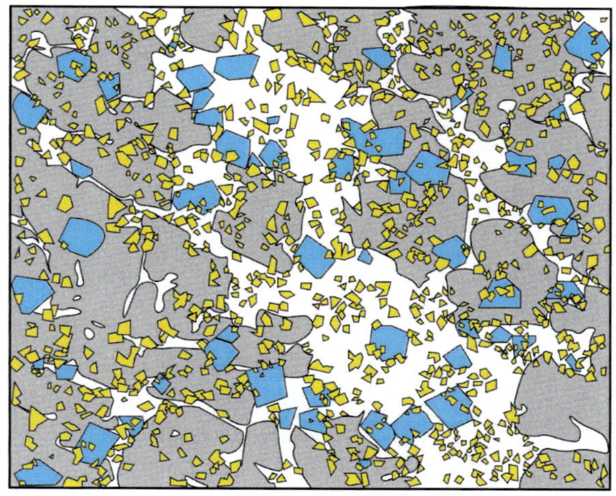

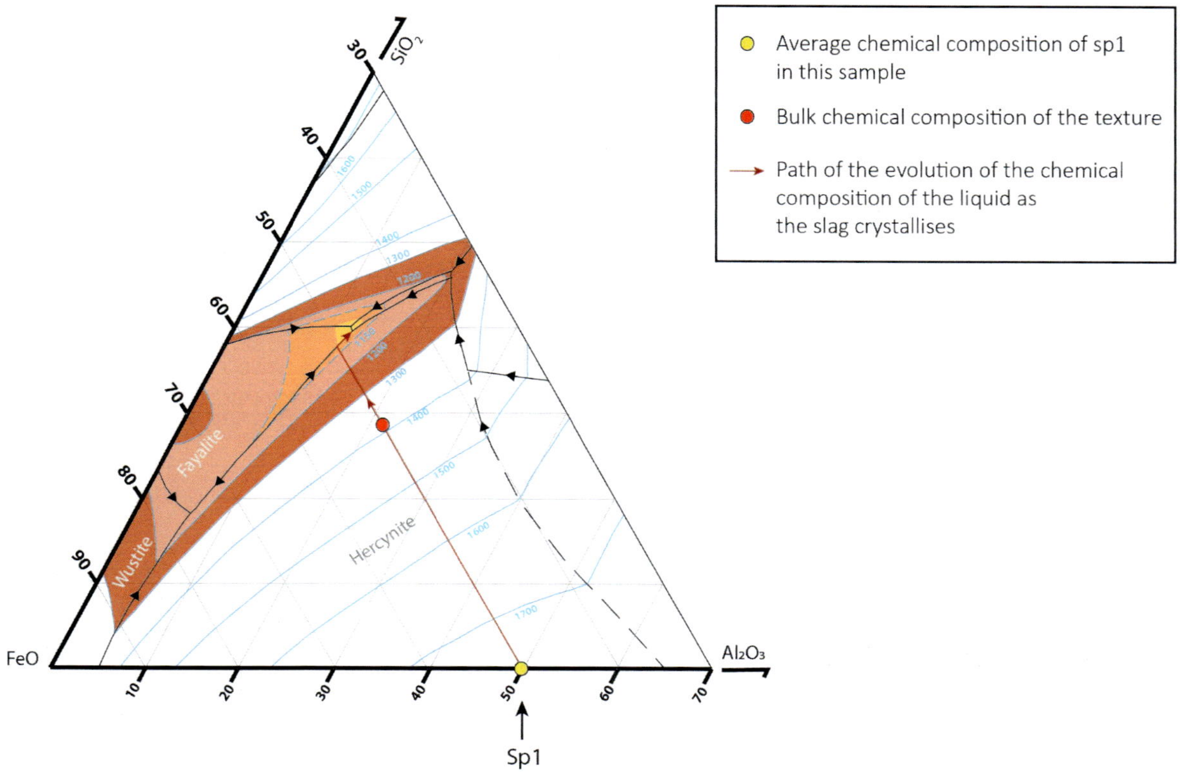

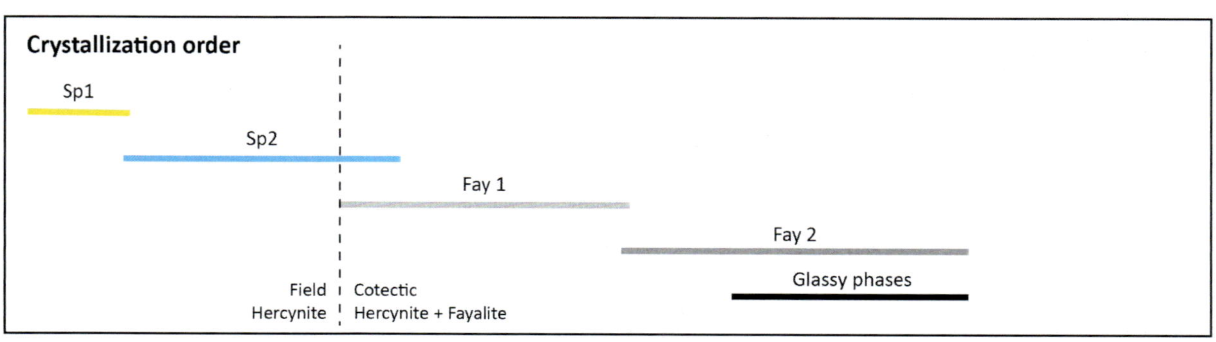

Figure 7.22: Summary sheet of **texture 4** (Example of BNY41006, Benavony). **Top**: A typical BSE image of texture 4, associated with a sketch of the texture. **Middle**: Crystallization path on the FeO_n-SiO_2-Al_2O_3 phase diagram (Kowalski et al., 2000). The chemical composition of the texture was measured by EDS on a large area of the BNY41006 sample. **Bottom**: Reconstruction of the crystallization order of the minerals.

Texture 1

MBR11005	Sp1	Sp1	Sp2	Sp2	Fay1
Si	0,4	1,16	0,38	0,53	30,66
Ti	2,6	2,27	6,88	10,57	0,31
Al	47,28	47,44	37,74	31,36	0,44
Cr	0,19	0	0,08	0	0
Fe	48,92	49,33	53,8	57,76	68,05
Ca	0,07	0	0,12	0,14	0,82
Mg	0,45	0,4	0,29	0,06	0,06
Mn	0,09	0,07	0,05	0,08	0,2
K	0	0	0	0,04	0
P	0	0	0	0	0,12
V	1,82	0,99	0,88	0,26	0
Total	101,82	101,67	100,22	100,81	100,65

Texture 2

BMK23002	Sp1	Sp1	Sp2	Sp2	Fay1	Fay1
Si	0,35	0,58	0,42	0,38	31,65	30,89
Ti	4,59	4,19	19,33	19,57	0,3	0,36
Al	46,38	47,19	16,79	16,9	0,24	0,36
Cr	0,27	0,1	0,1	0	0	0
Fe	48,84	49,02	61,25	60,79	66,84	66,46
Ca	0	0	0	0	0,25	0,26
Mg	0	0	0,1	0	0,59	0,59
Mn	0,14	0,12	0,15	0,18	0,51	0,41
K	0	0	0	0	0	0
P	0	0	0	0	0,3	0,71
V	0,95	0,52	0,83	0,68	0	0,07
Total	101,53	101,7	98,98	98,5	100,67	100,1

Texture 3

DMD82002	Sp1	Sp1	Sp2	Sp2	Fay1
Si	0,63	0,78	0,98	2,54	32,49
Ti	3,76	3,73	7,55	3,79	0,09
Al	41,48	40,38	20,13	10,02	0,34
Cr	0,73	1,07	0	0	0
Fe	52,32	50,81	70,41	81,46	66,99
Ca	0,00	0,07	0	0	0,28
Mg	0,32	0,32	0,14	0,18	0,21
Mn	0,10	0,08	0,06	0,08	0,23
K	0,00	0,03	0	0	0
P	0,00	0,00	0	0	0,79
V	1,86	2,03	0,78	0	0
Total	101,2	99,3	100,05	98,07	101,43

Texture 4

BNY41006	Sp1	Sp1	Sp2	Sp2	Fay1
Si	0,27	0,37	0,25	0,25	30,68
Ti	2,47	4,95	23,32	23,25	0,4
Al	52,06	47,67	12,04	12,19	0,24
Cr	0	0,12	0	0	0
Fe	45,27	45,89	61,31	60,78	67,65
Ca	0	0	0	0	0,24
Mg	0,26	0,12	0,08	0	0,23
Mn	0	0,19	0,32	0,33	0,6
K	0	0	0	0	0
P	0	0	0	0	0,19
V	1,21	1,26	1,4	1,22	0
Total	101,53	100,56	98,72	98,03	100,21

Figure 7.23: Selection of EDS analyses performed on main minerals representative of textures 1, 2, 3 and 4.

Texture 5 : Example of MTY12005 (Matavy, Bottom Furnace Slag)

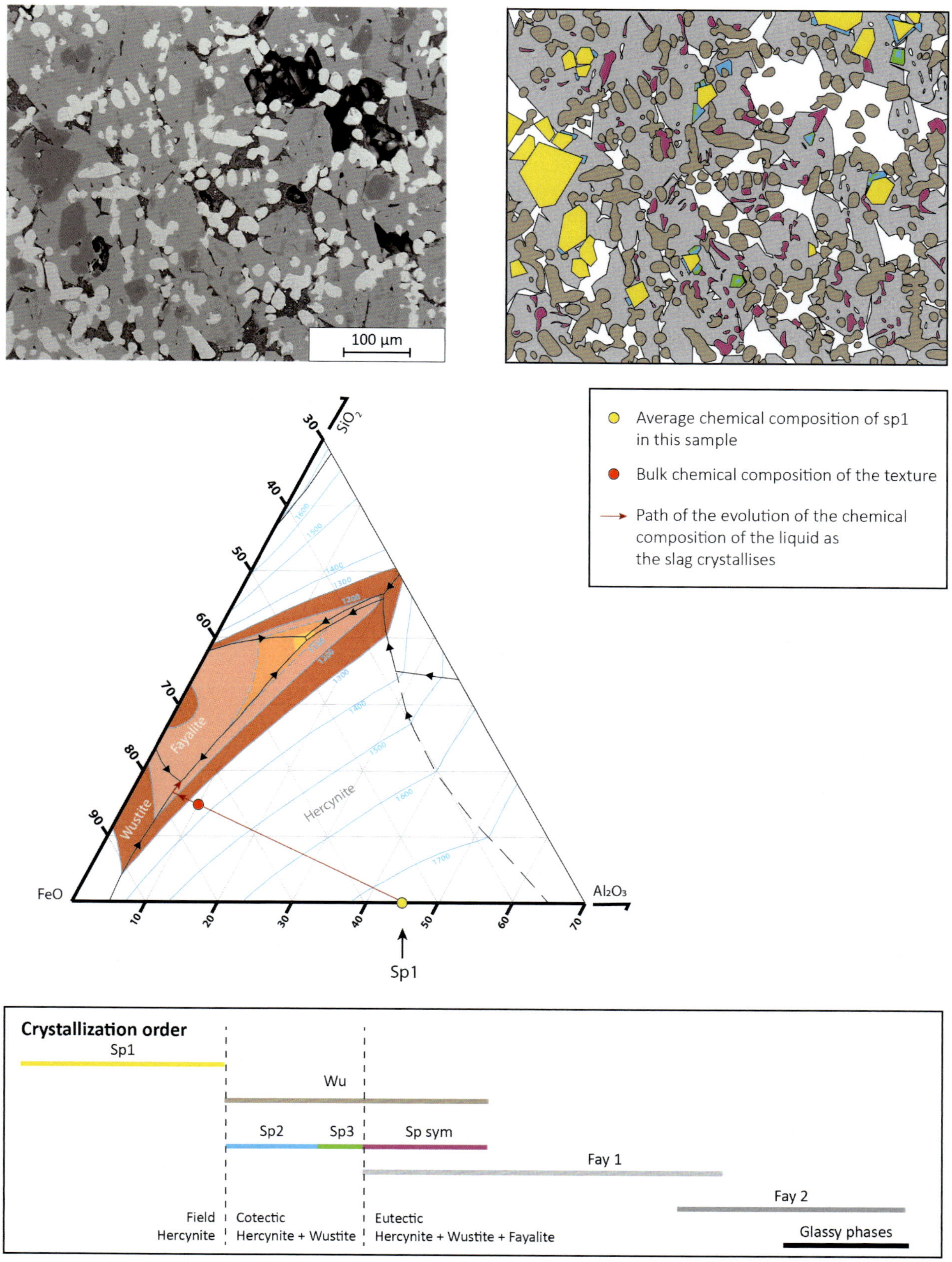

Figure 7.24: Summary sheet of **texture 5** (Example of MTY12005, Matavy). **Top**: A typical BSE image of texture 5, associated with a sketch of the texture. **Middle**: Crystallization path on the FeO_n-SiO_2-Al_2O_3 phase diagram (Kowalski et al., 2000). The chemical composition of the texture is the bulk XRF analysis of MTY12005. **Bottom**: Reconstruction of the crystallization order of the minerals.

Texture 6 : Example of DMD92003 (Ambodimadiro, Tapped slag)

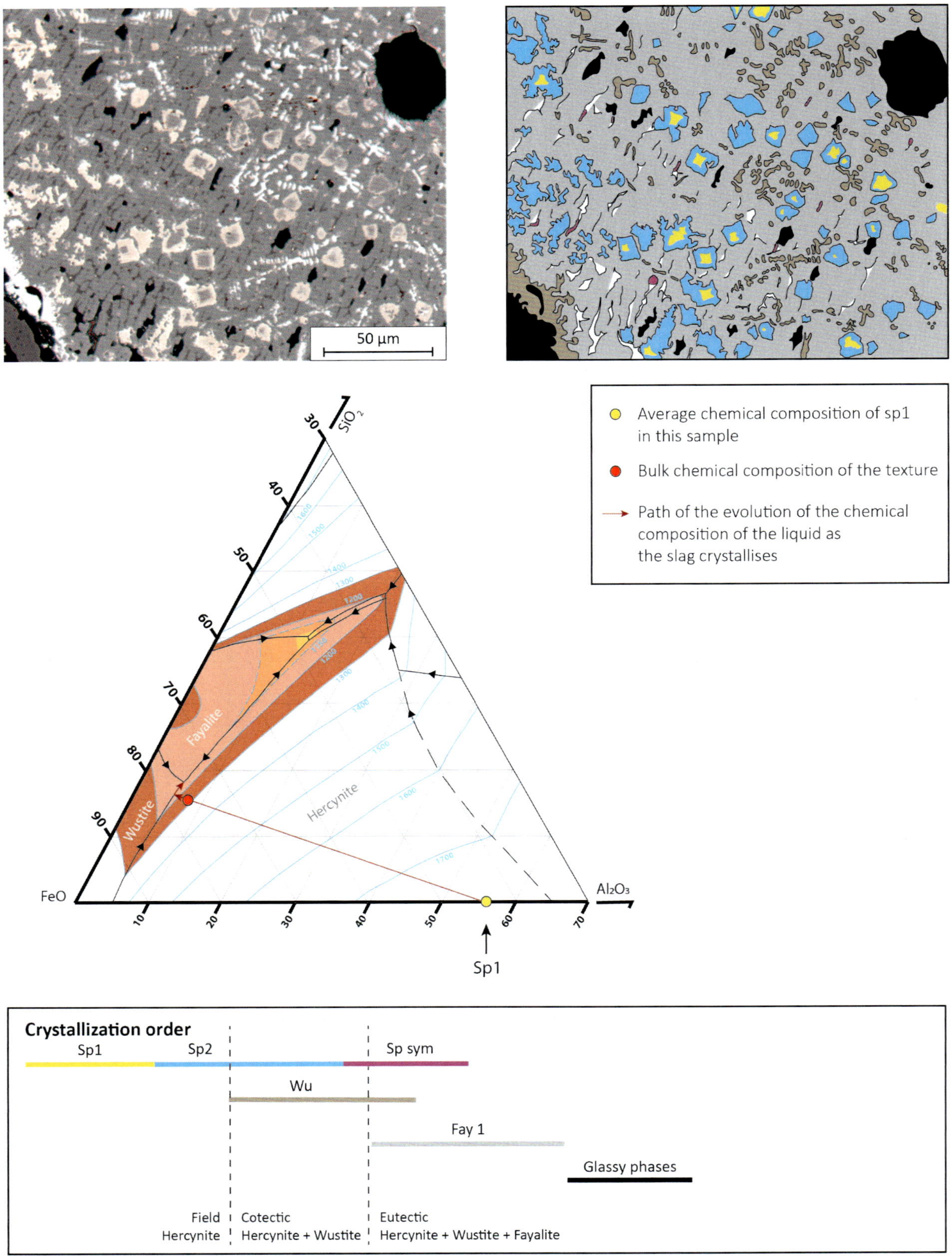

Figure 7.25: Summary sheet of **texture 6** (Example of DMD92003, Ambodimadiro). **Top**: A typical reflected light microscopy image of texture 6, associated with a sketch of the texture. **Middle**: Crystallization path on the FeO_n-SiO_2-Al_2O_3 phase diagram (Kowalski et al., 2000). The chemical composition of the texture is the bulk XRF analysis of DMD92003. No SEM-EDS analyses has been conducted on this sample. **Bottom**: Reconstruction of the crystallization order of the minerals.

Texture 7 : Example of BNY41030 (Benavony, Bottom Furnace Slag)

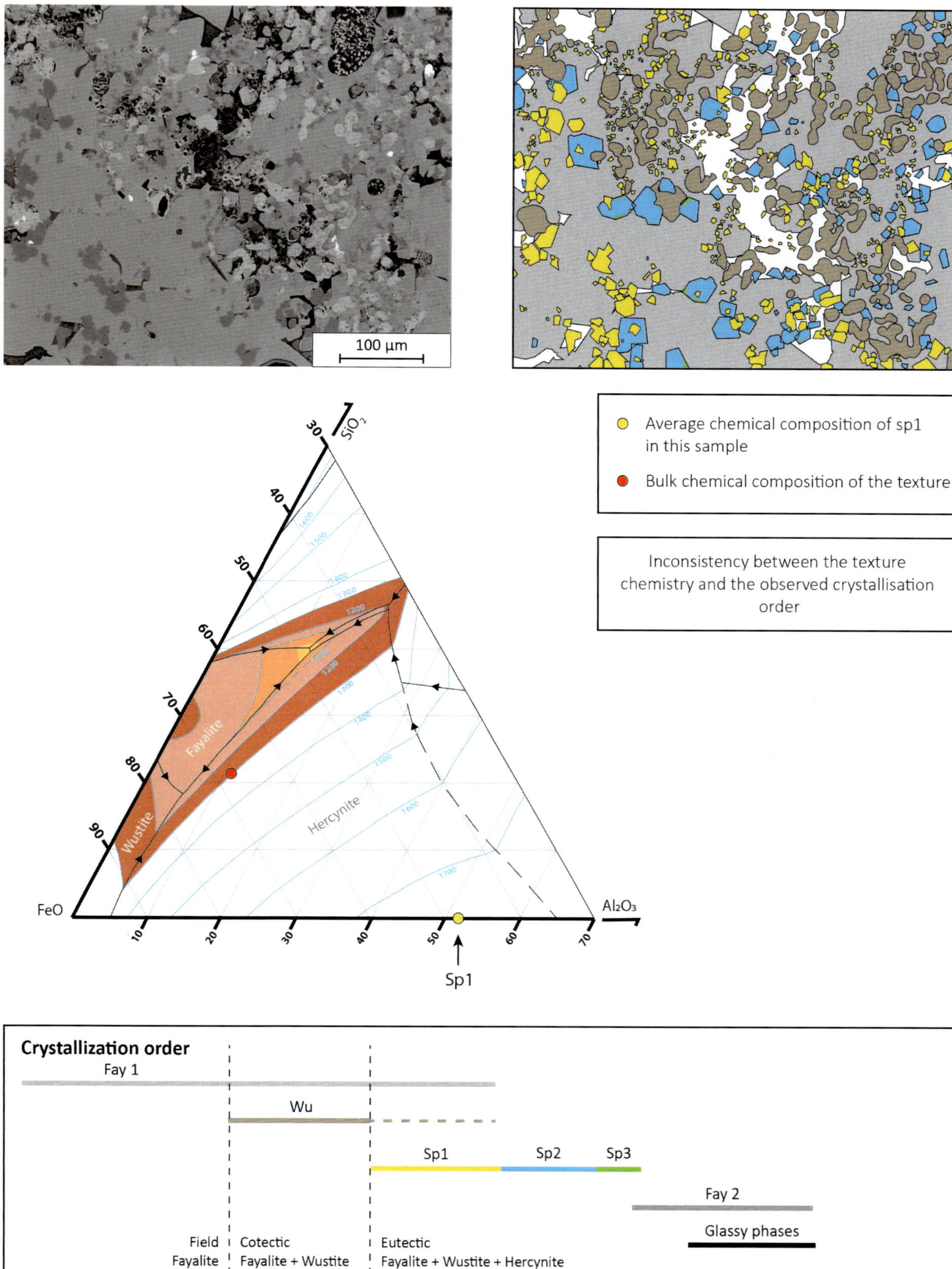

Figure 7.26: Summary sheet of **texture 7** (Example of BNY41030, Benavony). **Top**: A typical BSE image of texture 7, associated with a sketch of the texture. **Middle**: Crystallization path on the FeO_n-SiO_2-Al_2O_3 phase diagram (Kowalski et al., 2000). The chemical composition of the texture was measured by EDS on a large area of the BNY41030 sample. **Bottom**: Reconstruction of the crystallization order of the minerals.

Texture 8 : Example of MTY18002 (Matavy, Bottom Furnace Slag)

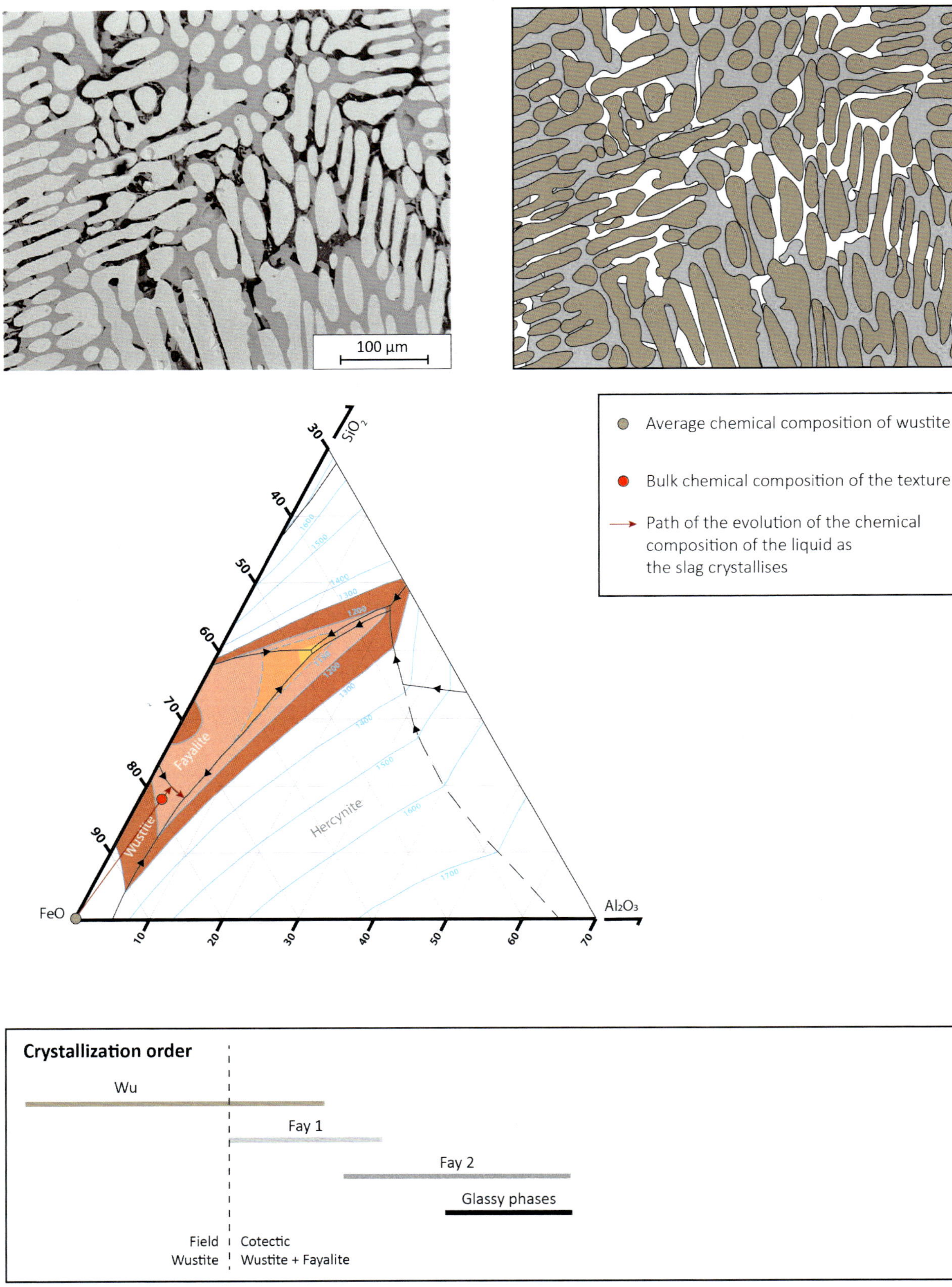

Figure 7.27: Summary sheet of **texture 8** (Example of MTY18002, Matavy). **Top**: A typical BSE image of texture 8, associated with a sketch of the texture. **Middle**: Crystallization path on the FeO_n-SiO_2-Al_2O_3 phase diagram (Kowalski et al., 2000). The chemical composition of the texture was measured by EDS on a large area of the MTY18002 sample. **Bottom**: Reconstruction of the crystallization order of the minerals.

Texture 9 : Example of BNY41033 (Benavony, Bottom Furnace Slag)

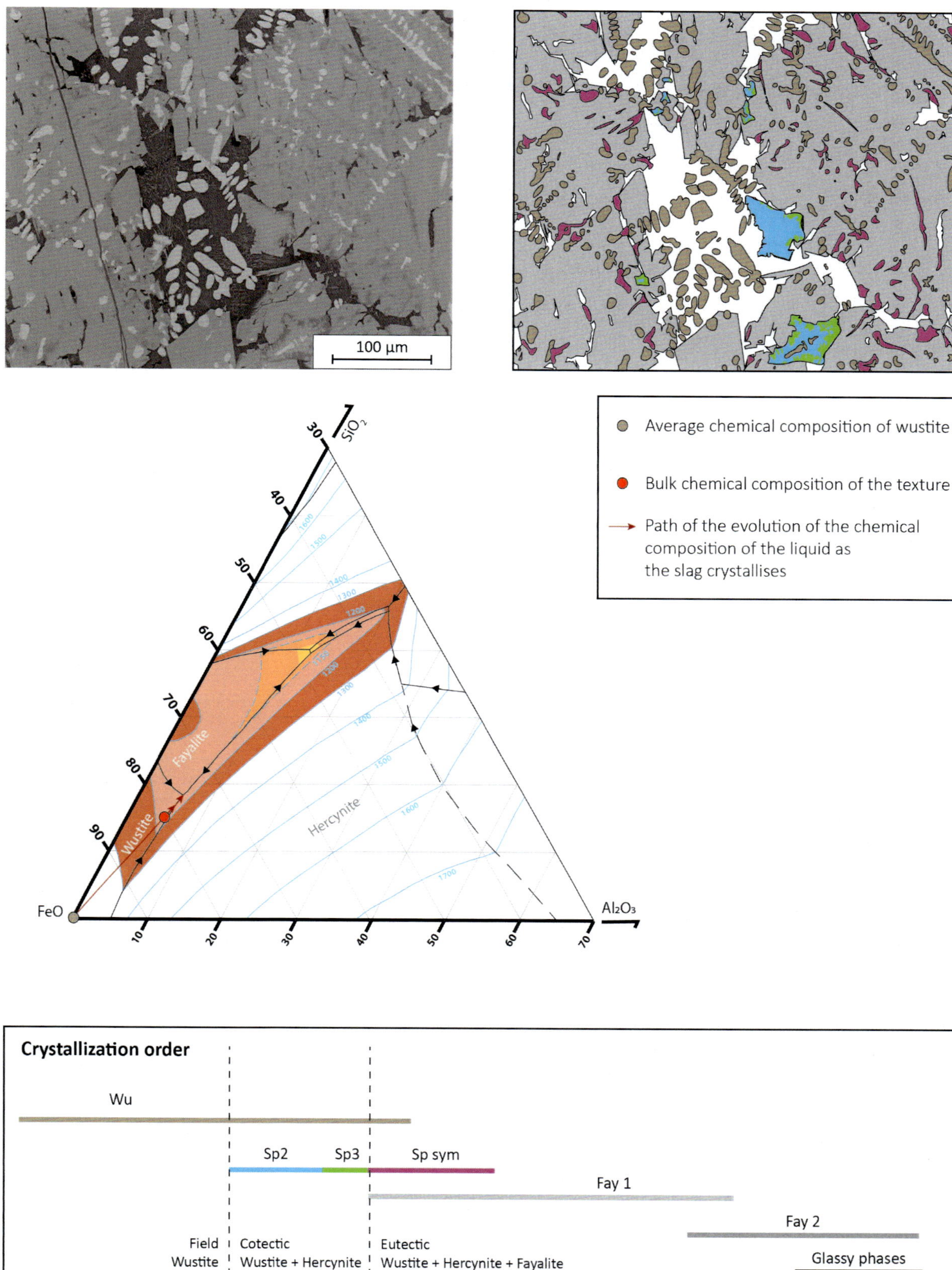

Figure 7.28: Summary sheet of **texture 9** Example of BNY41033, Benavony). **Top**: A typical BSE image of texture 9, associated with a sketch of the texture. **Middle**: Crystallization path on the FeO_n-SiO_2-Al_2O_3 phase diagram (Kowalski et al., 2000). The chemical composition of the texture was measured by EDS on a large area of the BNY41033 sample. **Bottom**: Reconstruction of the crystallization order of the minerals.

Texture 5

MTY12005	Sp1	Sp1	Sp2	Sp2	Wu	Fay1
Si	0,14	0,19	0,55	0,41	0,57	29,74
Ti	2,53	2,49	17,12	16,41	1,88	0,22
Al	44,19	46,87	16,17	17,16	0,93	0,48
Cr	2,52	0,16	0	0	0	0
Fe	49,49	49,08	64,3	64,29	94,21	67,63
Ca	0	0	0	0	0	0,21
Mg	0,12	0,11	0,16	0,12	0,08	0,26
Mn	0,69	0,6	0,18	0,12	0	0,54
K	0	0	0	0	0	0
P	0	0	0	0	0	0,41
V	1,98	1,79	1,01	1,24	0,26	0,03
Total	101,66	101,29	99,48	99,91	98,07	99,52

No EDS on Texture 6

Texture 7

BNY41030	Sp1	Sp1	Sp2	Sp 2	Sp3	Wu	Wu	Fay1	Fay1
Si	0,19	0,22	0,27	0,27	0,62	0,54	0,25	29,42	29,55
Ti	5,02	6,09	20,44	20,97	14,33	0,57	0,64	0,29	0,24
Al	44,37	42,48	13,3	13,93	26,33	0,34	0,26	0,26	0,3
Cr	0,5	0,22	0	0,18	0	0	0,05	0	0
Fe	47,93	50,44	62,97	61,2	58,23	96,29	96,95	67,54	67,91
Ca	0	0,05	0	0	0,09	0	0,03	0,27	0,27
Mg	0,42	0,37	0,13	0,15	0,15	0	0	0,68	0,58
Mn	0,27	0,29	0,37	0,33	0,33	0	0,04	0,84	0,81
K	0	0,04	0	0	0,05	0	0	0	0
P	0	0	0	0	0	0,07	0	0,24	0,2
V	1,66	1,02	0,61	1,78	0,32	0	0	0	0,06
Total	100,51	101,22	98,27	98,95	100,45	98,14	98,66	99,53	100,09

Texture 8

MTY18002	Wu	Wu	Fay1	Fay1	Fay2	Fay2	Leucite
Si	0,53	0,51	31,35	31,31	30,99	31,23	58,35
Ti	0,77	0,59	0	0,04	0,06	0,07	0
Al	0,57	0,41	0,08	0,15	0,15	0,11	24,28
Cr	0,63	0,83	0,05	0	0	0	0
Fe	97,61	96,81	63,03	63,44	63,87	63,82	1,92
Ca	0	0	0,42	0,39	3,46	3,25	0,19
Mg	0,16	0,08	3,35	3,47	0,28	0,27	0
Mn	0,1	0	0,06	0,05	0,06	0,05	0
K	0	0	0	0	0,1	0,06	20,56
P	0	0	0	0,13	0,25	0,17	0
V	0,39	0,33	0	0,13	0,05	0	0
Total	100,75	99,57	99,09	99,11	100,09	100,02	111,32

Texture 4

BNY41033	Sp2	Wu	Fay1	Fay1
Si	0,58	0,79	31,14	29,04
Ti	11,81	1,17	0,37	0,61
Al	28,63	0,59	0,28	1,18
Cr	0	0	0	0,05
Fe	57,91	96,87	66,26	66,67
Ca	0,05	0,14	1,84	0,6
Mg	0	0	0	0,17
Mn	0,15	0,09	0,41	0,45
K	0	0,08	0	0
P	0	0	0,29	0,33
V	0,25	0	0	0
Total	99,23	99,57	100,39	99,24

Figure 7.29: Selection of EDS analyses performed on the main minerals representative of textures 5, 7, 8 and 9.

Texture 10 : Example of BMK42004 (Bemanevika, Bottom Furnace slag)

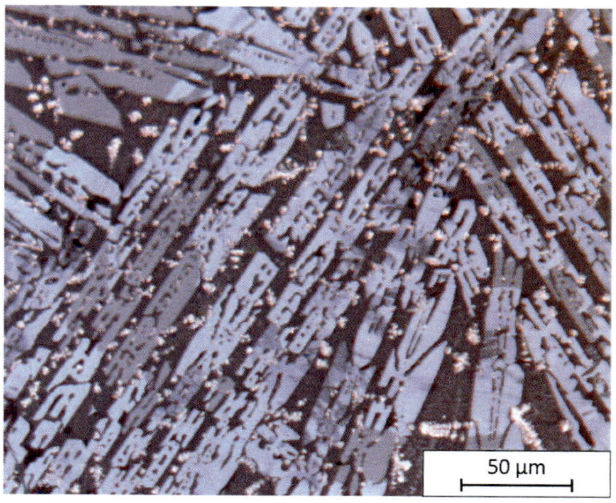

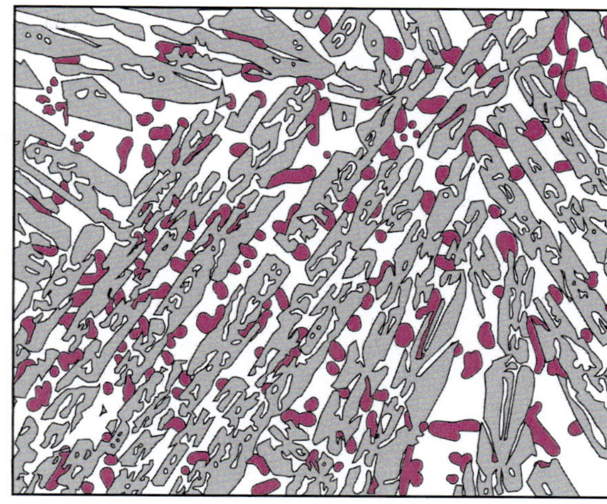

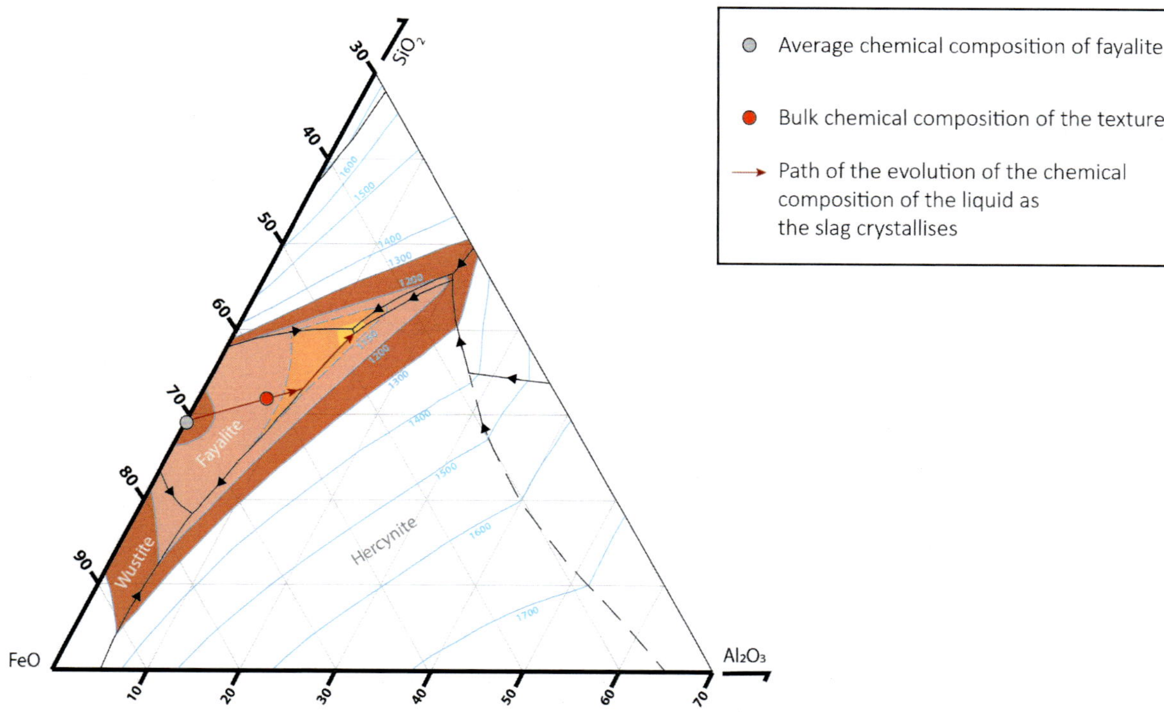

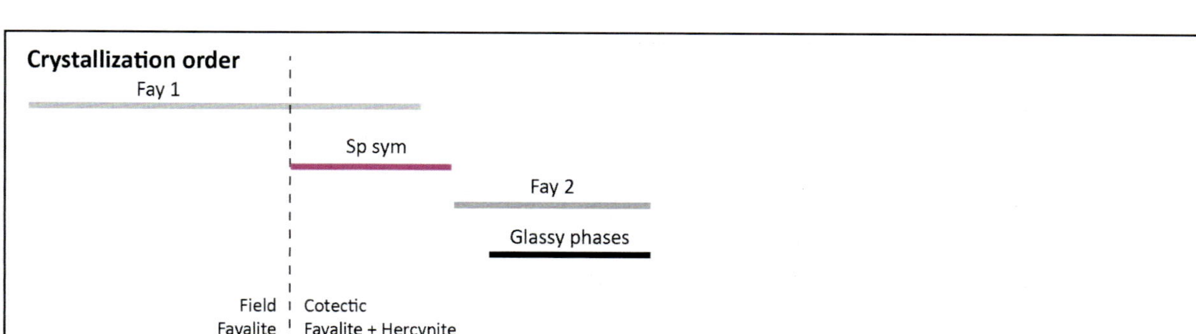

Figure 7.30: Summary sheet of **texture 10** (Example of BMK42004, Bemanevika). **Top**: A typical reflected light microscopy image of texture 10, associated with a sketch of the texture. **Middle**: Crystallization path on the FeO_n-SiO_2-Al_2O_3 phase diagram (Kowalski et al., 2000). The chemical composition of the texture was measured by EDS on a large area of ISH20001, which shows a small area with texture 10. No SEM-EDS analyses were conducted on BMK42004. **Bottom**: Reconstruction of the crystallization order of the minerals.

Texture 11 : Example of MTY18002 (Matavy, Bottom Furnace Slag)

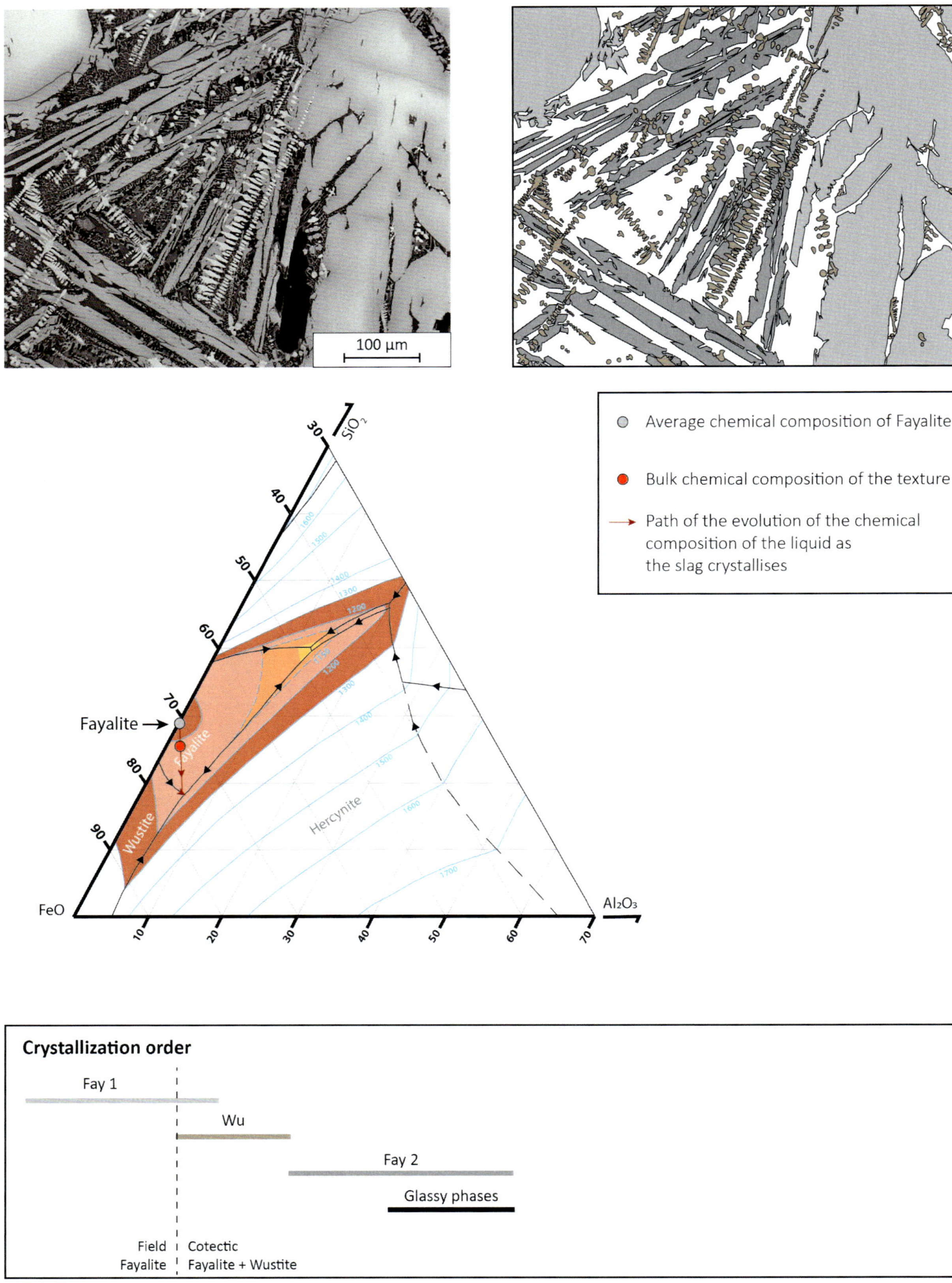

Figure 7.31: Summary sheet of **texture 11** Example of MTY18002, Matavy). **Top**: A typical BSE image of texture 11, associated with a sketch of the texture. **Middle**: Crystallization path on the FeO_n-SiO_2-Al_2O_3 phase diagram (Kowalski et al., 2000). The chemical composition of the texture was measured by EDS on a large area of the MTY18002 sample. **Bottom**: Reconstruction of the crystallization order of the minerals.

Texture 12 : Example of MBR14010 (Amboronala, Bottom Furnace Slag)

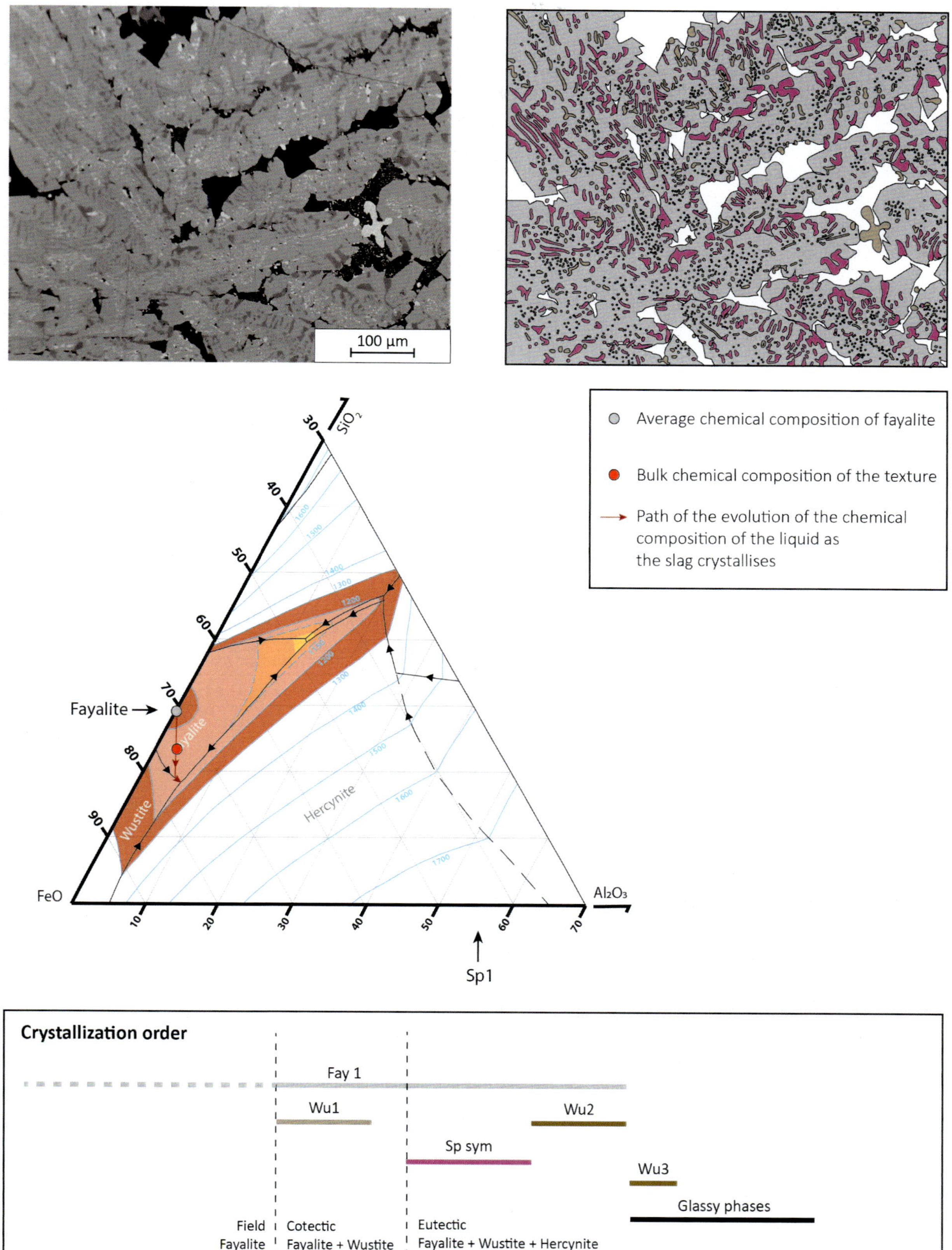

Figure 7.32: Summary sheet of **texture 12** Example of MBR14010, Amboronala). **Top**: A typical BSE image of texture 12, associated with a sketch of the texture. **Middle**: Crystallization path on the FeO_n-SiO_2-Al_2O_3 phase diagram (Kowalski et al., 2000). The chemical composition of the texture was measured by EDS on a large area of the MBR14010 sample. **Bottom**: Reconstruction of the crystallization order of the minerals.

Texture 13 : Example of MBR13006 (Amboronala, Bottom Furnace Slag)

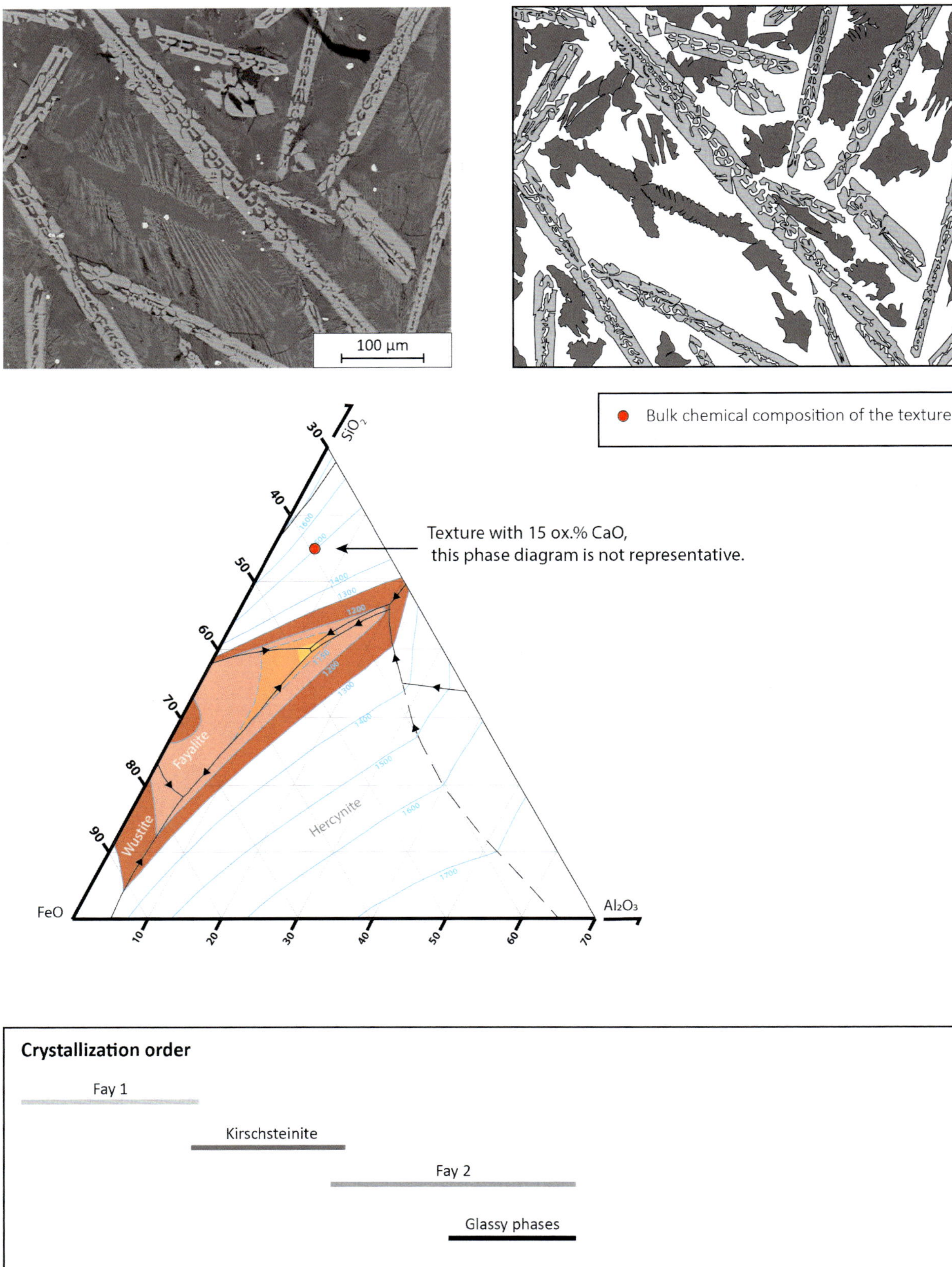

Figure 7.33: Summary sheet of **texture 13** Example of MBR13006, Amboronala). **Top**: A typical BSE image of texture 13, associated with a sketch of the texture. **Middle**: Crystallization path on the FeO_n-SiO_2-Al_2O_3 phase diagram (Kowalski et al., 2000). The chemical composition of the texture was measured by EDS on a large area of the MBR13006 sample. **Bottom**: Reconstruction of the crystallization order of the minerals.

No EDS on Texture 10

Texture 11

MTY18002	Wu	Wu	Fay1	Fay1	Fay2	Leucite
Si	0,72	0,66	30,84	31,19	30,95	51,45
Ti	1,33	1,1	0,05	0,06	0,05	0,18
Ni	0	0	0,09	0,15	0,1	0,25
Al	0,24	0,34	0,14	0,17	0,17	19,17
Cr	0,19	0,33	0,05	0,04	0	0
Fe	96,29	96,64	67,49	66,93	67,08	8,32
Ca	0,18	0,15	0,7	0,69	1,52	0,51
Mg	0	0	0,82	1,2	0,32	0
Mn	0	0,08	0,11	0,07	0	0
Na	0	0	0	0	0	0,75
K	0,18	0,11	0,03	0	0,11	16,75
P	0	0	0,06	0,11	0,15	0,7
V	0,38	0,49	0	0	0	0
S	0	0	0	0	0	0,91
Total	99,51	99,88	100,39	100,6	100,45	99

Texture 12

MBR14010	Sp sym	Sp sym	Wu	Fay1	Fay1
Si	0,38	0,52	0,32	31,72	31,39
Ti	4,33	4,75	16,72	0,14	0,21
Al	43,91	42,88	10,88	0,11	0,15
Cr	0	0	0	0	0
Fe	52,69	53,33	69,78	69,01	69,08
Ca	0	0	0,07	0,74	0,72
Mg	0	0		0,18	0,16
Mn	0	0	0,08	0	0
K	0	0	0	0	0
P	0	0	0	0,08	0,11
V	0,18	0	0,69	0	0
Total	101,5	101,47	98,79	101,97	101,82

Texture 13

MBR13006	Fay1	Fay1	Kirchteinite	Kirchteinite
Si	32,69	32,54	45,14	45,34
Ti	0	0,1	0,94	0,98
Al	0	0,09	4,49	4,99
Cr	0	0	0,11	0,04
Fe	56,08	55,85	25,05	25,48
Ca	4,77	4,91	21,79	20,83
Mg	5,82	5,74	2,32	1,59
Mn	0,13	0,13	0,1	0,11
K	0,06	0,13	0,19	0,68
P	0	0,11	0,48	0,49
V	0,07	0	0,17	0,21
Total	99,62	99,59	100,77	100,73

Figure 7.34: Selection of EDS analyses performed on main minerals representative of textures 11, 12 and 13.

7.4 Relationship between slag chemistry, mineralogy and cooling rate

7.4.1 Link between chemistry and mineralogy

The samples studied have a high chemical variability. We already mentioned that the chemistry of a cooling liquid directly impacts the texture of the solidified material (see Texture sheets). Hence, samples with similar chemical compositions tend to have similar textures (Figure 7.37).

Samples enriched in iron oxides are more likely to have wustites. Indeed, the XRD analyses detected wustite in samples with iron content above 60 wt.% Fe_2O_3 (Figure 7.35). XRD analyses are consistent with microscopy observations: samples with high iron oxides content usually show a texture 5.

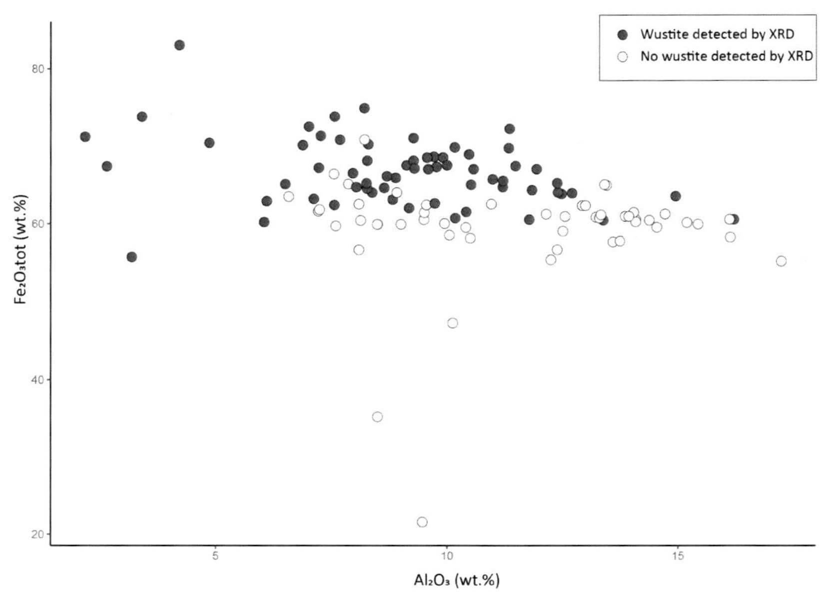

Figure 7.35: Al_2O_3-Fe_2O_{3tot} bivariate diagram for all studied slag, showing the correlation between high iron content and the presence of wustite. The data are bulk XRF compositional analyses.

Similarly, samples with slightly higher alumina content tend to have a texture 1 or 2, i.e. textures with large sp1 and sp2 spinels. These samples correspond mainly to samples from Bemanevika and Benavony (Figure 7.38 - C), which are enriched in TiO_2. It can be seen that for equal iron content the samples from these two sites do not have wustite whereas the slags from the other sites do (Amboronala, Ambodimadiro and Matavy). It could be because, during the crystallization process, the spinels mobilize a lot of iron. The liquid thus becomes too depleted in iron to be able to crystallize wustite grains.

In Figure 7.37, we can see some inconsistencies between the XRF bulk composition of some slag samples, and therefore where they are plotted on the phase diagram, and the main texture observed in these samples. For example, there is a sample classified with texture 8, and therefore with wustites crystallising first, which is inconsistently plotted in the fayalite domain. It is due to the fact that slags are heterogeneous materials that can have several textures within the same piece of slag. The chemical composition of a specific slag sample is therefore not typical of a texture but a mix of several textures. Furthermore, although we try to make the XRF bulk analyses as representative as possible, it is always possible given the heterogeneity of our materials that some slags contain a lot of entrapped sand grains from the furnace wall. The chemical compositions seem indeed to be drawn towards the SiO_2 pole.

7.4.2 Impact of the cooling rate on slag texture and mineralogy

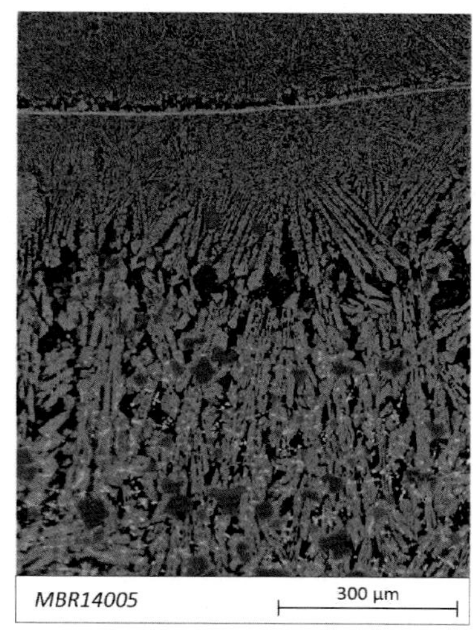

Figure 7.36: BSE image of a texture gradient in a tapped slag (MBR14005, Amboronala) due to a variation of the cooling rate. At the top we can see a white line related to the interface between two flows of tapped slag, and therefore corresponding to the cooling surface. The further away from this line, the more developed the grains are. The cooling rate of the liquid decreases when moving away from the cooling surface.

The cooling rate is the second essential criterion for understanding textures. It influences the size of the minerals and therefore the textures. Textural gradients and changes in mineral size can be observed as one moves away from the cooling surfaces (Figure 7.36). This trend is particularly visible in the tapped slag samples, but also to a lesser extent in bottom furnace slags close to the cooling surfaces.

We note that some textures are specific to tapped slag and in particular close to cooling surfaces. These surfaces are identifiable in the TS because a weld line can be observed at the interface the different slag flows (Figure 7.36). This line is mostly composed of wustite. Textures 3 (without wustite) and 6 (with wustite) are visible in the upper part of the slag flows, near the weld lines. These texture are specific to fast cooling rates (Figure 7.37 and Figure 7.38 - A).

The appearance of the spinels is very different in these two textures compared to other textures with lower cooling rates (Textures 1, 2 or 5 for example). There is no separation into two distinct grains of sp1 and sp2 compositions as in the other textures. The core of the spinels is aluminous and shows titanium content around 3-5 ox.% TiO_2. Then the composition evolves into titanium-rich and aluminium-depleted sp2 spinels. Chromium is concentrated in the cores of these spinels, associated with the titanium-poor and aluminium-rich compositions. What is important to note is that these two compositions are found in the same grain. Moreover, the closer a grain is to the weld line, the more titaniferous its core is.

This spinel texture is actually typical of a high cooling rate. It seems that titaniferous sp2 spinels are not phases that crystallize at high temperature. In most of the observed cases, these spinels crystallize in second or third position. Only in the case of very

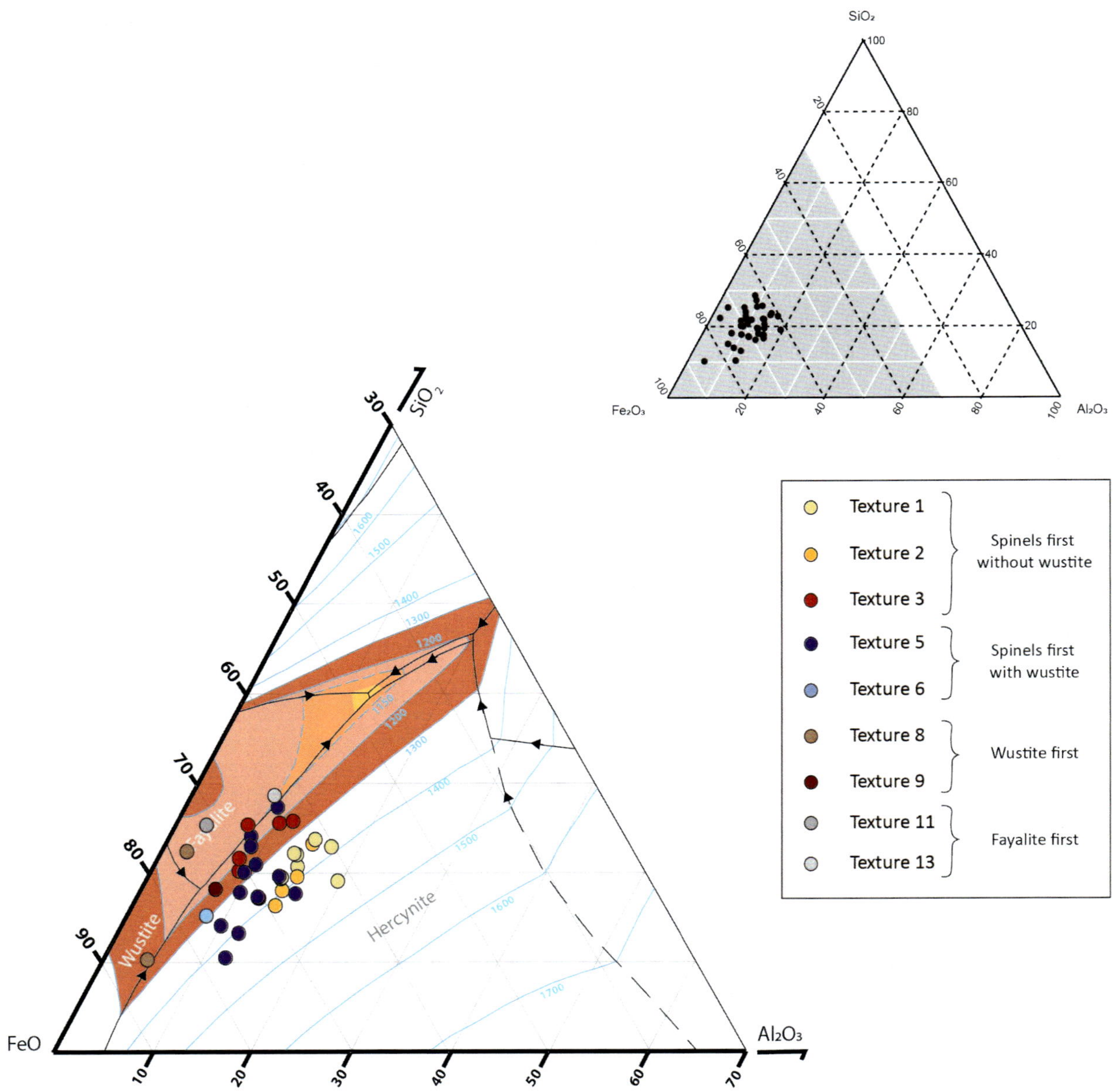

Figure 7.37: *FeO-SiO_2-Al_2O_3* ternary phase diagram showing the variability of the bulk chemical compositions of the slags according to their texture. The predominant texture is shown on this diagram. Several samples are composed of juxtaposed areas with different textures. This could explains some inconsistencies between the chemistry and the texture.

rapid cooling are these spinel compositions observed to crystallize among the first phases.

When the liquid is no longer very hot, it is less liquid and more viscous. The diffusion of chemical elements is less strong because they are less mobile. Minerals that crystallise therefore integrate the chemical elements available in their immediate environment. Here, spinel is forced to integrate titanium. The cooling rate therefore impacts not only the size of the grains, but also their composition.

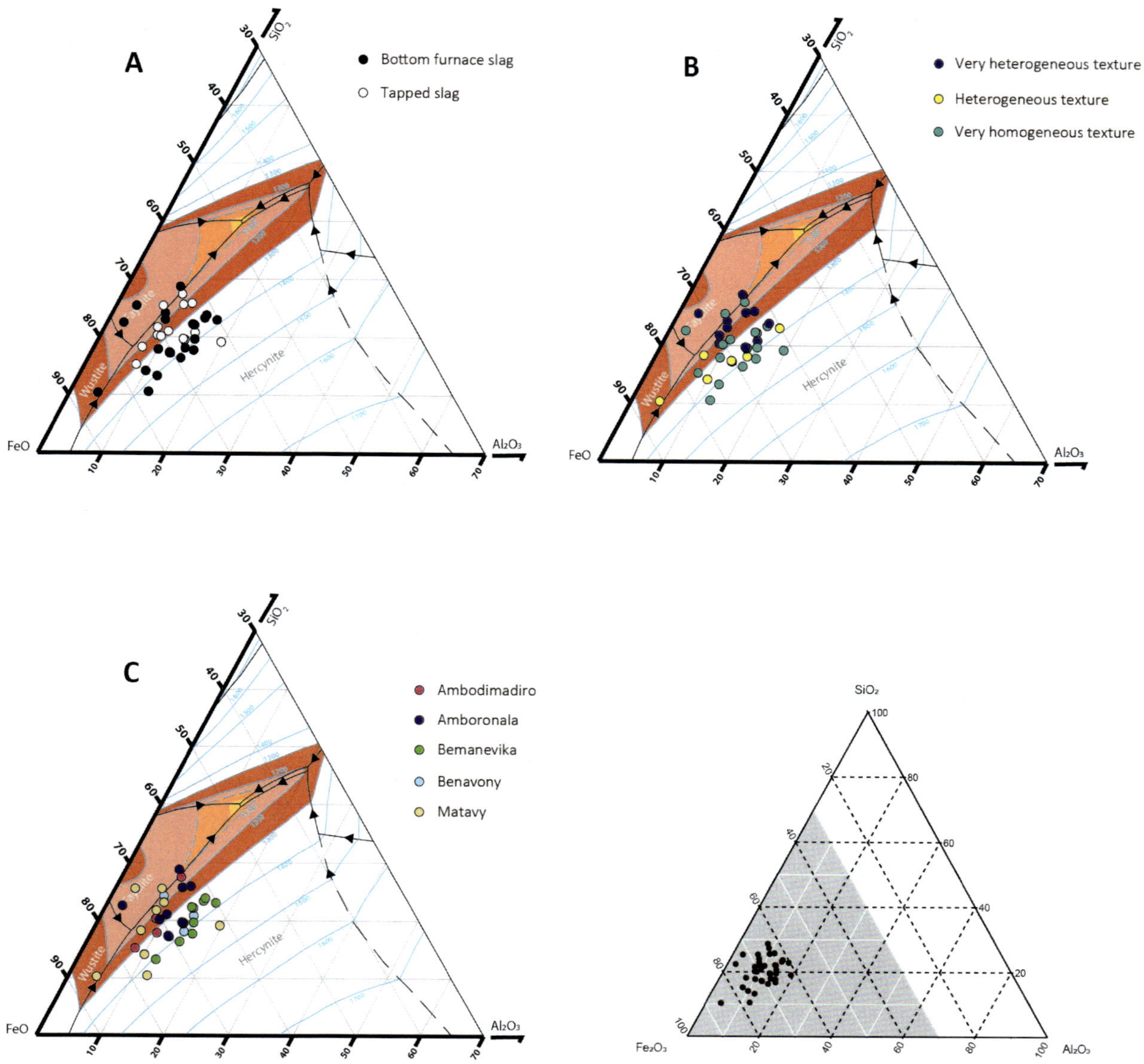

Figure 7.38: *FeO-SiO$_2$-Al$_2$O$_3$* ternary phase diagram showing the variability of the bulk chemical composition of the slags according to their morphology (**A**), their textural heterogeneity (**B**) or the smelting site from where they come from (**C**).

7.4.3 Considerations on the homogeneity of liquid slag

Ores are not homogeneous materials, not only from one fragment to another, but also within the same grain. During the smelting process, the first liquid that forms therefore doesn't necessarily have the same chemical composition as the second.

It is often assumed that when the slag forms at the bottom of the furnace, the material is sufficiently hot and liquid to gradually homogenise, by convection or diffusion. Indeed, it has been shown that some large pieces of slag are made of a single bloc of homogeneous material, and not of successive layers (Humphris et al., 2009). Some studies have also shown that successive flows in tapped slags can have the same chemical composition (Serneels, 1993).

However, in the slag from northeastern Madagascar, there is significant textural heterogeneity within many samples (Figure 7.18). 22 of the 41 samples observed by optical microscopy have several textures and cannot be considered as homogeneous. The transition is not gradual or linked to a cooling gradient as described above. There is a clear separation between two areas with radically different textures (Figure 7.39). These are two liquids of different composition that have cooled next to each other without mixing.

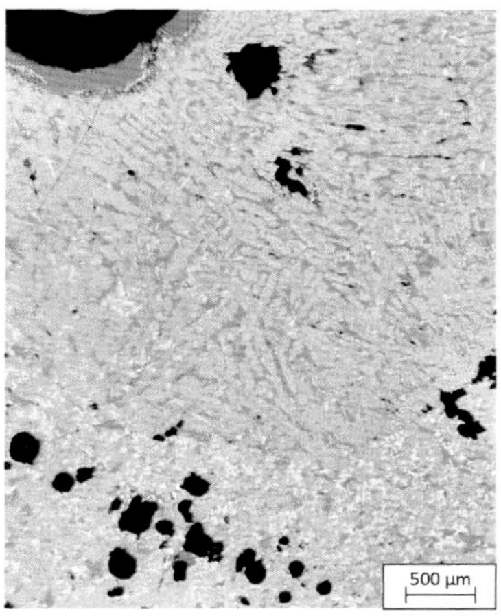

Figure 7.39: Reflected light microscopy image of BNY45005 showing two juxtaposed textures with a clear boundary.

We can observe that samples with a heterogeneous texture tend to have a lower melting temperature than homogeneous samples (Figure 7.38). This result is counter-intuitive, one would expect the opposite. One explanation could be that slag pieces with low melting temperatures and heterogeneous textures were formed during a smelting operation with low temperatures in the furnace. The temperature was high enough for the slag to fluidise, but not enough to create a hot liquid that could homogenise. On the contrary, slag pieces with a high melting point were formed at high temperatures and the liquid could homogenise.

However, these interpretations must be qualified. On the one hand, the phase diagrams are too simple to correctly estimate the melting temperatures of a slag. Slags are not only composed of SiO_2, Al_2O_3 and FeO, but also of TiO_2, CaO, MgO, which also impact on melting temperatures.

Moreover, the melting point of a slag piece only gives the minimum temperature which must have been reached during the operation to form this specific slag, and not the temperature actually reached in the furnace. Slag with low melting points could have been in a furnace that reached high temperatures and therefore be very liquid.

In general, the number of samples observed is too small to draw definitive conclusions. The stratification of samples with heterogeneous textures and the respective proportion of textures should be studied. This would allow the study of the evolution of the chemistry of the liquids and to understand the temperature changes in a furnace during a smelting operation.

Iron Working in Madagascar | 8

To better understand the arrival and diffusion of iron metallurgy in Madagascar, historical, ethnographic and archaeological sources were re-examined with an archaeometallurgical approach.

A few archaeological studies focusing specifically on iron metallurgy had already been carried out in some regions of Madagascar, in particular the highlands around Antananarivo (Gabler, 2005; Radimilahy, 1988) and in the Betsileo country (Coulaud, 1973), and in the south of the island (Radimilahy, 1988). Numerous archaeological survey reports also mention the presence of metallurgical remains (Crossland, 2001; Parker Pearson, 2010; Griffin, 2009; Dewar and Wright, 1993; Radimilahy, 1988; Vérin, 1986 and others). We have therefore sought to differentiate between primary production sites and consumption sites on the basis of this documentation (see differentiation smithing/smelting slag in Chapter 2). All the iron-using populations master at least basic smithing techniques and are able to carry out basic repairs, but some populations only consume iron without producing it.

Only few archaeological excavations have been carried out on these sites and no furnace structures have been excavated, with the exception of our work in the Northeast. The diversity of Malagasy iron production techniques is therefore not yet known. For these reasons, it is in most cases impossible to link the description of smelting slag with a specific technical tradition.

The historical documentation is very rich from the time of the first European explorers. However, the descriptions do not go into detail on the issues of iron metallurgy until the 18th century. The villages or regions mentioned in these texts could be located using maps from the 19th century (Cartothèque of the University of Bordeaux-Montaigne).

This work has sought to take into account as many sources and previous research works as possible. This chapter therefore corresponds to the current state of knowledge on Malagasy iron metallurgy and partly reflects the lack of surveys in some regions of Madagascar. The western half of the island, in particular around Mahajanga, would deserve to be more deeply investigated.

Tables containing all the historical sources used to write this overview can be found in the Appendix (Figure C.1).

8.1 The exploitation of iron ores in Madagascar

8.1.1 Availability of iron ore in Madagascar: Iron ore deposits

Iron is abundant in Madagascar and it has been exploited since early periods. From Fianarantsoa to Antananarivo, the Precambrian basement contains magnetite-rich quartzite, with an iron content that can locally reach 40% (Lacroix, 1922, vol.2:72-74). The quartzite is a very hard rock, impossible to mine with simple means, but the superficial layer has undergone severe weathering leading to the formation of a red tropical soil, a kind of laterite. The iron hydroxides concretions can be collected easily. After washing they become a good raw material for smelting by the bloomery process.

Also from other rock types, such as basalts as in the northeastern part of the island, the alteration process leads to the formation of iron-rich concretions near the surface. In many areas of the island, the conditions are favourable enough and exploitable concentrations of iron-rich material can be found. From the modern point of view, these have no economic value but at the scale of pre-modern industry, they were valuable. From the 17th century, early observers noticed the smelting of iron and the abundance of ore:

> "Iron of excellent quality abounds in the central provinces, around the capital, where it is found near the surface, and so rich is the ore in one of the mountains, Ambohimiangavo, that it is called the iron mountain. The ore is so abundant at the surface, that the soil has seldom been penetrated more than a few feet in depth, so that at present no idea can be formed of the riches of the country in this valuable metal".
>
> William Ellis (1858:264)

8.1.2 Iron ore extraction

Mining techniques in Madagascar are relatively well documented from the 19th century (Breton, 1898; Oliver, 1886; Hastie, 1817), but no archaeological mining remains have been recovered so far. Lateritic ferruginous concretions appear to have been the privileged ore. These ferruginous concretions and the associated lateritic substratum were collected in pits that were no more than 4-5 metres deep (Breton, 1898:687; Oliver, 1886:Vol.2, p.88). The miners focused on extracting the loose parts of the lateritic formation, probably as during the pre-colonial periods. The lack of archaeological evidence, such as galleries, suggests that the ore was obtained by surface collection. Breton describes that the task of collecting the ore is generally carried out by women, 'with their hands when the ore is soft, or with a wooden or iron instrument' (Breton, 1898:687). The ore is then washed to keep only the pisolites before being crushed to form small grains ranging from 1 to 3 centimetres in diameter (Ellis, 1838:306; Vacher, 1899:543–544). There is no mention of ore roasting in the written sources nor in the archaeological records.

8.2 Chronological overview of iron production in Madagascar

The origins of iron metallurgy in Madagascar are still poorly known. The traditional narrative about the introduction of iron production refers to King Andriamanelo, considered to be at the origin of the Merina royal lineage (16th century CE).

In the *Tantara ny Andriana eto Madagascar* (1873), which relates the History of the kings of Imerina according to Malagasy manuscripts and oral sources, it is told that Andriamanelo, by teaching his subjects how to produce iron and make it into spear heads, could defeat the Vazimba people and set up a new kingdom (Mullens, 1875; Callet, 1974). In fact, archaeology has shown that iron was already used several centuries earlier at many places around Madagascar (Serneels et al., 2021; Crossland, 2001; Parker Pearson, 2010; Griffin, 2009; Gabler, 2005; Radimilahy, 1988; Dewar and Wright, 1993). The traditional account should not be taken literally. However, it has undoubtedly a symbolic dimension which may also reflect the relationship between the development of a more efficient iron industry and the rise of the centralized Merina kingdom.

Based on recent fieldwork and previously published data, we propose a general overview of the early development of iron consumption and production. For the last centuries, the written record provides a substantial amount of data, both on techniques and about organisation of the production.

8.2.1 Iron consumption before the 11th century

The populating history of Madagascar is poorly delineated and still heavily debated (Campbell, 2019; Douglass et al., 2019; Radimilahy and Crossland, 2015; Dewar and Wright, 1993). Cutmarks on bones can be interpreted as very early traces of human activity which put back human occupation to 2000 BCE, but they remain very controversial (Gommery et al., 2011; Macphee and Burney,

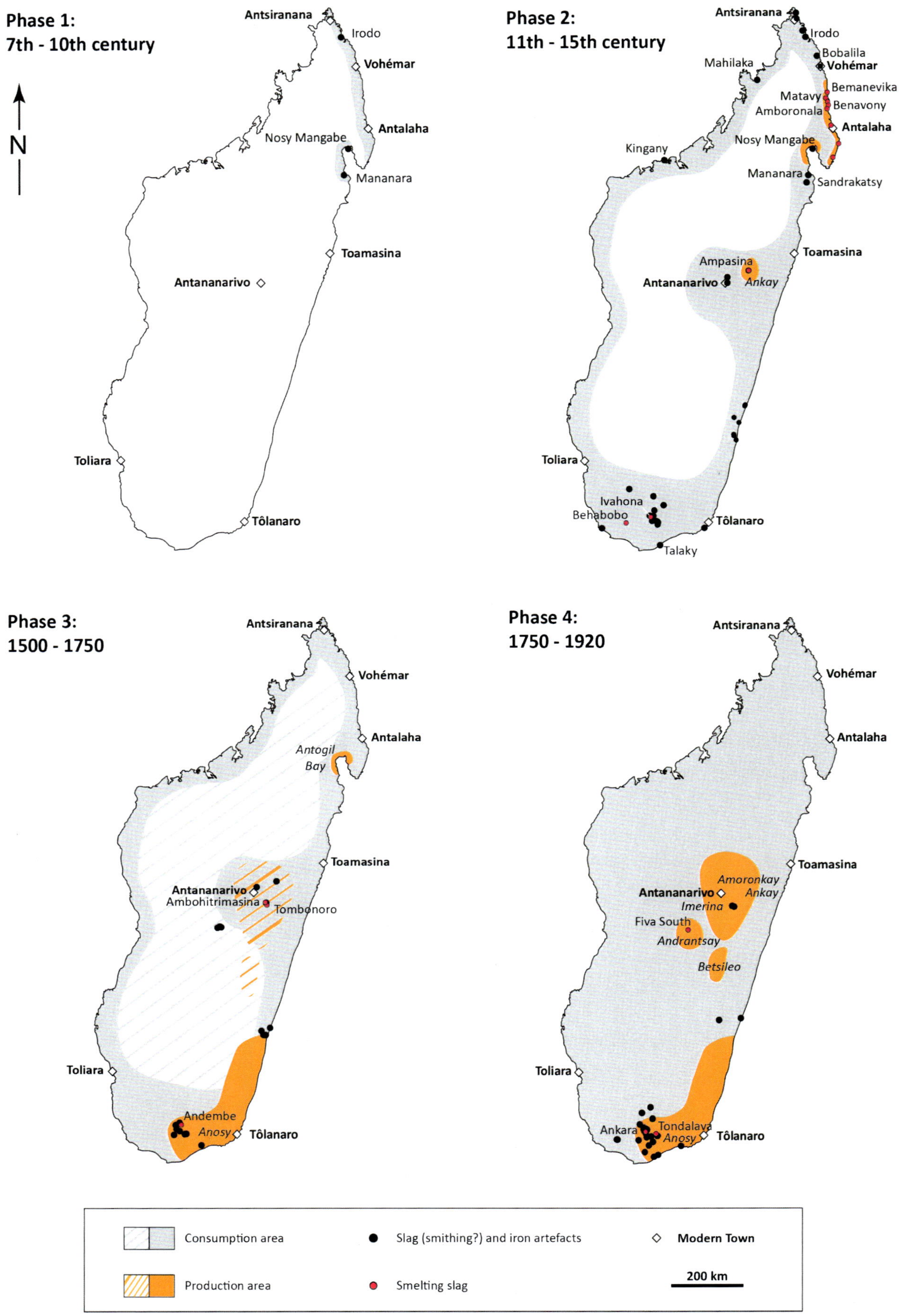

Figure 8.1: Map representing the spread of iron consumption and production in Madagascar from the first attested permanent human settlements (7th century) to 1920. The compiled data are from the archaeological records (Douglass et al., 2019; Crossland, 2001; Parker Pearson, 2010; Griffin, 2009; Gabler, 2005; Radimilahy, 1988; Dewar and Wright, 1993; Wright et al., 1993; Wright and Fanony, 1992; Rasamuel, 1984; Heurtebize and Vérin, 1974 and others) and from written sources (Breton, 1898; Lacombe, 1840; Ellis, 1838; de Flacourt, 1661). For the detailed list, see Appendix Figure C.1.

1991). Some authors support the opinion that these cutmarks were made with iron tools (Macphee and Burney, 1991) but this interpretation is not credible because no human society was using iron in time periods as early as 2000 BCE. The stone artefacts from Anja Cave suggest the presence of people already during the second millennium BCE (Dewar et al., 2013). But no evidence of use of iron is associated with the finds from these sites.

Much evidence, whether archaeological, linguistic or genetic, supports the hypothesis that, by the second half of the first millennium CE, permanent settlements had developed on Madagascar (Pierron et al., 2017; Beaujard, 2003; Allibert, 2008). By the 7th century, the island population is connected to the Indian Ocean trading network since Islamic ceramics are found on several settlements (Rakotoarisoa and Allibert, 2011; Dewar and Wright, 1993; Wright and Fanony, 1992; Vérin, 1986). After 1000 CE, there are signs of a significant increase of the population size and a spread of the settlement pattern.

On archaeological sites dated before the 11th century CE, the use of iron is attested only in northeastern Madagascar (Figure 8.1. Phase 1). Excavations in Nosy Mangabe (Wright, 1992; Vérin, 1986) and near Mananara (Wright and Fanony, 1992) unearthed pieces of slag, probably from smithing. Moreover, at several places, soapstone sherds were found, indirectly signalling the use of iron tools for stone extraction and carving. The absence of conserved iron artefacts elsewhere in Madagascar may be due to unfavourable burial conditions, but reflects more probably the actual absence of iron.

No traces of primary iron production have been observed for these early periods in Madagascar, including in the recently surveyed Northeast. It would therefore seem that iron was supplied through overseas contacts. Iron artefacts or bars could have been provided by most of the populations involved in the Indian Ocean trading network as the smelting of iron was already an established technique in Persia, southern and southeastern Asia. Even along the eastern coast of continental Africa, Bantu populations mastering the smelting of iron were already settled by the 5th century CE (Craddock, 1995).

8.2.2 Iron between the 11th and the 15th century

From the 11th century onwards, archaeological evidence of iron consumption is numerous all around the island in the coastal areas (Parker Pearson, 2010; Radimilahy, 1998; Dewar and Wright, 1993;Wright and Fanony, 1992), as well as some in the Central highlands (Gabler, 2005; Crossland, 2001; Rasamuel, 1984 – Figure

8.1. Phase 2). The use of iron spreads during this period, mostly along the coasts but also in the central part of the island.

Our recent fieldwork in the northeast of Madagascar (Chapter 4) brought to light the remains of significant iron production. They are well-dated by excavations and radiocarbon measurements. This production is contemporary to the settlements of the so-called *Rasikajy* population. We refer to this particular technical tradition as **MDG01**. It is for now the oldest iron production identified in Madagascar (Serneels et al., 2021). Early iron production has also been identified and studied in the central highlands, in Ankay, by Gabler (2005 – Figure 8.1. Phase 2). The furnaces were probably small structures with clay lining and a superstructure partially made of stones. More field work is still needed to present a convincing reconstitution of the furnace architecture. The iron production remained at a small scale and was poorly standardized and mastered. Gabler also assumes a seasonal activity, since smelting sites are often located on riverbanks, which are flooded during the rainy season. In many respects, the situation is similar in Ankay and the northeast: a poorly controlled process and a small-scale production intended for local needs.

It is however not clear what process could have prompted people to start a local metallurgical production, rather than keep on relying on imports. It could be because of an increasing demand in iron linked with demographic growth, the new arrival of people with metallurgical know-how, or a reorganisation of the Swahili trading network. For now, population mobility in the western Indian Ocean is too poorly understood to support one hypothesis over another.

8.2.3 Iron between 1500 and 1750

The first European navigators pointed out the broad use of iron objects by the Malagasy populations, albeit in small quantities (Barbosa, 1516:54 and Lopez, 1585:147 in COACM I). Engravings from this period show people using iron weapons, such as scenes of whale hunting with iron harpoons off the island of Sainte Marie (de Bry, 1598). The written sources focus on the description of armament which consists of a large spear with an iron tip, a large knife and sometimes arrows. Fish hooks, spades for agriculture or simple carpenters' tools are also occasionally mentioned (Boothby, 1646:93 and Everard, 1686:415 in COACM III). The most complete list is given by Flacourt:

> "[...] by smithing implements, such as axes, hammers, anvils, knives, hansards, spades, which they call Fanghali, razors, tweezers for plucking hair, grills for roasting meat, hooks for drawing it out of the pot,

> hatchets or trapdoors for wounding their enemies, and all kinds of javelins, darts, small darts, and great knives for cutting the throat of oxen".
>
> Translated from Etienne de Flacourt (1661: 73-74)

Blacksmithing became a common craft across Madagascar (Boothby, 1646:93-94 and Everard, 1686:415 in COACM III; de Flacourt, 1661:451). The smithing slag pieces found at numerous archaeological sites support the widespread use of iron (Parker Pearson, 2010; Gabler, 2005; Crossland, 2001; Radimilahy, 1988 - Figure 8.1. Phase 3). Smithing hearths are described as small structures associated with piston bellows and simple tools are used, like a big stone as an anvil:

> "I also saw a blacksmith making spear-heads in a strange way; he had for tools, so far as I could see, only two stones, which served him, one as an anvil and the other as a hammer, and his bellows were formed of two bamboos about three-quarters of a yard long, and about as big as a man's leg or thigh, which were pierced at the end with small holes through which air escaped over a small fire placed near them. This air was expelled by small plates or discs similar to butter-bats; as for the bamboos, they had their upper part closed by a sheep or goat skin with the hair outside, and resembled the swabs used to clean cannons; these bats [or pistons], being alternately raised and lowered, expelled the air onto the fire".
>
> Richard Boothby (1646 in COACM III: 93-94)

A later source indicates that the tuyere is a drilled stone:

> "At the bottom of the two tree trunks, two small bamboos are sunk into a stone with a single opening in front of the fireplace".
>
> Translated from Louis Catat (1893:41)

Unfortunately, smelting techniques are not detailed in early written sources, but it seems that this metallurgical know-how spread across Madagascar (Figure 8.1. Phase 3). Some rare textual allusions, as well as archaeological slag discoveries, attest the development of primary iron production in Imerina, i.e. the area around Antananarivo, and in the southern regions (Drury, 1729 in COACM IV:359; Hamond, 1640 in COACM III:10; Parker Pearson, 2010; Radimilahy, 1988). Flacourt described a smelting process observed in the region of Fort Dauphin (today Tôlanaro) in the South-East. Unfortunately, his description is too vague to reconstruct accurately the architecture of the furnace. He describes a small hearth with clay walls reinforced with some stones. No superstructure is described, so it is probably an open hearth. An original superstructure, such as the conical structure of the MDG03 technique (described below), would have been noticed and mentioned by Flacourt. The slag is discharged outside the furnace. Piston bellows are used. Only small quantities of ore are smelted and therefore only small amounts of metal are produced. It is therefore not a technique meant for intensive production, but probably rather for local consumption.

> "The smiths of this country take about a basket of ore such as they find it, they pound it and throw it into the fire between four stones lined with greasy earth and blow with their bellows, made in the shape of a wooden pump, and by dint of blowing, at the end of an hour they find their ore melted, which they make flow: then they heat it and beat it so much that they form a fonze which means a bar of the weight of three or four pounds".
>
> Translated from Etienne de Flacourt (1661:146-147)

8.2.4 Iron between 1750 and 1920

After 1750, iron consumption spread across the whole island and basic smithing was mastered almost everywhere (Figure 8.1. Phase 4). Smelting is well attested in the Central Highlands around Antananarivo and in the Southeast. It might have been practised at smaller scale in other parts of the island.

For this period, Madagascar has a very rich historical record concerning iron production. Numerous written sources, sometimes illustrated with pictures and drawings, are available. Unfortunately, the documents are not always clear and some confusion occurred. It appears that at least two different smelting techniques were used at the same period but in different regions (Vacher, 1899). On the contrary, the archaeological record is limited to a few surface observations, few test pits and finds of probable smithing slag (Parker Pearson, 2010; Griffin, 2009; Crossland, 2001; Radimilahy, 1988). No well-preserved furnace has been excavated. Only rubbed pits and stone tuyeres are reported, but it is not enough evidence to present a convincing reconstitution of the smelting process.

Barrel-shaped furnace used in the Central Highlands (Merina) – Technical tradition MDG02

A photograph published by Lacroix (1922:170) shows a smelting furnace still in use at the beginning of the 20th century in Amoronkay, east of Antananarivo (Figure 8.3). Similar furnaces are described in the literature, both in written sources (Breton, 1898 686-693; Oliver, 1886; Mayeur, 1785; Vacher, 1899:544) and ethnographic investigations (Radimilahy, 1988:27-39). This smelting technique will be referred to as **MDG02** in the following pages.

The furnace is circular in plan and barrel-shaped in section. The structure is a little less than one metre high and 60-70 centimetres in diameter. The superstructure is built with small stones and the inside is lined with a layer of clay a few centimetres thick. The use of slag blocks instead of stones is reported in the frame of an intensive production site (Breton, 1898). The furnace is built against a slight slope with, downslope, an opening made of three large stones allowing the tapping of slag, the extraction of the bloom and hence, the reuse of the furnace. Upslope, at ground level, a single tuyere is installed. The tuyere can be made of "clay with big grains" according to Breton or of schist stone, called *didy*, according to ethnographic record (Radimilahy, 1988). The tuyere is connected by two pipes to a pair of vertical piston bellows which must be about 1.20 metres high.

	Total mass of materials	Mass of iron	Iron content in ore
Ore loaded in the furnace	200 kg	140 kg	70 % Fe_{tot}
Bloom after smelting	70-80 kg		
Iron after bloom smithing	35-40 kg		
Yield	30%		

Figure 8.2: Calculation of the yield of a smelting operation carried out in a furnace of the MDG02 technical tradition, according to the description by Breton (1898).

According to Breton (1898), 180 to 220 kilograms of washed and crushed lateritic ore are loaded into the furnace at the beginning of the smelting process, with alternating layers of charcoal. The ore is arranged as a ring in the furnace so that a column of charcoal can be installed in the centre of the furnace. The ore appears to be loaded only once:

> "The first step is to put burning charcoal at the bottom of the furnace, then a layer of ore, but only in a ring, and so on, alternating layers of ore and charcoal so that there is a continuous column of coal in the centre through which the wind can blow. Without this precaution, some furnaces have burst".
>
> Translated from Breton (1898: 689)

After approximately seven hours, 70 to 80 kilograms of iron mixed with slag are removed from the furnace. This dirty bloom is broken up and sorted to remove as many impurities as possible, leaving only 50% of the initial amount. Breton therefore describes a technique with a yield close to 30% (Figure 8.2). It is then heated up in a hearth similar to those used for smithing to be purified from slag inclusions and compacted. Breton reports that for a specific village named Marorangotra, between 700 and 800 kilograms of iron were sent each week to supply Antananarivo.

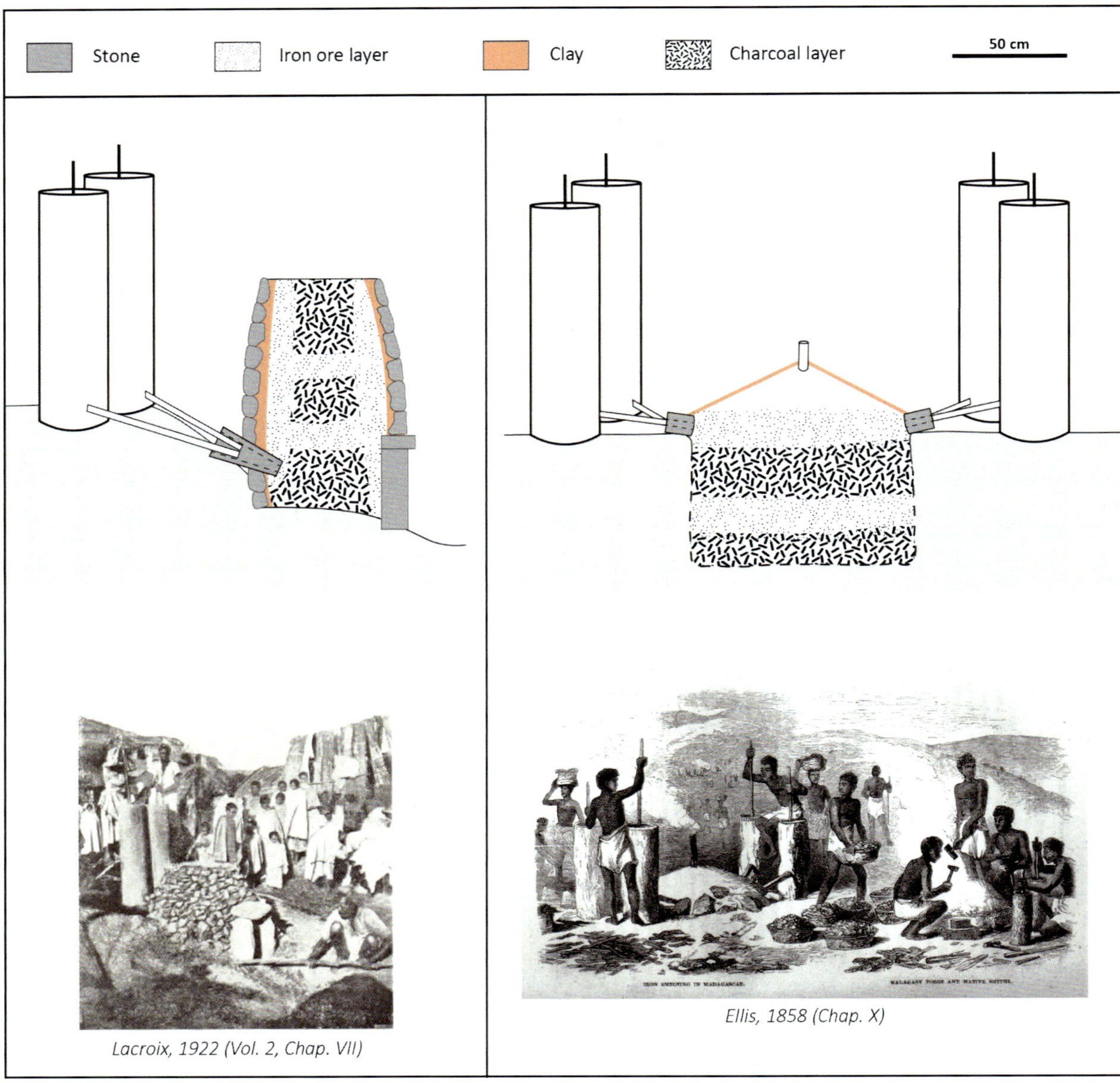

Figure 8.3: Reconstitution of two Malagasy smelting furnaces used during the 18th-19th century. (**Left**) The barrel-shaped furnace used in the Central highlands (Radimilahy, 1988; Lacroix, 1922; Breton, 1898). (**Right**) The covered pit furnace used in the South (D'Escamps, 1884; Sibree, 1873; Lacombe, 1840; Ellis, 1838). (From Morel and Serneels, *under review*)

The covered pit furnace used in the South (Anosy) – Technical tradition MDG03

The most famous image of iron smelting in Madagascar is an engraving published by William Ellis in his book *Three visits to Madagascar during the years 1853-1854-1856* (Ellis, 1858 - Figure 8.3). According to the author, this technique, called here **MDG03**, was implemented in Amoronkay to supply the capital Antananarivo with iron. However, Ellis probably never saw these furnaces personally, but only reported observations of previous authors. Indeed, Ellis had already published a very similar engraving and description twenty years before in his *History of Madagascar* (Ellis, 1838). At that time, Ellis had never set foot on Madagascar and relied only on the accounts of at least four English missionaries and in particular the work of David Griffiths (Campbell, 2012) or Robert Drury (Bazargan, 2017; Drury, 1729). The engraving probably comes from one of those texts.

Figure 8.4: Comparison of engravings of smelting furnaces in the four main 19th century written sources.

The two versions of the engraving in Ellis' two books are slightly different. However, both are almost identical with other engravings from other written sources (Ellis, 1838/Lacombe, 1840 and Ellis, 1858/Sibree, 1873 - Figure 8.4). It is not yet clear who the authors of the original engravings are and whether they actually observed a Malagasy furnace themselves.

In the 1838 version, it would even seem that Ellis mixed the MDG02 and MDG03 techniques into one furnace with high stone walls covered by a conical structure (Figure 8.5). The accuracy of his descriptions can hence be doubted, as did Jones in 1839 (Valette, 1966) who wrote to Ellis:

> "The whole account of the smelting of iron is very incorrect on the three pages. I have been at Moronkay six times and I spend a whole day to examine the whole process from the beginning in the mine to the end in the blacksmith forge; and I have not seen any process like the one described in the History whoever gave the account to you".
>
> David Jones (1839, in BAM 45)

Leguével de Lacombe (1840:161-162) gives a detailed account of a similar furnace in the Antarayes region, located north of Tôlanaro, which was out of the control of the Merina power (Figure 8.1. Phase 4). Lacombe's testimony seems quite reliable and he probably observed himself the metallurgical processes. D'Escamps (1884: 448-449) and Sibree (1873: 218-220) describe a similar technique without specifying where or by which population this kind of furnace was used. The MDG03 technique is clearly attested in the south of Madagascar, but the written sources are too vague to exclude other parts of the island.

Author	Iron ore		Furnace				Air supply		Duration of the smelting process	Smelting tradition
	Extraction	Preparation	Shape	Size	Building materials	Slag tapping	Bellows	Tuyeres		
Copalle, André 1825					Stones plastered with clay	Yes	2 Piston bellows		6 hours	MDG02?
Ellis, William 1838	Pits	Washed/crushed	Pit furnace, Conical cover	Heigth (walls): 90-120 cm, Diameter: 180 cm, Depth: 30-60 cm	Stones withouth mortar, plastered with clay	No	Piston Heigth: 150 cm			Mix between MDG02 and MDG03
Leguével de Lacombe 1840			Pit furnace, Conical cover		Clay		Piston Heigth: 100 cm, Diameter: 30 cm	Stone		MDG03
Ellis, William 1858		Washed/crushed	Pit furnace, thick clay cover	Depth: 90 cm	Stones plastered with clay	No	Piston			MDG03
Sibree, James 1873		Crushed	Pit furnace covered with a dome of stones and clay		Clay and stones	No	Piston Diameter: 30 cm			MDG03
Breton 1898	Pits	Washed/crushed	Ogival structure	Heigth: 80-90 cm, Diameter: 70 cm	Stones/slag plastered with clay	Yes	Piston Heigth: 90 cm, Diameter: 20 cm	Coarse-grained refractory clay	7-8 hours	MDG02
Vacher 1899	Pits	Washed	Ogival structure	Heigth: 100 cm, Diameter: 60 cm	Lump of soil plastered with clay	Yes	Piston	Sandstone	5-6 hours	MDG02
			Pits	Diameter: 100 cm, Depth: 100 cm	Lower part of stone slab	No	Piston			MDG03?

Figure 8.5: Comparative table of historical sources describing iron smelting processes in Madagascar (For the complete list of sources, see Appendix, Figure C.1).

By comparing the testimonies, we could reconstruct the MDG03 furnace implementations (Figure 8.5). Many authors repeat, sometimes word for word, Ellis' (Oliver, 1886; Wills, 1885) or Lacombe's (D'Escamps, 1884) descriptions. We decided not to rely on these texts for our interpretation, since it is highly probable that these authors never observed furnaces by themselves. The furnace is a circular structure with a diameter around 1 metre, covering a pit 60 to 90 centimetres deep. The superstructure of the furnace is unusual: a few stones are installed around the pit at ground level and a shallow conical clay structure closes the pit. The engravings (Figure 8.3) represent a small chimney at the apex of this cone whose purpose is to evacuate the excess of gas within the furnace. The confined structure of the furnace undoubtedly creates strong reducing conditions. Apparently, the slag cannot be tapped outside the furnace. Two pairs of piston bellows are installed face to face. The pit is filled up with alternating layers of charcoal and crushed lateritic ore and then closed with the cover. Smelting lasts only a few hours and once completed, the bloom is heated a second time to remove all impurities (Sibree, 1873). The metal is then compacted and shaped into bars.

> "The Malagasy forges are very different from ours; their bellows especially are very curious and of the greatest simplicity; they consist of two tree trunks pierced from one end to the other except for a small portion at the

> lower end which forms the bottom, and above which is a hole. These cylinders are about a foot in diameter and three and a half feet in length; they resemble two pumps which are held together by means of a mortise cut in the length of one of them; two iron pipes about a foot long and an inch in diameter are placed a few inches above the bottom in the holes of which I have just spoken. The two pipes, as they approach each other, enter round holes that are made in the stones that form a masonry structure consolidated with clay. This hearth is shaped like a Chinese hat; in the middle there is an iron pipe, wider than the first ones, through which the smoke comes out; each pump has a piston lined with oakum that the blower, placed in the middle, holds in each hand, and that he makes go alternately; these bellows produce a lot of wind".
>
> Translated from Leguével de Lacombe (1840:161-162)

Origin of the Malagasy iron smelting techniques

The MDG02 and MDG03 techniques share a few characteristics and show quite different features (Figure 8.6 and Figure 8.5). These technical characteristics can be compared with other technologies from the shores of the Indian Ocean.

Both use similar raw materials. Lateritic ore similar to the ones available in Madagascar are widespread in the tropical areas all around the Indian Ocean and have been used widely in Africa as well as in Asia, but not in Persia. Charcoal is the most common fuel for bloomery smelting and iron working. There is however no record of a specific choice of wood.

The air supply system is common to the two technical traditions MDG02 and MDG03. The vertical piston bellows are recognized as a typical technological borrowing taken from the Austronesian package. Cylindrical piston bellows are known only from the region of southeastern Asia (Marschall, 1968). The earliest document showing such bellows is a bas relief depicting the hero Bhima as a blacksmith from the temple of Candi Sukuh on Java built in the 15th century (O'Connor, 1985). The use of drilled stone to make a tuyere (Vacher, 1899; Lacombe, 1840) is not a common feature as tuyeres are mainly made of clay all over the world. This technical characteristic is common to both the techniques documented in the written sources (MDG02 and MDG03), and those documented archaeologically (MDG01 and the surveys carried out by Radimilahy and Gabler). This practice is not recorded from any other context around the Indian Ocean except Madagascar.

On the contrary, the design of the two types of furnaces is very different (Figure 8.3). It can hardly be traced to one single origin. The MDG02 barrel-shaped morphology as well as its sloping implantation are quite usual features for smelting furnaces. On the contrary, the shallow conical cover of MDG03 is not described in any archaeological or ethnographic record. Moreover, slag can be tapped outside of the MDG02 furnace whereas, for MDG03, slag is trapped in the pit and cannot flow outside. Hence, the main technical features of these two smelting techniques are too different to share a common origin. There is no technical continuity, one technique cannot be an evolution of the other.

	Central Highlands Technology *18th-19th century*	**Southeastern Technology** *18th-19th century*	**Northeastern Technology** *11th-15th century*
Tapping of slag	Yes	No	Occasional
Furnace architecture	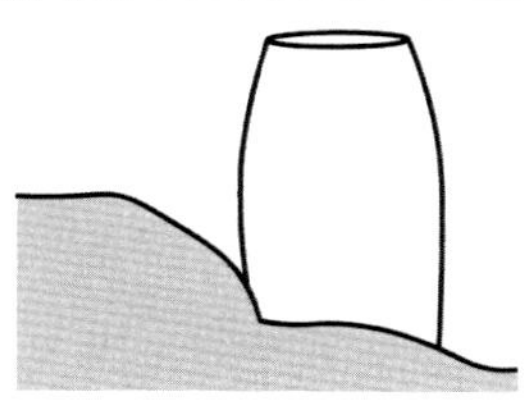 50 cm	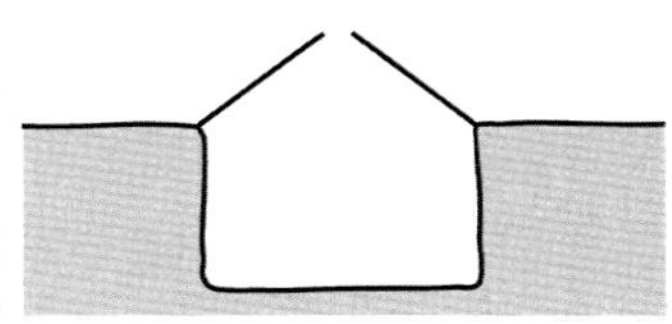 50 cm	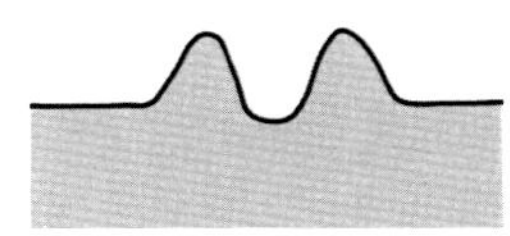 50 cm
Tuyere	One stone tuyere	Two stone tuyeres	One stone tuyere
Bellows	Austronesian piston bellows	Austronesian piston bellows	?
Production size	**Medium** 200 kg of ore → 40 kg of iron	**Medium ?** ? → ?	**Very small** 20 kg of ore → 1-5 kg of iron
Sources	Ethnographic records (Breton, 1898; Lacroix, 1922)	Ethnographic records (Ellis, 1838; Lacombe, 1940)	Archaeological records (Serneels et al., 2019)

Figure 8.6: Comparative table of the main technical features of the three known Malagasy smelting technologies (From Morel and Serneels, *under review*).

It is highly speculative to interpret these data in terms of technology transfer, but some assumptions can be discussed. The Central Highlands barrel-shaped furnace (MDG02) seems to be the result of a technological transfer, maybe sometime before 1750. The traditional narrative is that the Merina kings brought the metallurgical know-how when they conquered the Highlands around 1750 (according to the *Tantara ny Andriana eto Madagascar*). It is said that the local populations, the Vazimba, did not know how to produce iron. Archaeological work has demonstrated since, that, on the contrary, there are traces of iron production before 1750 in the Malagasy Highlands (Gabler, 2005). This myth probably means that the Merina arrived with a new technical tradition (MDG02) and set up a massive production that supplanted the Vazimba technical tradition. The Merina came from the eastern part of the island of Madagascar. Even if their origins outside Madagascar are not known, an Austronesian origin cannot be excluded. There are unfortunately only limited archaeological data are available about iron smelting furnaces in the Austronesian area, and we cannot make a direct comparison with an Autronesian technical tradition. This hypothesis would, however, be consistent with the linguistic similarities observed in the vocabulary related to metallurgy between Malagasy and Malay (Radimilahy, 1988; Dez, 1965).

The southern covered pit furnace (MDG03) could be the improvement of the *Rasikajy* furnace (MDG01). The low efficiency of the *Rasikajy* open bowl furnace leads to poor reducing conditions. This problem can be solved by covering the pit, as in MDG03. The intermediate technique would be the one described by Flacourt in the south of the island. Indeed, it was probably also small open furnaces with tapped slag. This hypothesis would be consistent with the traditional narrative suggesting that the *Rasikajy* would have moved down to the south of Madagascar from the 15th century (Grandidier and Grandidier, 1908).

The context of the development of the Merina proto-industry

The 18th century saw the emergence of the Merina state in the region around Antananarivo. The Merina established a centralized state that controlled the exploitation of natural resources as well as commercial relationships with the Europeans, including the slave trade. The economic growth and the progress in rice-growing techniques generated sufficient surpluses to lead to demographic growth and the development of a full-time specialized craft industry, of which iron metallurgy was one (Campbell, 2005).

King Radama I implemented a policy of proto-industrial development to ensure industrial self-sufficiency. He rejected in 1825 the British free trade alliance and adopted autarkic policies. His priority was to garner indigenous resources and to learn European technology, notably in armament, to support his expansionist policy. The Merina thus negotiated with Europeans to benefit from their iron working know-how. European blacksmiths could settle in Madagascar only under the condition that they trained Malagasy apprentices (Campbell, 2005; Ellis, 1838:197).

The Merina had the upper hand on iron production to control armament. Moreover, as the growth of their power was built on the development of rice cultivation, iron tools were essential to maintain efficiency of agriculture production. They intensified the iron production by creating state-controlled workshops for smelting and smithing, which operated simultaneously to support their high iron needs (Campbell, 2005). Labour was conducted by workers forced under the orders of the royal power and who could not leave under the threat of the death penalty (Oliver, 1886; Ellis, 1838:197). A proto-industrial iron production thus emerged in Madagascar under Merina rule, using the previously described MDG02 furnaces. In parallel, there was local production to meet the daily metal needs of the population. The smelt was carried out by non-specialized workers who had other activities during the year. There is no text that documents these techniques or the furnace shape. It is not known whether they are technical traditions derived from MDG02 or implemented before the arrival of the Merina.

> "In the provinces remote from the capital charcoal is burnt, and iron is worked by the chiefs and their people, or by native labourers, for their own advantage; but in Imèrina and in Antsihànaka, all the iron obtained is for the service of Government; hence five or six hundred men are constantly employed by the order of Government in burning of charcoal for the foundries in the provinces and the smitheries at the capital. The only return these men receive in the shape of compensation for their labour is exemption from certain taxes levied on other members of the community. The charcoal-burners, as well as the miners and founders, are, however, a sort of Government slaves; they live in the forests, or near the places where the ore is found, and they dare not leave their occupations on pain of death."
>
> Samuel Pasfield Oliver (1886, Vol.2:89)

Figure 8.7: Blast furnace from the Mantasoa industrial complex of Jean Laborde (Credits: Vincent Serneels). Cast iron flowed through the furnace opening on the right. The opening on the left was connected to the hydraulic blower.

A modern iron production from the 19th century: Laborde's blast furnace

In the framework of this proto-industrial policy of the Merina rulers, a trial to take advantage of the European technology took place. Armament was a royal priority and an emphasis was put on weapon production. Queen Ranavalona I, succeding Radama I, established in 1828 several factories in the vicinity of Antananarivo, including a musket factory and cannon foundry at Ilafy under the direction of the Frenchman Jean Laborde.

To meet expansion needs, the factory was moved to Mantasoa in 1837, 40 kilometres from Antananarivo. Laborde developed an industrial complex producing cast iron, as well as glass, soap, gun powder, etc. (D'Escamps, 1884). These products were intended to be delivered to Antananarivo. A blast furnace, still visible today, was built in 1843 to cast artillery cannons and produce metal for the manufacture of muskets. The presence of numerous lakes in the surrounding area enabled the construction of aqueducts to power the blast furnace (Campbell, 2005).

8.2.5 Traditional iron working at the beginning of the 21st century

Traditional Malagasy iron smelting significantly decreased at the beginning of the twentieth century. Local production could not compete with European iron production. The last recorded metallurgical practices were located in Amoronkay, in the region of Merikanjaka (Radimilahy, 1988).

The iron used today in Madagascar comes from modern blast furnace production and hence from imports (Coulaud, 1973). Most of the everyday life objects are made by recycling metal pieces from cars or other tools. Today's blacksmiths therefore have the role of recycling or repairing daily objects and making simple objects, such as machetes, spades, axes or small knives. A preliminary ethnographic survey was carried out by Walker Chrisoël Jaony and Hervé Mangany Totobemahefa as part of this project in the fall of 2021[1]. They interviewed nine blacksmiths in order to observe their practices, their workshops and to define their identity. This survey would need to be replicated and refined, but still records current metallurgical practices in the SAVA region[2].

1: Unpublished report: Jaony, W. C., Totobemahefa, M. H. Rapport de prospections métallurgiques - octobre 2021.

2: There are 23 adiministrative regions in Madagascar. The SAVA region is one of the regions in the North-East and includes the areas of Sambava-Antalaha-Vohémar-Andapa.

Figure 8.8: Early 20th century postcards featuring Malagasy highland blacksmithing.

Two types of practices can be distinguished among interviewed blacksmiths. Most of them did not inherit a know-how that had been passed down over several generations. Their father or they became blacksmiths out of economic necessity. It would however seem that once the smithing activity is established in the family, it is passed on from father to son. They are self-taught craftsmen who learn specific techniques according to the evolution of the demand. This category of blacksmiths does not master complex skills. They mainly repair tools or make simple objects, such as spades or knives, from pieces of iron that already have a shape close to the desired final product. Their forge is a simple hearth, sometimes set in a raised wooden trough or directly in the sand

(Figure 8.9.a). Their bellows are manual and are activated by a bicycle wheel. They have a variety of simple tools: an anvil, a hammer, pliers, punches and a water tank. They were asked if they knew how the forefathers made iron in this region. They had no idea of the ancient metallurgical practices.

On the contrary, some blacksmiths claim that they have inherited their smithing skills for generations and that it is a gift that is passed on from father to son. They master a wider range of technical skills and are able to produce a greater variety of objects of better quality. For example, they can make a machete from a crowbar. The fireplaces are installed on the ground (Figure 8.9.b). A cylindrical stone tuyere is installed between other stones or a low clay wall. The shape of the tuyere is very similar to the one from the *Rasikajy* smelting technique. Some blacksmiths even use archaeological tuyeres. Two metal pipes are inserted in a V shape into the tuyere opening and connected to a pair of piston bellows. These may be made of wood or sheet metal.

Figure 8.9: Photographs of different smithing workshops still operating nowadays. (a) Example of a non-traditional smithing workshop (Ambondrona-Antalaha, SAVA region). (b) Example of a traditional smithing workshop (Ankarango-Bekony, SAVA region). (c) Example of a smithing workshop near Antananarivo (Analamanga region).

These forges are simple workshops but are very similar to those described in the written sources (Boothby, 1646; de Flacourt, 1661; Ellis, 1838) or those visible on some early 20th century postcard photographs (Figure 8.8). It is interesting to note that these blacksmiths keep in memory old metallurgical practices. They are able to describe ancient furnaces that are compatible with what is known about iron metallurgy in Madagascar:

> "Yes, he remembers the techniques to extract iron from the ore using a forge. Firstly, four bellows must be installed on the four different sides surrounding the hearth (left, right, front and back), which must be put in the middle, to blow the fire. Thanks to these four bellows, we can have a very high temperature that will make the ore liquid. He pointed out that the iron-rich stones that the ancient inhabitants used to extract iron were collected in a village where slag was found. To extract the iron, these stones must be put into the fire. But before doing so, one must not forget to create channels under the fire directed to four different sides. When these stones are heated, they will become liquid and flow through these four channels. This is what will become the iron. But some of the stone that has been put into the fire will become slag, like the slag that has been discovered so far"
>
> Translated from the transcript of the interview with Mr. Bazy (Ankarango-Bekony, SAVA) conducted by Walker Chrisoël Joany.

There are also blacksmiths still operating around Antananarivo. No survey has been carried out among them, but their workshop is very similar to the ones of the traditional blacksmiths of the SAVA region (Figure 8.9.c).

General Conclusions 9

The study of the MDG01 *Rasikajy* iron metallurgy in northeastern Madagascar has led to the definition of a technical tradition that had not yet been described in the archaeological record. It is a simple smelting technique: the charcoal and lateritic pisolitic concretions are loaded in a small bowl furnace dug directly into the sandy substratum, without any clay-lining. A single short cylindrical tuyere, which may be made of clay or stone, is set horizontally into a wall built with sand. This tuyere is connected to bellows (Chapter 5). The same archaeological remains are found all over the study area which leads us to the conclusion that a single technical tradition with some local variations was implemented for about 400 years (11th to 14th century).

The metallurgical remains and particularly the study of the slag show that the technique was poorly mastered and that production was far from being standardised (Chapter 6). It would appear that some smelting operations failed to produce metal or only produced very little. The metallurgists were not specialised craftsmen, but rather craftsmen producing iron from time to time to meet local needs. If iron production had been carried out by specialised craftsmen, production would indeed be much more standardised and controlled.

On the other hand, it is also likely that the very architecture of the furnace, i.e. a simple open pit built of loose sand, made smelting particularly difficult to control.

From the point of view of the history of techniques, the question of the origin of this unusual smelting technique arises. Indeed, we know that before the 11th century iron was already known and used by the *Rasikajy* in this area of Madagascar (Vérin, 1986). However, no primary production site has been identified so far for such early periods. From the 11th century onwards, metallurgical production appears without any change in the material culture (Chapter 8). The emergence of iron metallurgy on the northeastern coast of Madagascar cannot therefore be linked to the arrival of a new population with metallurgical know-how that replaced the pre-existing population. A change in material culture and in particular ceramics would otherwise have been observed. It is more relevant to look for similarities with the technical traditions that existed around the Indian Ocean at that time.

Iron metallurgy around the Indian Ocean

The *Rasikajy* smelting technique may have been learned from another population of the Indian Ocean during commercial contacts. From 500 CE onwards, most of the regions around the Indian Ocean mastered iron smelting technology. At the time of the *Rasikajy*, at the beginning of the second millennium, there is no doubt that all these populations knew how to produce iron, sometimes with complex techniques.

As already discussed before (Chapter 2), the way in which a metallurgist makes iron depends on the way he has learned to. A smelting technique is adapted to the surrounding raw materials and there may be occasional technical innovations, but in most cases, one reproduces a know-how transmitted either by one's parents or by another population through cultural contacts. At the current state of the research, the MDG01 *Rasikajy* technique is the oldest Malagasy smelting technique. It is then unlikely that this technique was transmitted by another contemporary Malagasy population. The *Rasikajy* were however in contact via the Indian Ocean trade with other foreign cultures, in particular the Swahili coast and the Middle East, as well as with Austronesia through several waves of migration. They may have been also in contact more sporadically with India and China. On the other hand, it is rather unlikely that the *Rasikajy* had any contact with continental Southeast Asia.

- **Swahili Eastern Africa**: The East African coast around the $10-11^{th}$ century CE had regular contact with Madagascar (Radimilahy, 1998). Numerous iron artefacts have been uncovered during archaeological excavations, mainly knives, arrowheads, spears and agricultural tools such as spades or hoes (Chittick, 1974). Iron was therefore a relatively common commodity. However, no metallurgical production appears to have taken place on the Swahili coast during this period. Iron slag pieces are often mentioned in the literature, but only in small quantities (Bauzyte, 2019; Pradines, 2019; Sinclair et al., 1988; Horton, 1984; Chittick, 1984). Based only on the archaeological reports and on discussions with the concerned researchers, I interpret these slags as smithing remains mainly. I do not think there was any metallurgical production on the Swahili coast at this time, or only very locally to meet specific needs.
  In the hinterland, however, there was probably metallurgical production, as the Bantu people mastered iron metallurgy (Childs, 1991). The Swahili coast could therefore easily have been supplied with iron. The smelting furnaces described in the archaeological and ethnographic record have very diverse morphologies and technical specificities (Childs, 1991

Schmidt and Avery, 1978). Some of the technical traditions appear to involve small-scale furnaces, sometimes interpreted as bowl furnaces (Iles and Martinón-Torres, 2009; Brown, 1995; Cline, 1937).

- **Middle East**: Iron metallurgy in the Middle East is particularly poorly known. It is known that Persia was renowned for the quality of its metal production and in particular for the quality of its weapons (Allan, 1979). No furnaces have yet been satisfactorily documented. It is however unlikely that iron was produced in the Arabian Peninsula because there is no iron ore available.

- **Austronesia**: Traces of iron primary production, including large slag heaps, are documented in Sulawesi and Borneo (Harrisson and J. O'Connor, 1969). No archaeological furnaces have been accurately described until now. There is an engraving in the ethnographic literature showing a smelting furnace in Borneo (Bleekrode, 1857; Schwaner, 1853). This engraving has been reproduced several times and slightly modified. On the different versions, the furnaces have either a circular or a rectangular base, which drastically changes the interpretation in terms of history of techniques (Figure 9.1). No bowl furnace is recorded for the moment for this period.
  It is known, however, that there is a link between Austronesian and Malagasy metallurgy, since piston bellows are implemented in the MDG02 and MDG03 technical traditions (O'Connor, 1985; Marschall, 1968). However, we have no archaeological evidence that these bellows were used by the *Rasikajy*.

Figure 9.1: Comparison of two 19th century engravings illustrating the same technical tradition in Borneo. In the first case **(a)** there are two circular furnaces side by side (Bleekrode, 1857), while in the second **(b)** there is a single rectangular furnace (Schwaner, 1853).

- **India**: Iron metallurgy in India is ancient and dates back to the first millennium BCE. India is famous for its steel production and used to export its metal. Various technical traditions have been used during this long metallurgical history (Tripathi, 2014; Park and Shinde, 2013; Buchanan, 1807), including crucible steel techniques (Girbal, 2017; Park and Shinde, 2013). Little is known about the technology around the 10th century CE, because research focuses mostly on the early metallurgy. Some authors mention pit furnaces with no superstructure (Hegde, 1973), but parallels with the MDG01 tradition are difficult to draw.

- **China**: The practice of iron metallurgy in China is also ancient, the earliest dates for Chinese iron objects dating back to the first millennium BCE. Chinese metallurgy is best known for the production of crucible steel, the earliest traces of which date back to the 5th century BCE (Wagner, 2008; Han, 2000). As early as the 3rd century BCE, in some parts of China, iron was a relatively common metal and metalworking was reserved for specialised workers. Production was even organized and controlled by the state. China is also famous for developing blast furnace technology at the end of the first millennium BCE.
  During the period we are interested in (1000 and 1500 CE), China thus had already a long metallurgical tradition. A wide variety of furnaces has been used over the centuries, including bowl furnaces (Li et al., 2019). However, the literature focuses mainly on very old techniques or on ethnographic records from the 20th century. For the medieval period, little is known about production techniques. We could not find convincing parallels with the *Rasikajy* technique.

- **Continental Southeastern Asia**: The iron production techniques documented for this region of the Indian Ocean and this period are radically different from what is observed in Madagascar. The furnaces described are linear with generally an elongated rectangular shape. These furnaces are from a different technological lineage and it is unlikely that there was a transfer of technology to Madagascar (Leroy et al., 2020; Juleff, 2009).

The state of knowledge on iron production techniques in the Indian Ocean in the $10^{th}-11^{th}$ century CE, and in particular on production structures, is still poor and would deserve more combined field and laboratory work. It is nevertheless possible to try to compare some technical specificities with what we have observed for the MDG01 technique.

Origins of the *Rasikajy* iron smelting technique

We will here look at some specificities of the *Rasikajy* technical tradition and see if similarities can be found with other technical traditions known in the archaeological or ethnographic records around the Indian Ocean.

First of all, *Rasikajy* metallurgists use lateritic ores while other iron rich rocks are available in the immediate environment, in particular black sands. The use of lateritic ores is relatively common, especially in Africa (Cline, 1937). However, the fact that they do not use black sands is indicative. Indeed, if someone learns to produce iron with black sand, it is unlikely that he will try to use a red stone (lateritic ore). These are two too visually different materials. There are a number of technical traditions around the Indian Ocean that use black sands as iron ore (Killick and Miller, 2014; Brown, 1995; Buchanan, 1807). The non-use of black sands by the *Rasikajy*, although it could have been an abundant and easily extracted ore, supports the absence of any technology transfer between northeastern Madagascar and other metallurgical district where black sand is smelted, in particular India and South Africa.

The tuyere is a very specific feature of the *Rasikajy* technique and in particular the fact that it is drilled in a stone. In the vast majority of the other documented technical traditions around the world, the tuyeres are made of clay and are moulded, or sometimes drilled while the clay is not yet fired. It is clearly more complex to drill a stone than to mould a tuyere with clay. On the other hand, it is likely that stone tuyeres can be reused a greater number of times. Today's blacksmiths in northern Madagascar still use stone tuyeres that they collect from ancient metallurgy sites.

The *Rasikajy* made softstone artefacts already before the 11th century. These pots were polished thanks to some turning techniques. When iron production techniques were introduced, the *Rasikajy* already had the technical knowledge to drill a stone with a rotary tool to make a tuyere.

Currently only two other places or cultures in the world with stone tuyeres are known: Namibia (Sandelowsky and Pendleton, 1969) and the Vikings. A direct technological transfer from these parts of the world is very unlikely, and the development of stone tuyeres is more a coincidence. Moreover, in these two cases, the discoveries of stone tuyeres are rather sporadic whereas in Madagascar this specificity is widely spread in all the island and can be found in other later technical traditions (Radimilahy, 1988; Breton, 1898), including modern smithing techniques.

Finally, simple bowl furnace structures dug directly into the sand substratum are very sparsely documented in the archaeological and ethnographic records. This kind of metallurgical remains are indeed fragile and damaged structures difficult to find and to study. As seen previously, there are still mentions of such furnaces, especially in eastern Africa. However, these are systematically more elaborate furnaces, notably with clay linings or a structure built to facilitate the tapping of the slag.

Thus, we could not find a convincing parallel in the recorded technical traditions. This may be because the technical tradition at the origin of the *Rasikajy* tradition is not yet documented.

Nevertheless, the simple hearth dug in the sand may be reminiscent of a smithing hearth. Most smithing hearths documented for eastern Africa are simple hearths dug in the substratum with a tuyere placed on the ground and connected to bellows (Brown, 1995). The hypothesis we propose for the origin of the MDG01 *Rasikajy* technique would be an adaptation of smithing techniques into an iron smelting technique. The *Rasikajy* mastered smithing techniques before smelting, and could have adapted a smithing hearth into a bloomery furnace by adding lateritic iron ore, which is commonly used in Africa. Through their commercial contacts, and possibly some occasional migrations, they must have had an idea of how to make iron: basically putting some ore in a furnace with charcoal. In the 11th century, for some unknown reasons, the *Rasikajy* felt the need to produce their own metal and developed a basic technique from their basic metallurgical know-how.

Reflections on iron consumption

Most of the human societies from the second millennium CE, or even earlier, have based their economic stability and development on the use of iron. Iron is not only an necessary metal for weaponry, and therefore necessary for defending a territory, but it is essential for the development of efficient agriculture. Agricultural tools made of iron are much more efficient than those made of bronze or stone. Iron tools increase agricultural productivity and therefore a larger population can be fed. However, agricultural tools wear out relatively quickly, especially when they are used on dry terrain or pebbly soils. The tool gradually erodes until it becomes unusable. It is clearly observed in modern societies where tillage tools are still widely used: the iron blade wears out and decreases in size within a few years. These objects must therefore be regularly renewed. It requires a regular production of iron, and the more mouths to feed, the more important the production has to be. Iron tools can of course break, but in this case the metal can be repaired or recycled into a new object, which does not reduce the quantity of available metal.

For this reason, a demographic increase is often linked to an improvement in agricultural techniques, which can be linked to an increase in metallurgical production. In the 18th century in Madagascar, during the development of the Merina royalty in the region of Antananarivo, new rice-growing techniques allowed for a demographic increase and were correlated with the development of intensive metallurgical production (Campbell, 2005). There are of course also exceptions. Some societies have developed successful agriculture without iron technology, for example in South America.

There are also sedentary societies with low iron consumption. Their subsistence economy is based less on agriculture and land use, but rather on fruit harvesting, animal husbandry, hunting and fishing. In this situation, iron tools wear out much less quickly. Hunting weapons are indeed mainly throwing weapons, i.e. arrows or spears, whose iron blades wear out very little. The same applies to fishing tools such as hooks or harpoons. These objects wear little, they are more likely to be broken or lost. People also usually carry with them a large knife, or machete, which can have various functions. This object is probably the one that wears out the most rapidly, but it still represents a small amount of metal compared to the tillage tools described above.

The *Rasikajy* seem to be a typical population with a low iron consumption. Indeed, only few iron objects have been found during archaeological excavations, and most of them were uncovered in tombs (Gaudebout and Vernier, 1941; Vérin, 1986). The unfavourable burial conditions for the preservation of iron are partly responsible for the scarcity of iron artefacts in the archaeological records, but the *Rasikajy* were also probably not using of lot of iron tools in their daily life. Moreover, when the first Europeans visited Madagascar, they described only a small range of iron objects used by the Malagasy populations. These are mainly knives, throwing weapons, fishing tools and possibly other everyday utensils such as meat cooking hooks. However, there is no text describing specifically the objects used at this period on the northeastern coast.

The extraction of chlorite schist must have been the most iron consuming activity for the *Rasikajy* since the tools wore out relatively quickly.

The study of the metallurgical waste in our study area has shown that the *Rasikajy* carried out a small-scale iron production. We estimate that between 75 and 200 tons of iron were produced over the whole period (11th – 14th century), which represents small quantities of metal. If we assume that approximately 20'000 people were living along the northeastern coast of Madagascar at that time, this production would correspond to less than 1 kg of available metal per person per year. Despite a low iron consumption, these quantities are not sufficient enough to supply the whole region with metal, in particular because iron is needed to make chlorite schist carving tools. Hence, iron was not an exported material but the production was rather intended for local needs, and the missing necessary iron must have been imported.

Production campaigns were probably organised sporadically rather than on a regular basis. It is possible that when metal stocks started to get low, smelting operations were carried out. Only the biggest settlements, such as Benavony, which must have had a larger population, seem to have had slightly larger and therefore more regular production.

The low consumption of iron therefore allows us to infer the subsistence economy of the *Rasikajy*. We can also assume that if so little metal was produced, it is because the *Rasikajy* did not have to produce large quantities of weapons. They were therefore not or rarely at war either with a population living in the hinterland or with an oversea population. On the contrary, their metal supply seems to have been based partly on imports and therefore on peaceful relations with outside populations.

Appendix

A

Archaeological Data

A.1 Smelting sites location and description

The slag heaps were located during the 2017-2021 survey campaigns. The level of accuracy of the descriptions depends on whether excavations or test pits were carried out or whether the site was only quickly visited. Estimates of slag volumes and masses are mainly made on the basis of surface observations, by comparing with the heaps that had been excavated and weighed.

Sites	Location details	Number of heaps	Slag heap	GPS		Type of intervention
				X	Y	
Inosy	Antaimby/Anjala	5	INY110	39L402480	8467504	Visit
			INY120	39L402468	8467491	Visit
			INY130	39L402463	8467483	Visit
			INY140	39L402469	8467481	Visit
			INY150	39L402451	8467479	Visit
	Masomamangy	2	INY210	39L400901	8462497	Visit
			INY220	39L400898	8462478	Visit
	Lac Inosy	1	INY310	39L406183	8466889	Visit
Mahanara	Ambodimanga	6	MHN110	39L403566	8457063	Visit
			MHN120	39L403562	8457066	Visit
			MHN130	39L403555	8457062	Visit
			MHN140	39L403549	8457065	Visit
			MHN150	39L403546	8457058	Visit
			MHN160	39L403581	8457040	Visit
Bemanevika	Andranginalo	4	BMK510	39L406626	8444025	Survey
			BMK520	39L406609	8444019	Survey
			BMK530	39L406613	8444013	Survey
			BMK540	39L406603	8444003	Survey
	Tsaratanana	4	BMK110	39L405694	8442228	Survey
			BMK120	39L405700	8442213	Survey
			BMK130	39L405699	8442204	Test pit
			BMK140	39L405717	8442174	Survey
	Tanambao-Lex	5	BMK210	39L406144	8441236	Survey
			BMK220	39L406149	8441258	Survey
			BMK230	39L406162	8441341	Test pit
			BMK240	39L406184	8441317	Survey
			BMK250	39L406194	8441299	Survey
		2	BMK410	39L408160	8437505	Survey
			BMK420	39L408154	8437346	Test pit
Sambava	Antaimby	5	ATB1	39L407968	8423790	Survey
			ATB2	39L407928	8423836	Survey
			ATB3	39L407892	8423884	Survey
			ATB4	39L407912	8423866	Survey
			ATB5	39L407921	8423840	Survey
Matavy		19	MTY1	39L408663	8417706	Survey
			MTY2	39L408659	8417729	Survey
			MTY3	39L408681	8417769	Survey
			MTY4	39L408594	8417877	Survey
			MTY5	39L408563	8418211	Survey
			MTY6	39L408855	8418159	Survey
			MTY7	39L408945	8418249	Survey
			MTY8	39L408854	8418275	Survey
			MTY9	39L408845	8418293	Survey
			MTY10	39L408825	8418297	Survey
			MTY11	39L408881	8418300	Excavations
			MTY12	39L408867	8418322	Test pit
			MTY13	39L408899	8418311	Survey
			MTY14	39L408904	8418326	Survey
			MTY15	39L408916	8418334	Survey
			MTY16	39L408929	8418337	Survey
			MTY17	39L408912	8418361	Survey
			MTY18	39L408928	8418353	Survey
			MTY19	39L408966	8418346	Survey
Benavony		5	BNY410	39L410957	8406592	Excavations
			BNY420	39L410987	8406642	Survey
			BNY430	39L410816	8406526	Test pit
			BNY440	39L410975	8406693	Survey
			BNY450	39L410908	8406686	Survey

Figure A.1: Complete list of slag heaps in northeastern Madagascar, their location, their size and the estimated mass and volume of slag.

Size	Slag assemblage	Substratum	Dimensions			Volume (m3)	Mass of slag per unit of volume (kg/m3)	Total mass (T)	Report
			Length (m)	Width (m)	Thickness (m)				
Small	BFS?	Sand	3,0	3,0				0,9	
Small	BFS?	Sand						1	
Small	Mixed	Sand	4,0	2,0				0,9	Serneels et al., 2021
Small	Mixed	Sand	4,0	4,0				1	
Small	BFS?	Sand	4,0	3,0				1	
Unknown	Unknown	Sand						1	Serneels et al., 2021
Unknown	Unknown	Sand						1	
Unknown	Mixed?	Sand						1	
Unknown	Unknown	Sand						1	
Small	Unknown	Sand	3,0	2,0				0,7	
Small	Unknown	Sand	5,0	4,0				2	Serneels et al., 2021
Unknown	Unknown	Sand						1	
Unknown	Unknown	Sand						1	
Unknown	Unknown	Sand						1	
Small	BFS	Sand						1,5	
Small	BFS	Sand						1,5	Serneels et al., 2020
Small	BFS	Sand						1,5	
Medium	BFS	Sand						8	
Small	BFS	Sand						2	
Small	BFS	Sand						2	Serneels et al., 2019
Small	BFS	Sand	7,0	5,0	0,3	4,1		0	
Small	BFS	Sand						2	
Small	BFS	Sand						1,5	
Small	BFS	Sand						1,5	
Small	BFS	Sand	4,6	4,0	0,2	1,5		0	Serneels et al., 2019
Small	BFS	Sand						1,5	
Small	BFS	Sand						1,5	
Small	BFS	Sand						1,5	Serneels et al., 2019
Small	Mixed	Sand	10,5	4,5	0,4	7,4		0	
Small	Mixed	Sand						1,5	
Small	Unknown	Sand			0,2			1,5	Serneels et al., 2019
Small	Unknown	Sand			0,3			1,5	
Small	Unknown	Sand						1,5	
Small	Mixed	Sand						1,5	
Small	BFS	Sand						2,5	
Small	BFS	Sand						2,5	
Small	BFS	Sand						2,5	
Small	BFS	Sand						2,5	
Small	BFS	Sand						2,5	
Small	BFS	Sand						2,5	
Small	BFS	Sand						2,5	
Small	BFS	Sand						2,5	
Small	BFS	Sand						2,5	
Small	BFS	Sand						2,5	Serneels et al., 2018
Big	Mixed	Sand	13,0	13,0	0,9	60,0	650	39	
Small	BFS	Sand	6,0	6,0	0,3	4,2		2,8	
Small	BFS	Sand						2,5	
Small	BFS	Sand						2,5	
Small	BFS	Sand						2,5	
Small	BFS	Sand						2,5	
Small	BFS	Sand						2,5	
Medium	BFS	Sand	10,0	10,0	0,6	23,6		15,3	
Small	BFS	Sand						2,5	
Big	Mixed	Sand	10,0	9,0	0,6	30,0	1400	35	
Big	Mixed	Sand						20	Serneels et al., 2018
Big	Mixed	Sand	5,0	5,0	0,3	2,9		4,1	
Big	Mixed	Sand						20	
Big	Mixed	Sand						20	

Sites	Location details	Number of heaps	Slag heap	GPS		Type of intervention
				X	Y	
Amboronala		12	MBR110	39L410832	8398584	Test pit
			MBR120	39L410836	8398624	Test pit
			MBR131	39L410856	8398609	Survey
			MBR132	39L410856	8398609	Survey
			MBR141	39L410891	8398807	Excavations
			MBR142	39L410891	8398807	Excavations
			MBR143	39L410891	8398807	Excavations
			MBR210	39L410738	8399857	Test pit
			MBR220	39L410734	8399868	Test pit
			MBR240	39L410740	8399854	Test pit
			MBR411	39L411516	8398162	Survey
			MBR412	39L411516	8398162	Survey
Ambodimadiro		14	DMD810	39L409849	8397868	Survey
			DMD820	39L409874	8397865	Test pit
			DMD830	39L409852	8397868	Survey
			DMD840	39L409863	8397868	Survey
			DMD850	39L409913	8397863	Test pit
			DMD860	39L409928	8397858	Survey
			DMD910	39L409807	8398073	Survey
			DMD920	39L409814	8398069	Survey
			DMD930	39L409868	8398075	Visit
			DMD940	39L409886	8398103	Visit
			DMD950	39L409873	8398112	Visit
			DMD960	39L409866	8398114	Visit
			DMD970	39L409859	8398112	Visit
			DMD980	39L409884	8398117	Visit
Ambodipont Limite	Hill	1	APL100	39L409789	8390892	Survey
	Village	1	APL200	39L410360	8390220	Visit
	Andranovato	1	APL300	39L406243	8390699	Survey
	Antintezampako	5	APL410	39L409452	8391121	Survey
			APL420	39L409449	8391164	Survey
			APL430	39L409452	8391160	Survey
			APL440	39L409437	8391111	Survey
			APL450	39L409437	8391101	Survey
Ampanantova	Antaimby	5	PTV110	39L409782	8385848	Survey
			PTV120	39L409781	8385854	Survey
			PTV130	39L409753	8385824	Survey
			PTV140	39L409688	8385981	Survey
			PTV150	39L409722	8385981	Survey
	Antoaka	3	PTV210	39L403386	8386852	Survey
			PTV220	39L403386	8386852	Survey
			PTV230	39L403386	8386852	Survey
	Vohotsarivo-Betaimby	2	PTY310	39L398144	8387156	Visit
			PTV320	39L398164	8387146	Visit
Ambodipont Isahana		3	ISH110	39L410565	8382954	Survey
			ISH120	39L410815	8381974	Survey
			ISH130	39L410717	8382355	Survey
Ampaha	Ankarango	2	MPH110	39L408787	8371108	Visit
			MPH120	39L408780	8371110	Visit
		4	MPH210	39L408827	8370727	Visit
			MPH220	39L408829	8370715	Visit
			MPH230	39L408829	8370701	Visit
			MPH240	39L408813	8370702	Visit
		4	MPH310	39L408513	8370782	Visit
			MPH320	39L408513	837078	Visit
			MPH330	39L408512	8370788	Visit
			MPH340	39L408505	8370812	Visit
		2	MPH410	39L409339	8371209	Visit
			MPH420	39L409376	8371183	Visit
		1	MPH510	39L406798	8372725	Visit
		2	MPH610	39L411037	8372052	Visit
			MPH620	39L411030	8372080	Visit

Size	Slag assemblage	Substratum	Dimensions			Volume (m3)	Mass of slag per unit of volume (kg/m3)	Total mass (T)	Report
			Length (m)	Width (m)	Thickness (m)				
Small	TS	Sand	6,4	6	0,2	3,0		2,8	
Small	BFS	Sand	5,1	4,4	0,3	2,7	935	2,5	
Small	BFS	Sand	7,8	3,6				2,5	
Small	TS	Sand							
Small	TS	Sand	6,1	5,4	0,3	3,4	801	2,7	
Small	TS	Sand	4,2	3,9	0,2	1,3		1,2	Serneels et al., 2019
Small	BFS	Sand	2,6	1,7	0,4	0,6		0,4	
Medium	BFS	Sand	9,0	8,5	0,4	11,1	859	9,6	
Small	BFS	Sand	4,7	4,5	0,2	1,4	1159	1,6	
Small	BFS	Sand	4,8	4,2	0,4	3,2	913	2,6	
Small	BFS	Sand	6,0	3,0				2	
Small	TS	Sand	5,0	4,0				2	
Small	TS	Sand	10,0	4,0				4	
Small	TS	Sand	8,0	8,0	0,4	10,1		0	
Small	TS	Sand	8,0	5,0	0,2	3,1		0	
Small	BFS	Sand	3,0	2,0				1	
Small	TSs	Sand	2,0	2,0	0,3	0,5		0	
Small	BFS	Sand	4,0	4,0	0,2	1,3		0	
Big	Mixed	Sand	10,0	5,0				10	Serneels et al., 2019
Big	Mixed	Sand	8,0	4,0				10	
Small	BFS	Sand						2	
Small	BFS	Sand						2	
Small	TS	Sand						2	
Small	TS	Sand						2	
Small	TS	Sand						2	
Small	Unknown	Sand						2	
Small	BFS	Clay						1,5	2020
Small	Unknown	Sand						1,5	Serneels et al., 2019
Medium	Unknown	Clay						4	Serneels et al., 2020
Medium	Mixed	Sand	12,0	4,0				8	
Small	BFS	Sand						1,5	Serneels et al., 2020
Small	BFS	Sand						1,5	
Small	Mixed	Sand						1,5	
Big	Mixed	Sand						1,5	
Small	TS	Sand	6,8	3,7	0,4	4,0		0	
Small	TS	Sand	6,0	3,5	0,4	3,3		0	Serneels et al., 2019
Small	BFS	Sand	2,0	2,0	0,3	0,5		0	
Small	TS	Sand	3,0	3,0	0,3	1,1		0	
Small	TS	Sand	3,0	2,0	0,2	0,5		0	
Unknown	Unknown	Clay						1,5	
Unknown	Unknown	Clay						1,5	Serneels et al., 2020
Unknown	Unknown	Clay						1,5	
Medium	BFS	Clay	5,0	4,0				1,5	Serneels et al., 2021
Medium	BFS	Clay	5,0	4,0				1,5	
Small	BFS	Sand						1	
Small	BFS	Sand	5,0	5,0	0,3	2,9		0	Serneels et al., 2020
Small	BFS	Sand	3,0	3,0	0,2	0,7		0	
Unknown	Unknown	Clay ?						2	Serneels et al., 2021
Unknown	Unknown	Clay ?						2	
Unknown	BFS	Clay ?						2	
Unknown	BFS	Clay ?						2	Serneels et al., 2021
Unknown	BFS	Clay ?						2	
Unknown	BFS	Clay ?						2	
Small	BFS	Clay ?						2	
Small	BFS	Clay ?						2	Serneels et al., 2021
Small	BFS	Clay ?						2	
Small	BFS	Clay ?	4,0	2,0				1,5	
Unknown	BFS	Clay						2	Serneels et al., 2021
Medium	Mixed	Clay	4,0	4,0				3	
Small	Mixed	Clay ?	4,0	2,0					Serneels et al., 2021
Unknown	Unknown	Clay						2	Serneels et al., 2021
Unknown	Unknown	Clay						2	

Sites	Location details	Number of heaps	Slag heap	GPS X	GPS Y	Type of intervention
Ankavanana	Andrika	3	TLH110	39L419461	8355975	Survey
			TLH120	39L419467	8355989	Survey
			TLH130	39L419469	8355982	Survey
	Andripitra	3	TLH210	39L419450	8355677	Survey
			TLH220	39L419440	8355663	Survey
			TLH230	39L419442	8355626	Survey
	Masimdrano	3	TLH310	39L419126	8355217	Survey
			TLH320	39L418992	8355191	Survey
			TLH330	39L418744	8354799	Survey
	Ankavanana	3	TLH410	39L414347	8360393	Visit
			TLH420	39L414449	8360386	Visit
			TLH430	39L414403	8360369	Visit
Ambodikakazo		8	KKZ110	39L424727	8347401	Survey
			KKZ120			Survey
			KKZ130			Survey
			KKZ140			Survey
			KKZ150			Survey
			KKZ160	39L424531	8347255	Survey
			KKZ170	39L424518	8347310	Survey
			KKZ180	39L424516	8347310	Survey
Cap Masoala	Antsiragnamatso/ Ambohimahery	4	CPM110	39L442094	8300913	Visit
			CPM120	39L442105	8300910	Visit
			CPM130	39L442099	8300904	Visit
			CPM140	39L442092	8300911	Visit
	Marifinaritra/ Ampanavoana	2	CPM210	39L427256	8259733	Visit
			CPM220	39L427265	8259711	Visit
	Antsirabato	1	CPM310	39L408515	8237726	Visit
	TOTAL	147				

Size	Slag assemblage	Substratum	Dimensions			Volume (m3	Mass of slag per unit of volume (kg/m3)	Total mass (T)	Report
			Length (m)	Width (m)	Thickness (m)				
Unknown	Mixed	Sand	2,0	2,0	0,3			1	Serneels et al., 2020
Unknown	Mixed	Sand	4,0	4,0				2	
Unknown	Mixed	Sand						2	
Unknown	BFS	Sand	4,0	4,0				2	Serneels et al., 2020
Unknown	BFS	Sand						2	
Unknown	BFS	Sand						2	
Unknown	Mixed	Sand						2	Serneels et al., 2020
Unknown	Mixed	Sand						2	
Unknown	Mixed	Sand						2	
Unknown	Unknown	Sand						2	Serneels et al., 2021
Unknown	TS	Sand	2,0	2,0				1	
Unknown	Mixed	Sand						2	
Unknown	BFS	Sand						2	Serneels et al., 2020
Unknown	BFS	Sand						2	
Unknown	BFS	Sand						2	
Unknown	BFS	Sand						2	
Unknown	BFS	Sand						2	
Unknown	BFS	Sand						2	
Unknown	BFS	Sand						2	
Unknown	BFS	Sand						2	
Unknown	Mixed	Sand						2	Serneels et al., 2021
Unknown	Mixed	Sand						2	
Unknown	Mixed	Sand						2	
Unknown	Mixed	Sand						2	
Unknown	Unknown	Sand						2	Serneels et al., 2021
Unknown	Unknown	Sand						2	
Small	Mixed	Sand						2	
							Estimated Total Mass (T)	419	

A.2 Chronology of iron production - Radiocarbon dating

Site	Sector	Categorie	Pit	Layer	Depth	Sample	C14 age BP	2 sigma range CE		Report
								Lower	Upper	
Benavony (BNY)	503	Settlement	503			BNV01	1197	789	982	Serneels et al., 2018
	503	Settlement	503	Black disturbance		BNV02	1250	771	886	
	410	Slag heap	411	Bottom slag layer	-60 cm	BNV03	617	1319	1415	
	410	Slag heap	411	Grey layer	-65 cm	BNV04	1200	780	975	
	410	Slag heap	412		-65 cm	BNV05	701	1286	1390	
	410	Slag heap	414	Black layer	-30 cm	BNV06	697	1287	1391	
	410	Slag heap	414	Grey layer		BNV07	970	1035	1163	
	430	Slag heap	430		-28 cm	BNV08	1281	689	878	
	430	Slag heap	430		- 40 cm	BNV09	1211	773	970	
	430	Slag heap	430		-28 cm	BNV 10	594	1303	1409	
	430	Slag heap	430		-28 cm	BNV 11	1049	908	1025	
Matavy (MTY)	11	Slag heap	MTY11-S2	Furnace	-30 cm	MTV01	883	1160	1263	Serneels et al., 2018
	11	Slag heap	MTY11-S2	Just above white sand layer		MTV02	663	1301	1397	
	11	Slag heap	MTY11-S1		-10 cm	MTV03	821	1221	1279	
	11	Slag heap	MTY11-S1		-60 cm	MTV04	842	1210	1276	
	12	Slag heap	Trench		-15 cm	MTV05	784	1228	1294	
Amboronala (MBR)	110	Slag heap	110			MBRD 01	838	1164	1253	Serneels et al., 2019
	120	Slag heap	120			MBRD 02	699	1269	1382	
	140	Slag heap	MBR141	Bottom slag layer		MBRD 03	900	1042	1207	
	140	Slag heap	MBR141	Black disturbance		MBRD 04	918	1035	1165	
	140	Slag heap	Ore stock pile	Black disturbance	-3cm	MBRD 05	988	994	1151	
	140	Slag heap	MBR143	Bottom of the structure		MBRD 06	957	1022	1155	
	140	Slag heap	MBR143	Log in place in the structure		MBRD 07	995	992	1149	
	140	Slag heap	Around the anvil	Black disturbance		MBRD 08	885	1046	1218	
	210	Slag heap	210			MBRD 09	736	1251	1292	
	220	Slag heap	220			MBRD 10	689	1271	1386	
	230	Slag heap	230			MBRD 11	791	1216	1274	
	240	Slag heap	240			MBRD 12	866	1051	1225	
Ambodimadiro (DMD)	420	Slag heap	420			DMDD 01	865	1051	1225	Serneels et al., 2019
	450	Slag heap	450			DMDD 02	957	1022	1155	
Bemanevika (BMK)	130	Slag heap	130		-30 cm	BMKD 01	779	1221	1276	Serneels et al., 2019
	230	Slag heap	230	Bottom slag layer		BMKD 02	862	1054	1225	
	420	Slag heap	420		- 20 cm	BMKD 03	662	1281	1390	

Figure A.2: Table with details of charcoal sampling and dating results.

A.3 Slag dimensions

Site	Heap	Length	Width	Thickness	Mass
Amboronala	MBR120	150	150	70	1230
Amboronala	MBR120	130	130	90	1370
Amboronala	MBR120	170	120	75	1620
Amboronala	MBR120	110	110	35	640
Amboronala	MBR120	130	110	50	1200
Amboronala	MBR120	140	115	60	1120
Amboronala	MBR130	180	175	60	2000
Amboronala	MBR130	170	130	70	1140
Amboronala	MBR130	140	105	60	1280
Amboronala	MBR130	190	130	160	2730
Amboronala	MBR130	170	140	60	1560
Amboronala	MBR140	200	110	70	2150
Amboronala	MBR140	175	155	65	2350
Amboronala	MBR140	155	140	55	1400
Amboronala	MBR140	170	160	95	2260
Amboronala	MBR140	195	130	80	1830
Amboronala	MBR140	170	120	80	1540
Amboronala	MBR140	210	140	100	2870
Amboronala	MBR140	155	140	55	1400
Amboronala	MBR140	200	110	70	2150
Amboronala	MBR140	195	120	45	1020
Amboronala	MBR140	170	140	55	1270
Amboronala	MBR140	165	145	70	1580
Amboronala	MBR140	140	90	55	970
Amboronala	MBR140	160	105	60	1180
Amboronala	MBR140	100	80	25	320
Amboronala	MBR140	120	100	80	550
Amboronala	MBR140	170	160	50	2080
Amboronala	MBR210	150	150	60	1410
Amboronala	MBR210	120	120	70	1930
Amboronala	MBR210	180	140	50	2180
Amboronala	MBR210	160	140	60	2200
Amboronala	MBR210	150	140	70	2150
Amboronala	MBR210	130	120	70	1690
Amboronala	MBR210	150	120	40	1200
Amboronala	MBR210	140	110	70	1340
Amboronala	MBR210	150	130	50	1500
Amboronala	MBR210	170	120	60	1940
Amboronala	MBR210	150	130	60	1560
Amboronala	MBR210	150	120	60	1330
Amboronala	MBR210	150	100	60	1390
Amboronala	MBR210	110	100	60	1490
Amboronala	MBR210	110	100	55	1310
Amboronala	MBR210	150	140	60	1600
Amboronala	MBR210	160	130	50	1560
Amboronala	MBR210	150	140	60	1600
Amboronala	MBR210	160	150	60	2170
Amboronala	MBR210	150	130	60	1360
Amboronala	MBR210	150	120	70	2000

Figure A.3: Documentation of sizes and masses of complete bottom furnace slag pieces for Amboronala, Ambodimadiro, Bemanevika, Benavony and Matavy.

Site	Heap	Length	Width	Thickness	Mass
Amboronala	MBR220	160	150	60	3400
Amboronala	MBR220	230	150	70	2860
Amboronala	MBR220	180	130	40	1860
Amboronala	MBR220	170	140	60	2500
Amboronala	MBR220	180	150	50	2100
Amboronala	MBR220	190	160	70	3180
Amboronala	MBR220	180	140	60	2560
Amboronala	MBR220	150	140	60	2180
Amboronala	MBR220	170	130	70	2390
Amboronala	MBR220	160	130	70	2100
Amboronala	MBR220	190	160	80	360
Amboronala	MBR220	150	140	60	2060
Amboronala	MBR220	180	130	70	2340
Amboronala	MBR220	170	140	60	1780
Amboronala	MBR220	150	140	70	1970
Amboronala	MBR220	160	140	60	1680
Amboronala	MBR220	150	130	70	2130
Amboronala	MBR220	165	130	65	1740
Amboronala	MBR220	150	110	70	1620
Amboronala	MBR220	160	150	60	1640
Amboronala	MBR240	190	170	65	2480
Amboronala	MBR240	160	120	50	1520
Amboronala	MBR240	170	120	60	2270
Amboronala	MBR240	170	150	80	2670
Amboronala	MBR240	160	130	50	1780
Amboronala	MBR240	180	120	70	1570
Amboronala	MBR240	120	120	50	1970
Amboronala	MBR240	140	130	50	1890
Amboronala	MBR240	130	120	70	1730
Amboronala	MBR240	170	110	50	1410
Amboronala	MBR240	120	11	90	1460
Amboronala	MBR240	150	100	40	1520
Amboronala	MBR240	170	140	60	1860
Amboronala	MBR240	150	120	60	1900
Amboronala	MBR240	140	110	70	1550
Amboronala	MBR240	130	130	70	1720
Amboronala	MBR240	150	100	70	1320
Amboronala	MBR240	140	90	80	2050
Amboronala	MBR240	130	150	90	2180
Amboronala	MBR240	110	100	60	1420
Ambodimadiro	DMD820	180	210	120	3200
Ambodimadiro	DMD820	150	95	70	1030
Ambodimadiro	DMD820	120	100	55	590
Ambodimadiro	DMD820	140	100	60	710
Bemanevika	BMK130	155	120	50	1260
Bemanevika	BMK130	160	115	40	1300
Bemanevika	BMK130	175	160	90	1130
Bemanevika	BMK130	170	175	60	1450
Bemanevika	BMK130	185	175	75	1660
Bemanevika	BMK230	150	170	60	1350
Bemanevika	BMK230	160	130	70	1370
Bemanevika	BMK230	145	140	60	1060
Bemanevika	BMK230	140	130	45	1390
Bemanevika	BMK230	150	130	70	1130
Bemanevika	BMK420	185	190	60	2480
Bemanevika	BMK420	185	160	100	2390
Bemanevika	BMK420	210	185	60	3050
Bemanevika	BMK420	230	155	50	2110

Site	Heap	Length	Width	Thickness	Mass
Benavony	BNY410	300	100	100	5000
Benavony	BNY410	240	240	60	5000
Benavony	BNY410	170	150	65	1500
Benavony	BNY410	180	150	110	3500
Benavony	BNY410	200	170	60	3000
Benavony	BNY410	220	200	80	4500
Benavony	BNY410	250	200	120	4000
Benavony	BNY410	240	175	115	3800
Benavony	BNY410	185	160	90	2900
Benavony	BNY410	230	200	95	4400
Benavony	BNY410	170	160	60	1800
Benavony	BNY410	185	170	85	1500
Benavony	BNY410	230	200	70	3900
Benavony	BNY410	200	190	90	3750
Benavony	BNY410	215	185	110	3600
Benavony	BNY410	195	160	65	2600
Matavy	MTY11	190	180	90	2500
Matavy	MTY11	200	180	70	4500
Matavy	MTY11	185	180	90	2500
Matavy	MTY11	200	175	85	4200
Matavy	MTY11	215	200	75	4300
Matavy	MTY11	220	190	90	3100
Matavy	MTY11	180	165	90	2900
Matavy	MTY11	200	190	85	4400
Matavy	MTY11	190	190	80	2700
Matavy	MTY11	220	210	70	4000
Matavy	MTY11	200	175	80	2800
Matavy	MTY11	200	180	75	3200
Matavy	MTY11	190	185	85	2600

B

Mineralogical and Chemical Data

B.1 Loss of Ignition (LOI)

B.1.1 Example of a calculation: MBR11001 (Amboronala)

This sample has an iron oxide content measured by XRF of 63.9 wt.% Fe_2O_{3tot}. The sum of all other measured components is 34.2 wt.%. Ideally in a smelting slag all iron is under the form of FeO, and not Fe_2O_3. The data is thus converted to wt.%FeO and normalized to 100%:

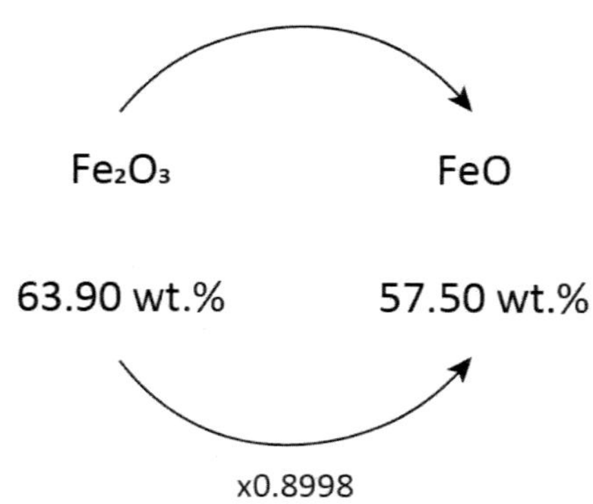

FeO	57.50 wt.%	57.50 wt.%	→ $FeO_{TOT\ norm}$
Other	34.20 wt.%	37.30 wt.%	
Σ	91.70 wt.%	100 wt.%	

A moisture-related mass loss of 4.8% was measured during the Thermogravimetric analyses. In goethite ($FeO.OH$, or $Fe_2O_3.H_2O$), which is the main iron oxide in slag, H_2O represents 10% of the mineral's mass. Fe_2O_3 therefore represents 90% of the goethite mass. 4.32% of Fe_2O_3 (3.89% of FeO) in the sample are therefore located in the alteration products and under the Fe^{3+} oxidation state.

$$FeO_{remaining} = 62.7 - 3.89 = 58.81\% \quad (B.1)$$

According to our calculations, 58.81 wt.% FeO are not in the alteration products.

During the heating protocol, the mass gain is due to the incorporation of oxygen because of the oxidation of iron. If all the iron in the sample was actually FeO, then there would be a mass gain of 5.88% to transform all the FeO into Fe_2O_3. In our case, LOI measurements show that the sample gains only 4.92% mass during heating. There is 0.96% in excess. Some of the iron was therefore already in the Fe^{3+} form.

A mass gain of 4.92% corresponds to 44.28 wt.% FeO as:

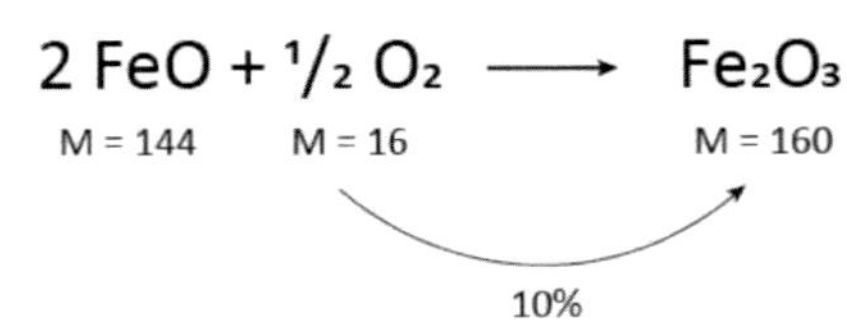

There is therefore 44.28 wt.% of actual FeO in the slag. The remaining iron was unreduced during the smelting process and is in the Fe^{3+} oxidation state.

	Alteration	Initial Fe^{2+}	Initial Fe^{3+}
FeO	3.89%	44.28%	14.53%
Fe_2O_3	4.32%		16.15%

B.1.2 Results of LOI calculations

	FeOtot (XRF) (wt.%)	FeO Alteration (wt.%)	FeO (Fe2+) (wt.%)	FeO (Fe3+) (wt.%)
MBR11001	62,7	3,9	44,3	14,5
MBR11002	62,5	6,5	39,2	16,9
MBR11003	58,8	2,7	48,0	8,2
MBR11005	59,1	9,0	38,6	11,5
MBR12001	74,0	5,7	46,6	21,7
MBR12003	64,1	9,1	32,0	23,0
MBR12004	56,7	1,8	40,9	14,0
MBR12005	58,1	1,5	39,3	17,2
MBR12007	72,5	2,2	50,2	20,1
MBR12009	68,7	5,5	41,6	21,6
MBR13001	69,5	4,5	41,3	23,7
MBR13002	63,7	6,3	27,2	30,2
MBR13006	57,4	3,6	35,7	18,1
MBR13007	60,2	3,2	41,3	15,7
MBR13009	58,2	3,8	37,9	16,5
MBR13010	64,6	5,7	36,9	22,0
MBR14002	56,9	2,9	44,0	10,0
MBR14003	53,7	5,7	37,2	10,8
MBR14004	62,6	9,6	32,3	20,6
MBR14005	56,8	1,6	49,1	6,0
MBR14006	66,0	3,1	47,3	15,6
MBR14007	65,1	2,9	52,1	10,1
MBR14009	57,1	2,7	41,0	13,4
MBR14010	66,5	4,0	52,1	10,3
MBR14011	70,0	7,4	41,9	20,8
MBR14012	60,1	11,3	28,6	20,2
MBR21001	69,0	8,7	51,8	8,6
MBR21002	63,9	5,7	42,3	15,9
MBR21005	62,9	4,7	42,6	15,7
MBR21008	63,3	3,4	48,2	11,8
MBR21010	59,1	7,1	37,9	14,1
MBR22001	60,6	9,6	30,2	20,8
MBR22002	61,1	9,6	30,8	20,7
MBR22005	65,8	3,6	35,7	26,5
MBR22006	61,8	12,4	26,3	23,1
MBR22008	61,1	8,4	37,9	14,7
MBR23001	72,1	3,2	57,6	11,3
MBR23002	61,6	2,9	43,7	15,0
MBR23003	62,9	12,6	27,0	23,3
MBR24001	66,1	7,7	25,7	32,7
MBR24003	63,3	2,3	47,3	13,6
MBR24004	59,3	9,0	30,6	19,7
MBR24005	69,2	17,2	27,0	25,0
MBR24006	67,5	4,9	43,4	19,2
MBR24007	67,5	15,5	23,9	28,1
MBR41001	63,9	11,2	32,1	20,6
MBR41002	62,8	11,3	34,7	16,9
MBR41003	67,1	17,4	36,7	13,0
MBR41005	58,2	8,8	38,3	11,0
MBR41006	64,7	14,3	33,1	17,3

Figure B.1: Calculation of the different oxidation states of iron from thermogravimetric measurements. The results are given in wt.%FeO for each sample.

	FeOtot (XRF) (wt.%)	FeO Alteration (wt.%)	FeO (Fe2+) (wt.%)	FeO (Fe3+) (wt.%)
BMK13001	58,7	13,5	32,4	12,8
BMK13002	59,3	10,1	35,6	13,6
BMK 13003	63,2	14,4	27,0	21,8
BMK13004	58,8	10,5	36,5	11,8
BMK 23001	54,3	12,2	31,8	10,3
BMK 23002	63,8	16,0	30,2	17,6
BMK 23004	58,4	7,8	35,0	15,6
BMK 24002	68,8	11,2	31,8	25,8
BMK 24004	59,3	5,7	39,5	14,1
BMK 24005	57,2	11,9	31,0	14,3
BMK 24006	56,1	19,8	35,0	1,3
BMK 24007	59,6	9,3	36,9	13,4
BMK 42001	61,3	13,3	30,2	17,8
BMK42002	59,5	11,7	32,0	15,7
BMK42003	61,2	22,7	25,7	12,9
BMK 42004	56,2	13,8	27,4	15,0
DMD 82001	69,5	2,3	45,6	21,6
DMD 82002	66,8	2,7	48,2	15,8
DMD 85001	33,0	14,3	14,4	4,2
DMD 85002	45,0	21,5	9,3	14,2
DMD 86001	66,8	6,3	44,3	16,2
DMD 92001	70,1	8,2	33,1	28,8
DMD 92002	58,4	6,2	42,8	9,3
DMD 92003	72,8	1,4	42,9	28,5
ATB001	60,9	15,1	26,4	19,4
ATB002	64,1	10,1	34,9	19,1
BNY50006	67,0	1,4	40,5	25,1
BNY50001	59,9	8,6	35,3	16,0
BNY 410001	59,3	1,5	52,1	5,8
BNY41002	59,6	9,5	33,1	17,0
BNY41005	64,3	3,1	46,2	15,0
BNY41006	60,0	11,7	33,3	15,1
BNY41008	60,5	10,5	30,6	19,3
BNY41029	63,7	6,6	42,0	15,1
BNY41030	61,8	8,1	36,5	17,3
BNY41033	65,2	4,6	46,2	14,4
BNY42001	59,6	7,4	41,8	10,5
BNY42003	66,9	7,3	38,2	21,4
BNY42004	60,5	28,7	22,1	9,7
BNY42006	55,6	7,7	34,7	13,2
BNY43007	61,7	5,3	39,6	16,8
BNY43008	59,7	22,1	30,0	7,6
BNY43009	58,8	6,0	39,2	13,7
BNY44001	60,5	9,6	33,6	17,4
BNY44002	58,8	1,6	48,9	8,3
BNY44004	61,6	3,9	44,2	13,5
BNY44005	64,6	7,0	39,6	18,1
BNY45001	62,5	1,5	47,3	13,6
BNY45002	59,5	7,5	37,2	14,8
BNY45004	55,5	2,5	46,2	6,8
BNY45005	61,2	6,8	41,0	13,4

	FeOtot (XRF) (wt.%)	FeO Alteration (wt.%)	FeO (Fe2+) (wt.%)	FeO (Fe3+) (wt.%)
MTY11001	68,8	1,1	50,9	16,8
MTY11002	63,0	1,9	44,2	16,8
MTY11004	54,6	1,5	46,1	7,0
MTY 11005	63,7	2,3	51,5	10,0
MTY11006	70,2	1,2	55,4	13,7
MTY11011	64,7	3,1	45,5	16,0
MTY11012	64,0	1,1	50,9	12,0
MTY12005	71,3	3,0	53,4	14,9
MTY12006	69,3	5,2	46,4	17,8
MTY12007	81,8	0,0	66,2	15,6
MTY18002	66,2	3,3	49,9	13,0
MTY18004	67,6	4,3	49,2	14,1
MTY18006	72,7	1,7	52,1	18,9
APL 10001	61,4	7,7	38,9	14,8
APL 10002	79,8	3,9	70,3	5,7
APL 10003	66,3	6,2	43,3	16,8
APL 30001	64,4	2,3	51,3	10,8
APL 30002	69,2	8,9	36,7	23,6
APL 30003	64,1	3,1	46,7	14,3
APL 30004	63,4	2,3	50,3	10,9
ISH 10001	63,4	7,3	41,0	15,1
ISH 10002	62,4	6,8	44,3	11,4
ISH 10003	67,0	3,2	56,8	7,0
ISH 20001	69,5	3,9	43,7	21,9
ISH 20002	60,4	3,9	43,7	12,8
ISH 20003	67,4	12,6	31,7	23,1

B.2 Processed XRD Data

The following tables are a compilation of the processing of the XRD spectra. The raw data are available on request.

B.2.1 Black sands and sandy substratum

Sample	Quartz	Ilmenite	Magnetite	Hematite	Amphibole	Garnet	Zircon	Rutile	Feldspath Na	Feldspath K	Muscovite
SBV610	x	x	x		x	x	x	x	x		
SBV611	x	x	x	x	x	x	x	x	x		
SBV620	x	x	x	x	x	x			x		
SBV621											
SBV630	x	x	x	x	x	x	x		x		
SBV631											
SBV640	x	x	x	x	x	x	x	x	x		
SBV641a	x	x	x	x	x	x	x	x	x		
SBV641b	x	x	x	x	x	x	x		x	x	
SBV642	x	x	x	x	x	x	x	x	x		
SBV650	x	x	x	x	x	x	x		x	x	
SBV651	x	x	x	x	x	x	x		x	x	
SBV652	x	x	x	x	x	x	x		x	x	
SBV660	x	x	x	x	x	x	x		x	x	
SBV661	x	x	x	x	x	x	x		x	x	
SBV662											
MBR Substratum	x								x	x	x

Figure B.2: XRD data after processing of black sand and sandy substratum samples for northeast Madagascar.

B.2.2 Iron ores

Sample	Comments	Quartz	Magnétite	Hématite	Goethite	Anatase	Kaolinite	Gibbsite
		SiO_2	Fe_3O_4	Fe_2O_3	FeO(OH)	TiO_2	$2SiO_2$, Al_2O_3, $2H_2O$	$Al(OH)_3$
MTY501	Lateritic block	x			x			x
MTY503	Lateritic block		x	x	x			x
MTY504	Lateritic block			x	x			x
MTY512	Lateritic block	x			x			x
MTY513	Lateritic block	x		x	x			
MTY514	Pisolites		x	x				
MTY515	Pisolites			x	x			
MTY516	Pisolites			x	x			
BMK501	Non washed pisolites			x	x	x		x
BMK502	Non washed pisolites	x		x	x			x
BMK503	Lateritic block	x		x	x	x	x	x
BMK510	Washed pisolites			x	?			
BMK511	Lateritic block			x	x		x	x
BMK512	Lateritic block			x	x			x
MBR14501	Pisolites		x	x	x		?	
MBR14502	Pisolites	x	x	x	x		?	
MBR14503	Pisolites		x	x	x		?	
MBR14504	Pisolites		x	x	x		?	
MBR14510	Pisolites	x	x	x	x		x	
MBR14511	Pisolites	x		x	x			
DMD501	Pisolites	x		x	x		x	
DMD502	Pisolites	x		x	x		x	
ATH501	Pisolites			x	x			
ATH502	Pisolites	?			x			
APL501	Pisolites				x			x
APL502	Pisolites				x			x
APL503	Lateritic block				x			x
APL504	Pisolites			x	x			x
APL505	Pisolites	x		x	x			x
APL506	Pisolites			x	x			x

Figure B.3: XRD data after processing of iron ore samples.

B.2.3 Smelting slag

Sample	Fayalite	Quartz	Wüstite	Goethite	Magnétite	Lepidocrocite	Spinel Alumineux	Pyroxenoid	Leucite	Anorthite	Rutile	Iron
Formule	Fe_2SiO_4	SiO_2	FeO	FeO(OH)	Fe_3O_4	FeO(OH)	$X^{2+}Al_2O_4$	$X^{2+}SiO_3$	$KAlSi_2O_6$	$CaAl_2Si_2O_8$	TiO_2	Fe
MBR11001	x	x	(x)		(x)		x					
MBR11002	x		(x)	x	(x)		x					
MBR11003	x	x	x		x		x					
MBR11005	x	x		x	x		x					
MBR12001	x	x	x	x	x		x		x			x
MBR12003	x	x	x	x	x		x		x			
MBR12004	x	x	x		x				x			
MBR12005	x	x	x		x		x					
MBR12007	x	x	x		x		x		x			
MBR12009	x		x		x		x		x			
MBR13001	x	(x)	x	(x)	x		x		x			
MBR13002	x	x	x	(x)	x		x		x			
MBR13006	x	x		(x)	x		x		x			
MBR13007	x	x	?	(x)	x		x		x			
MBR13009	x	x		(x)	(x)		x					
MBR13010	x	x	x				x					
MBR14001	x	xx			x		x					
MBR14002	x	x			x		x			x		
MBR14003	x	x			(x)		x			x		
MBR14004	x	x	x		x		x	x				
MBR14005	x	x			(x)		x					
MBR14006	x		x				x					
MBR14007	x	x	x				x					
MBR14009	x	x			(x)		x		?			
MBR14010	x	x	x		x		x					x
MBR14011	x	x	x		(x)				x			
MBR14012	x	x		x	x	x	x				x	
MBR21001	x	x	x		x		x					x
MBR21002	x	x			x		x				x	
MBR21005	x	x			x		x					
MBR21008	x	x		x	x	x	x					
MBR21010	x	x		x	x		x				x	
MBR22001	x	x	x	x	x		x					
MBR22002	x	x		x	x		x					
MBR22005	x	x	x		x		x				x	
MBR22006	x	x	x	x	x		x					
MBR22008	x	x		x	x	x	x					
MBR23001	x	x	x		x		x					x
MBR23002	x	x	x				x					
MBR23003	x	x	x	x	x		x					
MBR24001	x	x	x	x	x	x	x					
MBR24003	x	x	x		x		x					
MBR24004	x	x	(x)	x	x		x					
MBR24005	x	x	x	x	x	x	x					
MBR24006	x	x	x		x		x					
MBR24007	x	x	x	x	x	x	x					
MBR41001	x	x	x	x	x		x		x			
MBR41002	x	x		x	x		x		x			
MBR41003	x	x	x	x	x		x					
MBR41005	x	x			x		x					
MBR41006	x	(x)	x	x	x	x	x					
BMK13001	x	x			x	x	x					
BMK13002	x	x		x	x	x	x				x	
BMK 13003	x	x		x	x	x	x				?	
BMK13004	x	x		x	x	x	x				?	
BMK 23001	x	x		x	x	x	x		?			
BMK 23002	x	x		x	x		x					
BMK 23004	x	x		x	x		x					
BMK 24002	x	x	x	x	x		x	?				
BMK 24004	x	x			x		x					
BMK 24005	x	x		x	x	x	x	?	x			
BMK 24006	x	x		x	x	x	x					
BMK 24007	x	x			x		x					
BMK 42001	x		x	x	x	xx	x		?			
BMK42002	x			x	x	x	x					
BMK42003	x			x	x	xx	x		x			
BMK 42004	x	x		x	x	x	x		?			
DMD 82001	x	x	x				x					
DMD 82002	x	x	x				x					
DMD 85001	x	xxx		x	x		x					
DMD 85002	x	x		x	x	x	x					
DMD 86001	x		x	x	x		x					
DMD 92001	x	x		x	x		x					
DMD 92002	x	x	x		x		x					
DMD 92003	x		x				x					

Figure B.4: XRD data after processing of smelting slag samples.

Sample	Fayalite	Quartz	Wüstite	Goethite	Magnétite	Lepidocrocite	Alumineous Spinel	Pyroxenoid	Leucite	Anorthite	Rutile	Iron
Formule	Fe_2SiO_4	SiO_2	FeO	FeO(OH)	Fe_3O_4	FeO(OH)	$X^{2+}Al_2O_4$	$X^{2+}SiO_3$		$CaAl_2Si_2O_8$	TiO_2	Fe
ATB001	x	x	x	x	x	x	x	x				
ATB002	x	x	x	?	x		x	x				
BNY50006	x	x		x	x		x					
BNY50001	x	x			x		x					
BNY41002	x	x		?	x	x	x					
BNY41005	x	(x)	x		x		x					
BNY41006	x	(x)		x	x	x	x					
BNY41008	x	(x)		x	x	x	x					
BNY41029	x	(x)	x	?	x		x					
BNY41030	x			?	x	?	x					
BNY41033	x		x		x	x	x					
BNY42001	x	(x)			x		x					
BNY42003	x		x	x	x	x	x					
BNY42004	x			x	x	x	x					
BNY42006	x	x			x		x					
BNY43007	x	x	x		x		x					
BNY43008	x			x	x	x	x					
BNY43009	x	x		x	x		x					
BNY44002	x	(x)		x	x		x					
BNY44004	x	x	x	(x)	x		x					
BNY44005	x	x	x	(x)	x		x					
BNY45001	x		(x)		x		x					
BNY45002	x	x		x	x		x					
BNY45004	x	x		x	x		x					
BNY45005	x	x		x	x	x	x		x			
MTY11001	x	x	x				x					
MTY11002	x	x			?							
MTY11004	x	x			x		x					
MTY11006	x		x				x					
MTY11011	x	x	x		x		x					
MTY11012	x	x	x		x		x					
MTY12005	x	x	x		x		x					
MTY12006	x	x	x		x		x					
MTY12007	x		xx		(x)		x					
MTY18002	x	x	x		x				x			
MTY18004	x	x	x		x		x					
MTY18006	x	x	x		x		x					

xx Mineral detected in large quantities

(x) Mineral detected in small quantities

x Mineral detected

? Uncertain identification of the mineral

B.3 XRF Data

Sample		SBV610	SBV611	SBV620	SBV621	SBV630	SBV631	SBV640	SBV641a	SBV641b
		Before panning	After panning	Before panning	After panning	Before panning	After panning	Before panning	After panning	After panning
SiO2	(wt.%)	46,02	17,88	52,73	21,04	37,91	27,61	15,09	9,24	6,56
TiO2	(wt.%)	11,60	17,90	9,40	18,80	13,80	17,00	23,40	25,70	27,40
Al2O3	(wt.%)	4,97	3,67	5,33	4,84	4,80	4,34	3,19	2,49	2,31
Fe2O3	(wt.%)	29,10	42,90	24,20	45,50	34,30	42,00	49,40	54,10	56,50
MgO	(wt.%)	1,77	1,78	1,85	2,51	1,98	2,03	1,48	1,37	1,35
MnO	(wt.%)	0,38	0,55	0,33	0,62	0,46	0,55	0,62	0,67	0,71
CaO	(wt.%)	1,70	1,40	1,90	2,00	1,90	1,80	0,90	0,80	0,70
Na2O	(wt.%)	0,93	0,54	0,91	0,58	1,06	0,69	0,63	0,48	0,43
K2O	(wt.%)	0,83	0,23	0,97	0,26	0,67	0,42	0,35	0,14	0,06
P2O5	(wt.%)	0,28	0,40	0,23	0,39	0,31	0,36	0,50	0,54	0,54
SUM	(wt.%)	98,64	89,00	98,80	98,16	98,55	98,38	97,36	97,46	98,61
Ba	ppm	346	335	351	310	322	324	450	450	386
Ce	ppm	1564	2895	1217	2369	1739	2112	3363	3719	4095
Co	ppm	133	117	157	118	126	99	79	66	71
Cr	ppm	236	401	212	417	297	361	423	457	466
Cu	ppm	6	10	5	11	8	10	13	14	15
Hf	ppm	125	218	110	215	195	236	205	224	231
La	ppm	788	1487	633	1206	884	1067	1673	1826	1992
Mn	ppm	2915	5255	2615	5378	3772	4617	5416	5807	6080
Nb	ppm	189	327	160	316	231	290	393	427	439
Nd	ppm	600	1106	470	909	676	816	1269	1404	1547
Ni	ppm	27	36	25	38	29	33	34	34	36
Rb	ppm	19	6	21	6	15	10	9	4	5
Sc	ppm	23	38	22	42	27	34	35	36	37
Se	ppm	1	0	0	1	1	1	1	1	1
Sm	ppm	85	163	69	137	107	124	178	195	220
Sn	ppm	11	20	8	17	13	17	19	18	21
Sr	ppm	75	31	79	35	69	51	27	13	12
Th	ppm	478	817	363	686	508	610	955	1051	1153
V	ppm	439	790	364	734	524	650	964	1061	1086
W	ppm	863	621	956	627	707	528	341	242	308
Y	ppm	162	261	143	269	217	257	212	237	249
Zn	ppm	174	305	148	304	215	269	356	404	404
Zr	ppm	4461	7865	4099	7789	7030	8249	7549	7963	8326
SiO2:Al2O3		9,26	4,87	9,89	4,35	7,90	6,36	4,73	3,71	2,84
SiO2:TiO2		3,97	1,00	5,61	1,12	2,75	1,62	0,64	0,36	0,24
V:Cr		1,9	2,0	1,7	1,8	1,8	1,8	2,3	2,3	2,3

Figure B.5: Bulk chemical compositions of black sands samples from the shores of Sambava, Benavony and Bemanevika. Some samples were analysed before being washed (before panning) and some other after being separated from light minerals (after panning).

Sample		SBV642	BMK610	BMK611	SBV650	SBV651	SBV652	SBV660	SBV661	SBV662
		After panning	Before panning	After panning	Before panning	After panning	Sables noirs	Before panning	After panning	After panning
SiO2	**(wt.%)**	5,77	52,35	30,40	38,27	26,43	15,60	77,04	43,73	42,93
TiO2	**(wt.%)**	27,50	5,90	9,50	14,80	18,30	22,20	2,70	10,90	11,30
Al2O3	**(wt.%)**	2,13	4,76	4,63	4,83	4,58	3,72	6,21	7,99	7,87
Fe2O3	**(wt.%)**	57,30	30,10	48,60	34,90	42,50	50,30	7,50	26,50	27,20
MgO	**(wt.%)**	1,23	1,47	2,28	1,73	1,93	1,69	1,50	4,18	4,25
MnO	**(wt.%)**	0,71	0,35	0,59	0,46	0,56	0,64	0,13	0,47	0,48
CaO	**(wt.%)**	0,60	1,70	2,40	1,30	1,40	1,10	1,70	3,70	3,80
Na2O	**(wt.%)**	0,43	0,86	0,53	0,78	0,52	0,70	1,05	0,67	0,66
K2O	**(wt.%)**	0,04	1,11	0,52	0,68	0,40	0,21	1,64	0,65	0,61
P2O5	**(wt.%)**	0,57	0,07	0,09	0,34	0,41	0,49	0,07	0,16	0,17
SUM	**(wt.%)**	98,27	99,06	99,94	99,23	98,44	98,26	99,86	99,46	99,82
Ba	**ppm**	450	367	224	350	330	366	403	261	263
Ce	**ppm**	3669	211	254	2133	2704	3291	114	263	323
Co	**ppm**	69	167	115	111	129	99	199	198	229
Cr	**ppm**	480	105	158	285	370	436	82	299	314
Cu	**ppm**	14	3	8	7	7	11	-2	2	3
Hf	**ppm**	231	24	29	113	138	164	12	43	43
La	**ppm**	1814	113	139	1092	1393	1641	63	148	177
Mn	**ppm**	6295	2736	4843	3690	4810	5507	1019	4046	4132
Nb	**ppm**	460	60	93	235	296	357	47	185	192
Nd	**ppm**	1379	87	104	810	1028	1245	45	109	134
Ni	**ppm**	34	22	31	27	33	33	15	39	39
Rb	**ppm**	2	24	9	15	9	5	36	12	11
Sc	**ppm**	39	15	28	28	36	38	11	43	45
Se	**ppm**	1	-2	-3	1	1	1	0	-1	-1
Sm	**ppm**	198	9	15	116	146	175	9	19	21
Sn	**ppm**	20	7	9	13	15	17	3	9	13
Sr	**ppm**	5	74	42	45	33	19	94	70	68
Th	**ppm**	1035	63	76	645	789	958	28	71	87
V	**ppm**	1125	457	760	584	749	897	112	424	442
W	**ppm**	288	1072	626	664	747	526	1347	1248	1372
Y	**ppm**	235	51	73	170	214	238	37	125	105
Zn	**ppm**	422	120	210	211	282	330	61	236	246
Zr	**ppm**	8427	984	1166	4046	5076	5837	489	1588	1682
SiO2:Al2O3		2,71	10,99	6,57	7,92	5,77	4,20	12,41	5,47	5,45
SiO2:TiO2		0,21	8,87	3,20	2,59	1,44	0,70	28,53	4,01	3,80
V:Cr		2,3	4,4	4,8	2,0	2,0	2,1	1,4	1,4	1,4

Sample		SiO_2 (wt.%)	TiO_2 (wt.%)	Al2O3 (wt.%)	Fe_2O_3 (wt.%)	MgO (wt.%)	MnO (wt.%)	CaO (wt.%)	Na_2O (wt.%)	K_2O (wt.%)	P_2O_5 (wt.%)	SUM (wt.%)
MBR 14501	Pisolites	6,68	1,30	7,50	83,50	0,04	0,03	0,00	0,04	0,02	0,21	99,52
MBR 14502	Pisolites	8,42	1,00	5,63	82,80	0,04	0,03	0,00	0,05	0,02	0,30	98,48
MBR 14503	Pisolites	5,55	1,20	6,50	84,80	0,06	0,05	0,00	0,05	0,01	0,23	98,69
MBR 14504	Pisolites	8,02	0,90	7,15	82,40	0,08	0,11	0,10	0,05	0,04	0,32	99,38
MBR 14510a	Pisolites	14,75	1,10	9,75	77,30	0,07	0,03	0,00	0,05	0,04	0,25	103,49
MBR 14510b	Pisolites	8,49	1,10	7,63	79,80	0,07	0,03	0,00	0,03	0,03	0,25	97,63
MBR 14511a	Pisolites	7,88	1,30	8,26	79,50	0,10	0,03	0,10	0,09	0,08	0,18	97,71
MBR 14511b	Pisolites	8,10	1,30	8,30	79,90	0,09	0,03	0,10	0,08	0,08	0,18	98,35
MBR 14512a	Pisolites	8,30	1,20	8,28	79,50	0,09	0,03	0,10	0,08	0,04	0,25	98,07
MBR 14512b	Pisolites	8,15	1,20	8,28	79,70	0,08	0,03	0,20	0,10	0,05	0,25	98,25
DMD 501	Pisolites	16,28	1,60	12,43	69,50	0,05	0,09	0,00	0,04	0,01	0,10	100,47
DMD 502	Pisolites	15,62	1,60	12,95	68,90	0,05	0,09	0,00	0,05	0,01	0,08	99,69
DMD 503	Pisolites	35,20	1,30	11,30	47,90	0,22	2,35	0,00	0,06	0,08	0,07	101,08
BMK 501	Unwashed pisolites	7,60	3,70	43,44	46,70	0,02	0,48	0,10	0,05	0,01	0,26	102,67
BMK 502	Unwashed pisolites	5,10	3,20	41,60	49,00	0,02	0,41	0,10	0,04	0,01	0,27	100,11
BMK 503	Block	18,76	3,50	19,88	57,90	0,05	0,09	0,10	0,06	0,02	0,29	100,97
BMK510	Pisolites	6,14	2,10	10,48	77,90	0,15	0,15	0,20	0,10	0,13	0,46	98,14
BMK511	Block	17,31	3,80	21,04	55,80	0,09	0,09	0,10	0,03	0,02	0,30	98,86
BMK512	Block	3,85	2,30	26,37	64,40	0,05	0,28	0,10	0,02	0,01	0,36	98,21
MTY501	Block	3,98	1,20	12,97	78,20	0,05	0,10	0,00	0,02	0,01	0,47	97,21
MTY503	Block	0,33	5,10	60,02	30,40	0,00	1,07	0,00	0,02	0,02	0,17	97,39
MTY504	Block	0,28	5,10	64,48	26,90	0,00	0,25	0,00	0,01	0,01	0,16	97,37
MTY508a	Pisolites	1,56	1,20	8,66	85,70	0,06	0,09	0,00	0,34	0,01	0,33	98,34
MTY508b	Pisolites	1,55	1,20	8,68	85,80	0,07	0,09	0,10	0,18	0,01	0,34	98,40
MTY509a	Pisolites	1,71	1,40	8,35	86,00	0,05	0,10	0,00	0,05	0,01	0,33	98,39
MTY509b	Pisolites	1,84	1,30	8,34	85,50	0,06	0,09	0,00	0,07	0,01	0,33	97,94
MTY510a	Pisolites	1,59	1,20	8,77	85,30	0,05	0,09	0,00	0,10	0,01	0,33	97,85
MTY510b	Pisolites	1,54	1,20	8,73	85,10	0,06	0,09	0,00	0,11	0,01	0,33	97,56
MTY512	Block	6,17	1,20	10,58	79,90	0,06	0,08	0,00	0,03	0,02	0,28	98,65
MTY513	Block	6,53	1,50	8,98	80,00	0,14	0,10	0,10	0,02	0,02	0,36	98,14
MTY514	Pisolites	1,64	1,20	8,19	84,90	0,04	0,10	0,00	0,03	0,01	0,31	96,79
MTY515	Pisolites	2,12	1,20	7,46	86,00	0,04	0,07	0,00	0,01	0,01	0,39	97,62
APL 501	Pisolites	5,24	1,10	12,07	77,40	0,07	0,13	0,10	0,00	0,02	0,36	96,78
APL 502	Pisolites	4,50	1,20	8,43	80,40	0,06	0,05	0,10	0,00	0,02	0,37	95,41
APL 503	Block	6,84	1,10	9,84	76,90	0,07	0,26	0,10	0,00	0,02	0,57	96,08
APL 504	Pisolites	8,02	1,90	43,31	43,40	0,01	0,22	0,00	0,01	0,01	0,09	97,17
APL 505	Pisolites	10,59	1,60	13,74	70,40	0,05	0,03	0,00	0,02	0,01	0,13	96,73
APL 506	Pisolites	9,95	1,90	26,49	58,30	0,04	0,12	0,00	0,01	0,00	0,09	97,07
ATH501	Pisolites	11,24	1,00	7,68	75,80	0,09	0,04	0,10	0,00	0,05	0,51	96,68
ATH502	Pisolites	14,02	0,90	12,92	69,00	0,08	0,03	0,00	0,01	0,24	0,22	97,64
ISH 501	Pisolites	3,26	1,20	7,61	83,90	0,06	0,02	0,00	0,00	0,01	0,34	96,68
ISH 502	Pisolites	2,86	1,10	7,09	84,20	0,05	0,02	0,00	0,00	0,01	0,34	95,92
ISH 503	Pisolites	6,77	1,00	8,80	78,20	0,07	0,24	0,10	0,00	0,03	0,68	96,17
MHN501	Block	14,51	3,70	13,47	65,60	0,11	0,84	0,10	0,14	0,03	0,29	99,18
MHN502	Block	15,67	3,60	14,97	62,70	0,09	0,58	0,10	0,02	0,02	0,32	98,42

Figure B.6: Bulk chemical compositions of the iron ores collected in northeastern Madagascar.

Sample	Ba (ppm)	Ce (ppm)	Co (ppm)	Cr (ppm)	Cu (ppm)	La (ppm)	Ni (ppm)	Rb (ppm)	Sr (ppm)	V (ppm)	Y (ppm)	Zr (ppm)
MBR 14501	20	0	8	768	88	12	30	0	1	898	5	141
MBR 14502	23	1	10	871	91	10	33	0	1	844	4	100
MBR 14503	30	0	16	1099	120	9	26	0	1	959	7	118
MBR 14504	62	0	9	786	122	11	24	0	5	952	3	81
MBR 14510a	13	0	26	537	106	11	37	2	3	659	6	128
MBR 14510b	27	2	16	684	115	13	45	1	1	899	8	135
MBR 14511a	35	14	14	774	113	15	27	1	7	763	10	156
MBR 14511b	31	19	12	780	112	19	27	1	6	766	11	129
MBR 14512a	27	0	21	845	117	15	39	1	7	737	9	104
MBR 14512b	27	0	18	886	117	8	39	1	7	761	9	104
DMD 501	31	13	39	1737	125	6	50	0	1	1257	6	211
DMD 502	34	13	38	1471	115	8	52	0	1	1346	5	173
DMD 503	641	133	902	672	92	22	101	6	3	672	11	240
BMK 501	243	21	52	1096	153	13	27	0	7	1152	3	267
BMK 502	166	12	48	1502	162	11	30	0	6	1269	3	265
BMK 503	50	13	23	816	253	12	54	0	2	1525	5	322
BMK510	44	16	105	549	146	18	89	3	13	1833	10	122
BMK511	51	9	22	688	269	15	56	0	2	1421	4	277
BMK512	55	0	49	2264	282	14	46	0	5	1519	5	229
MTY501	37	7	45	477	112	7	76	0	1	988	3	76
MTY503	247	11	171	450	93	15	14	0	16	1000	1	301
MTY504	75	5	79	463	119	13	9	0	2	887	1	293
MTY508a	25	0	55	1570	117	14	67	0	1	1809	3	100
MTY508b	26	0	55	1579	117	8	67	0	1	1806	3	95
MTY509a	30	0	56	1600	128	8	65	0	1	1860	3	102
MTY509b	28	0	53	1624	127	9	66	0	1	1880	4	104
MTY510a	26	0	51	1676	125	12	64	0	1	1851	3	95
MTY510b	24	0	51	1674	124	11	64	0	1	1844	3	107
MTY512	16	3	67	894	202	14	85	0	1	1679	7	105
MTY513	22	0	71	1251	220	9	67	0	3	1858	4	99
MTY514	23	0	53	1449	111	12	64	0	1	1717	3	106
MTY515	20	0	38	1058	209	11	60	0	0	1572	5	91
APL 501	65	0	12	1449	213	8	32	0	2	916	3	138
APL 502	27	0	11	1524	176	10	28	0	2	848	2	128
APL 503	75	2	141	1806	163	11	95	0	2	1500	4	103
APL 504	45	13	236	827	77	18	62	0	2	551	4	120
APL 505	25	3	25	712	96	13	70	0	2	657	3	93
APL 506	39	16	98	665	93	16	64	0	3	584	3	114
ATH501	27	0	39	667	171	10	17	1	2	670	1	85
ATH502	54	50	25	1127	51	33	29	14	9	322	23	115
ISH 501	14	0	17	1181	202	8	60	0	1	1215	5	86
ISH 502	14	0	13	1032	186	11	49	0	1	1169	3	76
ISH 503	52	3	160	663	86	11	90	0	4	1713	5	97
MHN501	520	38	239	562	158	14	71	0	5	1797	5	250
MHN502	113	34	150	512	173	16	66	0	3	1904	4	282

Sample		SiO_2 (wt.%)	TiO_2 (wt.%)	Al_2O_3 (wt.%)	Fe_2O_3 (wt.%)	MgO (wt.%)	MnO (wt.%)	CaO (wt.%)	Na_2O (wt.%)	K_2O (wt.%)	P_2O_5 (wt.%)	SUM (wt.%)
MBR11001	TS	18,52	1,70	12,70	63,90	0,24	0,06	0,60	0,10	0,11	0,21	98,43
MBR11002	TS	18,91	1,70	12,47	63,80	0,24	0,06	0,60	0,11	0,11	0,21	98,47
MBR11003	TS	24,22	1,40	10,17	60,70	0,29	0,09	1,00	0,25	0,51	0,34	99,26
MBR11005	TS	24,05	1,40	9,49	60,50	0,21	0,13	1,50	0,31	0,30	0,34	98,46
MBR12001	BFS	12,09	1,50	8,21	74,90	0,40	0,05	0,50	0,13	0,46	0,37	98,88
MBR12003	BFS	21,52	1,30	7,97	66,50	0,74	0,04	0,60	0,21	0,80	0,30	100,22
MBR12004	BFS	30,47	0,40	3,18	55,70	0,65	0,02	1,60	0,36	1,46	0,19	94,18
MBR12005	BFS	28,42	0,80	6,05	60,20	0,57	0,04	1,40	0,32	1,20	0,28	99,48
MBR12007	BFS	17,95	0,30	3,40	73,80	0,67	0,02	1,10	0,23	1,36	0,18	99,20
MBR12009	BFS	21,09	0,60	4,86	70,40	0,58	0,03	0,60	0,18	0,75	0,22	99,52
MBR13001	BFS	16,93	1,20	7,68	70,80	0,48	0,04	0,60	0,13	0,60	0,30	99,02
MBR13002	BFS	20,87	1,10	8,89	65,90	0,58	0,05	1,00	0,18	0,80	0,34	99,93
MBR13006	BFS	27,07	1,00	7,59	59,70	0,67	0,05	1,60	0,30	1,27	0,31	99,76
MBR13007	BFS	24,95	0,90	7,21	61,60	0,63	0,05	1,30	0,26	1,12	0,29	98,50
MBR13009	TS	23,80	1,50	9,94	60,00	0,50	0,04	2,20	0,21	0,31	0,28	99,00
MBR13010	TS	20,39	1,50	9,77	67,30	0,30	0,04	0,70	0,14	0,16	0,18	100,66
MBR14001	BFS	65,34	2,00	9,47	21,60	0,15	0,04	0,40	0,17	0,21	0,12	99,76
MBR14002	BFS	26,07	1,50	10,05	58,50	0,27	0,08	0,70	0,35	0,63	0,25	98,59
MBR14003	TS	26,52	1,30	12,24	55,30	0,33	0,05	0,80	0,39	0,94	0,37	98,54
MBR14004	TS	18,61	1,60	12,40	64,00	0,34	0,13	0,70	0,17	0,20	0,33	98,72
MBR14005	TS	23,84	1,60	10,50	58,10	0,33	0,10	2,10	0,58	0,40	0,36	98,20
MBR14006	TS	18,13	1,10	9,59	67,00	0,19	0,06	1,10	0,29	0,29	0,34	98,37
MBR14007	TS	18,76	1,40	8,70	66,10	0,34	0,14	1,50	0,28	0,35	0,39	98,18
MBR14009	BFS	26,38	0,90	9,00	59,90	0,33	0,02	0,80	1,69	1,05	0,24	100,51
MBR14010	BFS	16,47	1,10	11,48	67,40	0,22	0,02	0,50	0,20	0,37	0,20	98,26
MBR14011	BFS	21,19	0,10	2,18	71,20	0,88	0,01	1,30	0,40	1,23	0,19	98,79
MBR14012	BFS	24,46	1,20	9,50	61,40	0,15	0,02	0,40	0,26	0,42	0,22	98,23
MBR21001	BFS	16,71	1,60	8,30	70,20	0,34	0,10	0,50	0,10	0,30	0,40	98,85
MBR21002	BFS	23,65	1,20	6,50	65,10	0,30	0,24	0,50	0,12	0,37	0,26	98,46
MBR21005	BFS	22,88	1,40	7,86	65,10	0,48	0,17	0,80	0,16	0,38	0,39	99,90
MBR21008	BFS	21,38	1,90	8,26	64,60	0,33	0,15	0,60	0,20	0,40	0,45	98,54
MBR21010	BFS	20,35	1,30	14,08	60,50	0,55	0,04	0,60	0,12	0,41	0,19	98,48
MBR22001	BFS	24,11	1,40	9,18	62,00	0,21	0,02	0,50	0,19	0,36	0,24	98,46
MBR22002	BFS	24,94	1,30	8,09	62,50	0,27	0,41	0,30	0,10	0,24	0,23	98,60
MBR22005	BFS	21,39	1,10	7,22	67,20	0,43	0,13	0,50	0,08	0,23	0,42	98,93
MBR22006	BFS	23,19	1,30	8,82	63,10	0,30	0,02	0,60	0,12	0,50	0,25	98,44
MBR22008	BFS	22,89	1,50	9,55	62,40	0,32	0,03	0,60	0,18	0,48	0,26	98,42
MBR23001	BFS	15,18	1,40	7,01	72,50	0,28	0,11	0,40	0,09	0,29	0,48	97,99
MBR23002	BFS	26,12	0,60	6,10	62,90	0,48	0,26	0,60	0,17	0,74	0,19	98,42
MBR23003	BFS	22,80	1,30	8,38	64,00	0,25	0,43	0,30	0,09	0,18	0,28	98,27
MBR24001	BFS	19,35	1,30	9,12	67,50	0,27	0,02	0,40	0,12	0,39	0,21	98,90
MBR24003	BFS	22,16	1,10	8,27	64,50	0,30	0,02	0,80	0,18	0,56	0,23	98,36
MBR24004	BFS	16,11	3,10	16,20	60,50	0,39	0,18	0,50	0,13	0,51	0,24	98,10
MBR24005	BFS	14,88	1,50	10,16	69,80	0,26	0,03	0,40	0,10	0,31	0,27	97,95
MBR24006	BFS	17,48	1,10	9,71	68,60	0,28	0,02	0,50	0,15	0,30	0,25	98,61
MBR24007	BFS	16,87	1,60	9,91	68,50	0,31	0,04	0,30	0,12	0,29	0,28	98,45
MBR41001	BFS	18,95	1,60	10,51	65,00	0,53	0,05	0,50	0,11	0,45	0,41	98,34
MBR41002	BFS	21,33	1,80	8,91	64,00	0,56	0,06	0,50	0,12	0,50	0,36	98,37
MBR41003	BFS	17,01	1,80	9,27	68,10	0,69	0,06	0,40	0,11	0,38	0,31	98,34
MBR41005	BFS	19,86	1,80	14,54	59,50	0,83	0,06	0,40	0,12	0,49	0,32	98,26
MBR41006	BFS	17,82	1,60	10,99	65,70	0,54	0,05	0,50	0,10	0,35	0,38	98,26
BMK13001	BFS	20,18	3,30	13,22	60,80	0,11	0,10	0,60	0,17	0,62	0,22	99,65
BMK13002	BFS	18,15	4,00	13,85	60,90	0,10	0,10	0,50	0,14	0,55	0,28	98,94
BMK 13003	BFS	15,62	3,60	13,45	64,90	0,12	0,11	0,50	0,10	0,31	0,24	99,30
BMK13004	BFS	18,61	3,50	14,08	60,20	0,12	0,10	0,70	0,13	0,55	0,23	98,57
BMK 23001	BFS	22,06	3,80	16,12	58,20	0,34	0,28	0,60	0,16	0,37	0,40	102,66
BMK 23002	BFS	15,25	3,30	13,40	65,00	0,18	0,18	0,30	0,08	0,14	0,39	98,45
BMK 23004	BFS	17,11	3,70	15,43	59,90	0,39	0,24	0,60	0,16	0,33	0,42	98,61
BMK 24002	BFS	12,32	3,30	11,34	69,70	0,21	0,20	0,50	0,07	0,11	0,44	98,43
BMK 24004	BFS	16,09	3,10	16,11	60,50	0,39	0,18	0,50	0,12	0,51	0,38	98,28
BMK 24005	BFS	20,20	3,70	12,50	59,00	0,53	0,29	1,20	0,18	0,76	0,44	99,07
BMK 24006	BFS	21,53	2,90	13,59	57,60	0,43	0,18	0,50	0,17	0,72	0,48	98,41
BMK 24007	BFS	18,32	2,90	14,04	61,40	0,28	0,18	0,70	0,13	0,47	0,43	99,17
BMK 42001	BFS	16,79	3,00	14,94	63,50	0,16	0,25	0,40	0,10	0,14	0,29	99,92
BMK42002	BFS	20,91	2,10	12,55	60,90	0,16	0,31	0,50	0,15	0,28	0,33	98,47
BMK42003	BFS	18,18	2,30	12,92	62,30	0,15	0,27	0,70	0,17	0,40	0,37	98,09
BMK 42004	BFS	22,21	2,30	13,74	57,70	0,16	0,25	0,80	0,19	0,49	0,36	98,53
DMD 82001	TS	17,98	1,30	7,26	71,30	0,31	0,09	0,60	0,13	0,18	0,27	99,61
DMD 82002	TS	19,37	1,50	8,27	68,10	0,19	0,11	0,60	0,08	0,08	0,33	98,82
DMD 85001	BFS	49,83	1,90	8,50	35,20	0,73	0,07	2,50	0,22	0,39	0,26	99,90
DMD 85002	BFS	38,58	1,80	10,12	47,20	0,22	0,08	0,40	0,17	0,26	0,25	99,37
DMD 86001	BFS	16,53	1,20	9,29	67,10	0,55	0,14	0,90	0,35	0,61	0,43	97,34
DMD 92001	TS	16,29	1,30	8,22	70,80	0,29	0,05	0,50	0,12	0,12	0,29	98,14
DMD 92002	TS	25,78	1,70	8,50	59,90	0,29	0,10	1,30	0,25	0,23	0,28	98,57
DMD 92003	TS	14,54	1,10	7,56	73,80	0,27	0,26	0,50	0,12	0,11	0,33	98,81

Figure B.7: Bulk chemical compositions of smelting slag samples collected in northeastern Madagascar (See more on the next pages).

Sample	Ba (ppm)	Ce (ppm)	Co (ppm)	Cr (ppm)	Cu (ppm)	La (ppm)	Ni (ppm)	Rb (ppm)	Sr (ppm)	V (ppm)	Y (ppm)	Zr (ppm)
MBR11001	51	5	27	1126	52	14	7	2	24	1154	10	188
MBR11002	54	2	21	1047	53	11	8	1	22	1088	10	190
MBR11003	87	8	36	789	79	12	10	9	62	1256	7	180
MBR11005	93	11	31	474	62	16	5	6	65	966	6	197
MBR12001	63	5	464	863	92	10	7	8	38	919	8	138
MBR12003	74	9	28	834	86	20	15	15	46	1024	15	144
MBR12004	148	5	26	522	140	12	39	33	169	333	4	100
MBR12005	125	10	28	561	114	12	26	23	120	594	6	123
MBR12007	112	-3	44	632	252	7	115	29	112	359	3	43
MBR12009	69	2	32	881	193	11	61	20	55	555	6	105
MBR13001	63	5	49	945	114	12	25	12	41	1008	8	128
MBR13002	85	6	27	811	97	9	25	17	58	853	8	129
MBR13006	150	7	29	595	69	12	10	26	142	697	8	147
MBR13007	129	6	28	569	105	12	24	24	117	663	7	133
MBR13009	51	8	26	715	100	10	29	5	111	771	10	229
MBR13010	58	11	27	481	77	13	14	3	38	751	6	186
MBR14001	61	67	68	469	39	45	22	4	32	581	12	589
MBR14002	130	16	26	447	61	14	9	13	71	767	10	224
MBR14003	149	6	28	740	77	14	6	17	88	1426	7	186
MBR14004	129	3	16	732	88	6	7	4	43	1109	8	177
MBR14005	140	11	30	849	41	12	4	8	135	1272	6	199
MBR14006	102	4	21	1043	91	15	12	5	56	1406	5	150
MBR14007	147	4	23	552	85	11	7	7	97	1072	5	166
MBR14009	228	14	28	423	52	11	5	25	95	798	6	108
MBR14010	74	2	26	853	84	13	16	8	34	1498	5	121
MBR14011	189	6	46	94	297	5	83	26	116	105	2	31
MBR14012	87	12	20	449	56	15	9	9	29	847	6	194
MBR21001	45	0	21	1452	100	10	18	4	29	1093	10	214
MBR21002	136	29	29	512	127	38	60	8	44	690	192	202
MBR21005	100	9	54	981	76	14	10	6	53	1102	7	196
MBR21008	80	18	15	832	59	21	7	8	40	1078	10	304
MBR21010	71	-6	29	1376	57	12	7	7	37	1386	5	129
MBR22001	54	16	18	532	58	16	7	6	34	1355	9	191
MBR22002	237	17	13	610	152	17	10	7	18	744	14	208
MBR22005	76	11	17	850	74	16	8	7	30	1051	8	143
MBR22006	70	6	11	508	72	14	7	9	38	1214	9	182
MBR22008	91	12	16	327	64	15	7	11	46	1049	13	191
MBR23001	51	0	28	920	66	9	8	5	30	1036	6	163
MBR23002	174	4	37	1060	192	15	19	12	55	761	6	97
MBR23003	212	2	22	812	80	15	8	4	15	993	10	170
MBR24001	65	20	17	373	85	18	10	8	32	1191	8	169
MBR24003	107	20	18	511	101	15	12	11	80	1154	11	144
MBR24004	121	14	18	471	400	12	15	23	80	953	8	135
MBR24005	61	12	13	445	116	15	15	6	22	1216	31	140
MBR24006	36	10	21	526	99	13	10	7	27	1217	7	126
MBR24007	42	28	17	410	85	40	20	4	25	1051	198	168
MBR41001	92	18	27	699	107	8	5	10	31	972	5	174
MBR41002	105	7	11	745	83	14	5	11	35	1037	5	189
MBR41003	72	6	57	398	120	12	7	7	30	1000	4	150
MBR41005	84	2	54	1025	80	7	5	10	35	1490	4	179
MBR41006	83	14	29	751	122	8	5	6	29	1024	5	163
BMK13001	147	33	19	739	150	23	7	16	50	1378	13	386
BMK13002	144	35	17	862	204	28	8	10	51	1725	14	378
BMK 13003	112	25	26	795	399	20	29	7	33	1498	11	337
BMK13004	122	24	19	840	160	21	8	11	51	1507	11	385
BMK 23001	227	25	25	689	92	23	15	10	56	1630	8	246
BMK 23002	200	34	21	389	119	17	9	4	22	1183	9	218
BMK 23004	157	24	21	653	68	16	6	7	60	1995	6	253
BMK 24002	153	21	40	325	125	13	23	4	39	1283	7	236
BMK 24004	155	11	52	977	99	16	11	15	77	2141	7	240
BMK 24005	240	27	32	428	95	18	12	22	117	1342	8	240
BMK 24006	194	26	41	566	94	18	11	28	76	1585	8	255
BMK 24007	127	15	53	644	115	15	19	10	79	1616	11	239
BMK 42001	238	35	30	852	88	17	6	3	63	1631	9	253
BMK42002	192	22	36	690	97	15	12	5	86	1104	9	277
BMK42003	172	20	27	742	143	10	10	8	113	1537	9	235
BMK 42004	194	0	25	807	123	8	8	9	125	1527	8	239
DMD 82001	35	13	35	571	117	14	15	2	29	714	5	215
DMD 82002	57	21	29	403	86	16	8	1	24	845	8	266
DMD 85001	55	53	41	830	65	40	27	6	115	762	9	582
DMD 85002	35	24	41	1023	161	23	43	4	27	740	8	406
DMD 86001	65	8	22	699	85	15	11	9	58	1062	6	166
DMD 92001	50	25	25	178	71	19	8	5	80	538	9	379
DMD 92002	38	16	32	783	124	17	38	2	20	996	6	222
DMD 92003	89	6	27	701	116	14	11	1	30	965	4	158

Sample		SiO_2 (wt.%)	TiO_2 (wt.%)	Al_2O_3 (wt.%)	Fe_2O_3 (wt.%)	MgO (wt.%)	MnO (wt.%)	CaO (wt.%)	Na_2O (wt.%)	K2O (wt.%)	P_2O_5 (wt.%)	SUM (wt.%)
ATB001	BFS	21,85	1,80	10,41	61,50	0,25	0,13	0,40	0,09	0,20	0,34	97,24
ATB002	BFS	20,79	1,60	8,64	64,60	0,25	0,10	0,50	0,09	0,19	0,37	97,30
BNY50006	BFS	16,15	1,50	10,57	67,00	0,52	0,05	0,30	0,09	0,22	0,35	97,05
BNY50001	BFS	20,34	2,70	11,78	60,50	0,32	0,16	0,40	0,11	0,32	0,31	97,22
BNY 410001	TS	22,58	2,30	10,96	62,50	0,31	0,14	1,30	0,18	0,42	0,34	101,30
BNY41002	TS	19,24	2,60	13,38	60,40	0,29	0,21	0,50	0,08	0,26	0,27	97,50
BNY 410004	TS	19,42	3,90	13,29	60,70	0,31	0,18	0,90	0,12	0,27	0,36	99,78
BNY41005	TS	15,69	3,00	11,20	64,70	0,34	0,10	1,40	0,10	0,28	0,27	97,39
BNY41006	BFS	16,07	4,20	13,93	60,90	0,30	0,20	0,90	0,15	0,38	0,34	97,72
BNY41008	BFS	15,25	4,10	14,72	61,20	0,31	0,20	0,70	0,13	0,29	0,33	97,62
BNY41029	BFS	17,10	2,30	11,84	64,30	0,26	0,15	0,70	0,07	0,25	0,30	97,55
BNY41030	BFS	16,53	3,30	12,99	62,30	0,26	0,38	0,60	0,07	0,18	0,30	97,23
BNY41033	BFS	15,89	2,50	11,21	65,50	0,27	0,26	0,70	0,09	0,25	0,30	97,30
BNY42001	BFS	17,20	2,90	15,19	60,10	0,26	0,22	0,30	0,07	0,18	0,27	97,04
BNY42003	BFS	13,99	2,00	11,94	67,00	0,25	0,42	0,70	0,05	0,11	0,39	97,18
BNY42004	BFS	17,32	3,00	13,33	61,10	0,38	0,36	0,50	0,13	0,52	0,29	97,22
BNY42006	BFS	23,54	2,40	12,38	56,60	0,30	0,23	0,90	0,16	0,43	0,26	97,48
BNY43007	BFS	23,84	1,60	7,56	62,40	0,25	0,08	0,80	0,14	0,38	0,25	97,48
BNY43008	BFS	17,68	2,50	14,37	60,40	0,35	0,35	0,80	0,09	0,27	0,29	97,43
BNY43009	BFS	21,88	2,40	10,40	59,50	0,34	0,11	1,20	0,20	0,64	0,31	97,23
BNY44001	BFS	17,58	3,60	12,14	61,20	0,43	0,39	1,10	0,08	0,31	0,27	97,43
BNY44002	BFS	23,30	2,30	8,49	59,90	0,59	0,06	1,80	0,27	0,68	0,25	97,96
BNY44004	BFS	20,31	2,50	9,73	62,60	0,46	0,04	1,30	0,17	0,37	0,21	97,96
BNY44005	BFS	19,34	2,30	8,26	65,20	0,43	0,05	1,00	0,14	0,32	0,25	97,54
BNY45001	BFS	21,89	1,90	7,11	63,20	0,49	0,08	1,70	0,24	0,51	0,25	97,62
BNY45002	BFS	23,12	2,70	8,13	60,40	0,55	0,22	1,40	0,19	0,48	0,26	97,71
BNY45004	BFS	26,45	2,30	8,09	56,60	0,62	0,09	2,10	0,29	0,62	0,25	97,67
BNY45005	BFS	21,90	2,40	7,23	61,80	0,73	0,07	1,60	0,22	0,76	0,29	97,30
MTY11001	TS	15,64	1,80	9,56	68,50	0,14	0,11	0,20	0,06	0,04	0,35	96,77
MTY11002	TS	23,81	1,60	6,57	63,50	0,20	0,12	0,60	0,17	0,13	0,37	97,27
MTY11004	TS	17,03	3,10	17,23	55,10	0,89	0,21	1,60	0,32	0,46	0,44	97,00
MTY 11005	TS	20,40	1,90	7,55	66,40	0,22	0,14	3,00	0,18	0,15	0,50	100,62
MTY11006	TS	17,09	1,60	6,88	70,10	0,13	0,11	0,30	0,04	0,07	0,53	97,06
MTY11011	BFS	16,06	1,90	12,38	65,20	0,39	0,13	0,40	0,14	0,28	0,38	97,73
MTY11012	BFS	21,68	1,80	8,04	64,70	0,17	0,12	0,40	0,05	0,09	0,37	97,73
MTY12005	BFS	13,12	1,70	9,27	71,00	0,31	0,12	0,40	0,13	0,23	0,44	97,09
MTY12006	BFS	13,42	1,70	10,47	68,90	0,30	0,14	0,40	0,29	0,34	0,43	96,79
MTY12007	BFS	9,98	0,40	4,21	83,00	0,47	0,04	0,60	0,18	0,47	0,26	99,88
MTY18002	BFS	23,81	0,30	2,65	67,40	0,98	0,03	1,20	0,35	1,43	0,23	98,55
MTY18004	BFS	15,09	1,40	9,99	67,50	0,48	0,14	0,80	0,18	0,49	0,48	96,91
MTY18006	BFS	9,75	1,80	11,35	72,20	0,35	0,22	0,30	0,09	0,18	0,33	97,00
APL 10001	BFS	19,64	1,80	12,14	62,60	0,38	0,14	0,50	0,15	0,29	0,35	98,30
APL 10002	BFS	13,91	0,20	1,24	79,00	0,70	0,04	0,80	0,33	0,47	0,28	97,05
APL 10003	BFS	21,16	0,60	5,27	66,80	0,66	0,05	1,30	0,29	0,94	0,28	97,51
APL 30001	BFS	17,29	1,80	11,33	65,10	0,84	0,08	0,50	0,06	0,24	0,26	97,77
APL 30002	BFS	14,96	2,00	9,45	69,00	0,49	0,06	0,30	0,03	0,03	0,32	96,95
APL 30003	BFS	20,70	1,90	8,38	64,60	0,63	0,13	0,30	0,07	0,13	0,31	97,36
APL 30004	TS	21,15	1,90	9,02	64,70	0,59	0,13	0,30	0,05	0,10	0,33	98,52
ISH 10001	TS	21,70	1,70	8,63	64,20	0,28	0,04	0,40	0,11	0,13	0,37	97,77
ISH 10002	TS	23,36	1,80	8,04	63,70	0,22	0,05	0,40	0,12	0,20	0,28	98,37
ISH 10003	TS	19,87	1,70	7,44	67,90	0,32	0,04	0,30	0,05	0,11	0,31	98,24
ISH 20001	TS	13,51	1,80	11,32	70,90	0,30	0,05	0,40	0,04	0,12	0,40	99,07
ISH 20002	TS	16,96	2,70	15,06	61,30	0,32	0,16	0,60	0,05	0,17	0,22	97,84
ISH 20003	BFS	13,12	1,90	13,05	67,90	0,33	0,13	0,60	0,02	0,09	0,37	97,77

Figure B.8: Bulk chemical compositions of smelting slag samples collected in northeastern Madagascar (See more on the previous pages).

Sample	Ba (ppm)	Ce (ppm)	Co (ppm)	Cr (ppm)	Cu (ppm)	La (ppm)	Ni (ppm)	Rb (ppm)	Sr (ppm)	V (ppm)	Y (ppm)	Zr (ppm)
ATB001	123	24	37	735	68	17	6	5	17	998	9	287
ATB002	92	40	37	191	96	20	7	5	25	549	29	232
BNY50006	70	3	41	1121	45	16	7	7	16	1225	8	174
BNY50001	135	51	34	702	65	34	6	5	30	1214	8	274
BNY 410001	117	41	42	728	41	36	6	7	85	1129	12	298
BNY41002	135	40	31	727	112	28	15	6	35	1088	10	274
BNY 410004	107	126	48	711	97	76	11	6	51	1343	16	430
BNY41005	86	37	104	846	75	30	10	5	67	1278	9	262
BNY41006	114	43	45	831	84	32	10	9	58	1611	10	329
BNY41008	116	42	48	1074	97	35	12	7	48	1880	10	320
BNY41029	108	30	37	750	52	21	6	7	25	1192	10	272
BNY41030	121	25	51	869	80	22	7	3	28	1494	9	306
BNY41033	113	47	35	866	84	31	7	5	50	1581	10	308
BNY42001	104	20	58	922	77	18	9	3	26	1422	9	409
BNY42003	155	20	34	1094	104	22	10	2	42	1483	9	229
BNY42004	149	15	47	749	101	16	12	9	45	1198	10	238
BNY42006	140	57	55	600	90	37	11	9	76	1048	12	281
BNY43007	132	51	43	320	83	30	6	8	63	652	9	227
BNY43008	204	41	44	932	139	22	12	6	46	1306	14	204
BNY43009	150	68	70	296	63	43	6	14	106	1025	11	267
BNY44001	161	47	71	692	106	33	12	7	45	1317	11	310
BNY44002	148	109	95	305	74	72	6	13	119	1284	14	474
BNY44004	121	87	39	249	80	51	5	8	77	1297	11	436
BNY44005	119	69	33	237	70	37	8	6	56	1297	9	336
BNY45001	136	123	136	312	64	67	6	11	111	657	15	483
BNY45002	157	103	41	493	70	64	7	10	87	788	13	462
BNY45004	165	137	48	286	45	81	4	13	138	726	16	551
BNY45005	161	120	28	619	74	73	6	17	86	982	15	530
MTY11001	33	19	118	1019	174	23	28	0	16	1606	6	193
MTY11002	52	28	68	213	193	23	26	1	50	702	6	194
MTY11004	61	12	60	2023	57	15	9	3	106	3121	9	282
MTY 11005	43	22	37	246	121	23	8	0	45	982	7	221
MTY11006	41	14	31	405	108	15	7	1	23	1188	7	179
MTY11011	42	13	111	1485	118	11	14	2	35	2142	7	187
MTY11012	39	22	49	840	130	18	12	1	28	1465	7	219
MTY12005	34	6	77	991	118	15	13	2	32	1975	7	155
MTY12006	50	-1	57	1247	120	14	11	5	35	2142	7	150
MTY12007	11	-5	92	1089	312	12	57	4	66	996	1	32
MTY18002	51	5	95	505	334	10	80	22	88	444	2	48
MTY18004	39	4	39	1121	117	11	12	6	41	1791	6	137
MTY18006	70	-1	61	1308	182	12	14	2	24	2265	8	161
APL 10001	56	1	25	1001	84	21	8	5	31	1550	8	205
APL 10002	71	5	50	144	320	8	99	8	61	87	2	35
APL 10003	78	29	31	494	253	10	51	18	98	410	6	97
APL 30001	64	13	18	979	87	16	7	7	33	1102	9	207
APL 30002	47	-4	17	1288	74	16	5	0	17	1463	11	200
APL 30003	85	3	15	664	78	14	6	3	25	988	6	220
APL 30004	80	5	60	816	51	14	8	3	19	1149	6	210
ISH 10001	41	6	19	476	79	16	8	2	24	1171	8	198
ISH 10002	38	5	20	527	97	11	5	2	27	936	5	213
ISH 10003	31	7	19	418	80	10	6	1	22	895	5	174
ISH 20001	64	8	18	743	83	17	9	4	17	1091	8	154
ISH 20002	126	15	36	739	50	20	4	4	27	1639	9	190
ISH 20003	77	10	20	794	162	17	16	3	16	1298	6	153

Sample		BNY41901	BNY 41902	MBR14901	MBR 14902	MBR 14903	MBR14910	MBR14911	MBR14912	MBR14913	MTY11904
		Substratum BNY410	Substratum BNY410	Substratum MBR140	Substratum MBR140	Washed substratum	From indurated wall	From indurated wall	From indurated wall	From indurated wall	From indurated wall
SiO_2	(wt.%)	86,58	79,53	78,22	78,61	77,87	86,65	85,61	82,76	85,66	74,19
TiO_2	(wt.%)	1,40	2,50	0,20	0,20	0,20	1,80	1,90	1,80	1,90	4,20
Al_2O_3	(wt.%)	5,53	5,59	11,10	10,75	10,43	7,56	8,15	8,12	7,88	10,79
Fe_2O_3	(wt.%)	4,40	6,60	3,10	2,50	3,50	2,90	2,70	6,70	3,10	10,30
MgO	(wt.%)	0,54	0,85	0,19	0,20	0,19	0,09	0,09	0,09	0,09	0,14
MnO	(wt.%)	0,06	0,10	0,02	0,02	0,02	0,03	0,02	0,02	0,03	0,07
CaO	(wt.%)	0,70	1,10	0,70	0,70	0,70	0,00	0,00	0,00	0,00	0,10
Na_2O	(wt.%)	0,84	0,97	1,92	1,93	1,93	0,15	0,12	0,13	0,13	0,10
K_2O	(wt.%)	1,70	1,71	4,12	4,06	4,00	0,08	0,06	0,05	0,05	0,03
P_2O_5	(wt.%)	0,08	0,13	0,04	0,05	0,04	0,03	0,03	0,03	0,03	0,08
SUM	(wt.%)	102,01	99,31	99,75	99,15	99,02	99,39	98,79	99,81	98,99	100,13
Ba	(ppm)	473	484	907	902	890	58	51	51	51	71
Ce	(ppm)	144	196	10	15	10	69	55	39	58	133
Co	(ppm)	162	52	20	37	37	91	71	78	78	39
Cr	(ppm)	48	75	47	49	58	145	143	174	170	76
Cu	(ppm)	4	11	2	1	2	3	3	5	3	70
Hf	(ppm)	15	25	1	1	1	10	11	10	12	9
La	(ppm)	81	100	11	10	12	38	23	27	25	50
Ni	(ppm)	10	12	6	5	6	25	24	23	25	37
Rb	(ppm)	44	50	99	100	97	2	1	1	1	0
Sr	(ppm)	87	104	162	166	165	4	3	3	3	2
V	(ppm)	64	102	54	67	75	153	138	282	157	336
Y	(ppm)	21	30	5	5	5	12	7	9	11	12
Zr	(ppm)	607	1066	100	88	99	488	523	463	540	525

Figure B.9: Bulk chemical compositions of sandy substratum samples.

C

Written sources

Author	Date	Title	Book	Reference to iron		Comments
				Pages	Details	
D'Albuquerque, Fernand		Tristan Da Cunha visite Madagascar avec sa flotte (1506)	COACM I	p.22	Mention of Malagasy soldiers armed with shields, sagaies, bows and arrows	
Barbosa, Duarte	1516	Description de Madagascar	COACM I	p.54	Mention of assegais with iron tips	
De Sousa, Balthazar Lobo	1557	Exploration de Madagascar	COACM I	p.101	Mention of the presence of iron ore in Madagascar	
Lopez, Ed.	1585	Description de Madagascar	COACM I	p.147	Mention of the presence of iron ore in Madagascar and description of the Malagasy iron weaponry	
?		Relation du Voyage de découverte fait à l'île Saint-Laurent, dans les années 1613-1614 par le capitaine Paulo Rodrigues Da Costa et les pères Jésuites Pedro Freire et Luis Mariano à bord de la caravelle Nossa Sanhora Da Esperença	COACM II	p.8	Mention of iron weaponry	
				p.11	Reference to iron spade	
				p.13	Reference to smithing work	
De Houtman, Cornelius	1595	?	COACM I	p.190	Mention of iron weapons in St. Augustine's Bay	
Lodewijcksz	1595	Histoire de la navigation Hollandaise aux Indes orientales	COACM I	p.202	Engraving with characters carrying iron weapons in the Bay of Antogil	
Van Neck	1598		COACM VI	p.255	Mention of iron weapons in St. Augustine's Bay	
Van der Hagen ?	After 1601	Relache sur la c'ote Est de Madagascar et à Antogil (Quatrième voyage des Hollandais aux Indes, avec trois vaisseaux, de 1599 à 1601, sous le commandement de l'amiral Et. Van Der Hagen)	COACM I	p.258	Reference to iron weapons in the Bay of Antogil	
Finch,William	1608	Observations faites dans la baie de Saint Augustin, à Madagascar	COACM I	p.415	Description of iron weapons in St. Augustine's Bay	
?	1609	Description véridique, complète et détaillée aussi bien qu'Historique et Chorographique, de l'île extrêmement riche, puissante et célèbre de Madagascar, autrement nommée Saint-Laurent, qui de nos jours est considérée comme la plus grande de toutes celles du monde	COACM I	p.441	Description of Malagasy iron weapons on the coast. He does not describe the central part of the island, which is not well known to Europeans.	
				p.442	Reference to iron weaponry	
De Bry, Théodore	1613	India Orientalis	COACM I	p.246	Engraving representing a whale hunt in Sainte Marie with the use of an iron hook	
Gomes, Jean	1620	Lettre du père Jean Gomes datée de Mazalagem	COACM II	p.342	Allusion to Malagasy people producing their own iron	
Purchas	1625	Description de Madagascar	COACM II	p.375	Reference to iron weapons	
Herbert, Thomas	1626	Madagascar et les îles Comores	COACM II	p.384-385	Reference to iron weapons	
Hamond, Walter	1640	Paradoxe montrant que les habitants de Madagascar sont, au point de vue des biens temporels, le peuple le plus heureux de la Terre	COACM III	p.10	Reference to the presence of exploitable iron ore in Madagascar and the existence of iron smelting techniques	
				p.57	Mention of iron weapons and the time needed to rebuild a stock of weapons	
Boothby, Richard	1646	Courte description ou découverte de la très fameuse île de Madagascar ou Saint-Laurent, en Asie, à proximité de l'Inde orientale (1644)	COACM III	p.74	Reference to iron weapons he brought back from Madagascar	
				p.91	Description of iron weapons	
				p.93	Mention of an iron hatchet used as a carpenter's tool for making pirogues	
				p.93-94	Detailed description of a smithing workshop	
				p.96	Mention of the abundance of iron ore in Madagascar	
Hunt, Robert	1650	Ile d'Assada (Ile de Nosy Bé), près de Madagascar	COACM III	p.264	Reference to iron primary production	

Figure C.1: Compilation of historical written sources from the 16th to the 20th century mentioning iron metallurgy. **Acronyms:** COACM: Collection des Ouvrages Anciens Concernant Madagascar (Grandidier, 1903); BAM: Bulletin de l'Académie Malgache; AAMM: Antananarivo Annual and Madagascar Magazine; NRE: Notes, Reconnaissances et Explorations.

Author	Date	Title	Book	Reference to iron		Comments
				Pages	Details	
Flacourt	1661	Histoire de la grande isle de Madagascar		p.73	Reference to iron smelting and smithing techniques in Madagascar,as well as general description of the different manufactured iron objects	
				p.74	Reference to iron carpenter tools	
				p.75	Reference to iron fishing tools	
				p.77	Description of kitchen furniture and equipment, including knives, hooks, etc	
				p.146-147	Brief description of a smelting technique and mention of the different ethnic groups that master this technique	
				p.451	Brief description of a smithing workshop	
Martin, François		Mémoires de Francois Martin, fondateur de Pondichéry (1665-1696)		p.123	Reference to iron working and to smithing tempering technique in Fort-Dauphin	
				p.161	Mention that smithing work is common	
				p.168	Mention that iron could be traded more because the different regions of Madagascar have only little contact with each other	
				p.257	Mention of a village with blacksmiths	
Everard, Robert	1686	Relation de trois années de souffrances dans l'île d'Assasa (Nosy Bé), près de Madagascar, au cours d'un voyage aux Indes en 1686	COACM III	p.415	Brief description of a smithing workshop and of the manufactured artefacts	Grandidier doesn't believe that Everard is accurate, there is too many inconsistencies according to him
Ovington, Jean	1690	Description de Madagascar et de l'ile d'Anjouan	COACM III	p.457	Mention that iron and steel are very important metals for the people of Madagascar, especially for weaponry	
				p.462	Mention that the ships of the island of Anjouan, which have trading contacts with the eastern African coast, do not have iron	
Drury, Robert	1729	Madagascar or Robert Drury's Journal during fifteen years Captivity on that Island	COACM IV	p.20	Mention that iron spades were used to calibrate fines	
				p.30	Mention that iron was a known but still "valuable" material	
				p.75	Reference to iron spades	
				p.158	Reference to iron spears	
				p.220	Reference to iron ore in Madagascar	
				p.221-222	Fines must be paid with spade	
				p.235	Some people know how to repair a gun	
				p.350	Reference to iron smelting	
				p.359	Reference to Merina smelting technique	
Edouard, Yves	1754	Relache dans la baie de Saint-Augustin d'une escadre anglaise	COACM V	p.256	Reference to iron spears	
Rochon, Alexis	1768	Voyage à Madagascar et aux Indes orientales		p.383	Reference to iron smelting and smithing techniques	Taken from Flacourt according to C. Radimilahy
				p.402	Reference to iron ore extraction close to Fort-Dauphin	
Mayeur, Nicolas	1775	Voyage dans le Nord de Madagascar			Description of ore extraction	
Mayeur, Nicolas	1785	Voyage au pays d'Ancove, par le pays d'ancaye autrement dit des Baizangouzangoux	BAM 12(2)	p,28	Brief description of iron smelting and of the manufactured iron artefacts	
Hastie	1817	Diary	BAM 1(2)	p.248	Description of ore extraction, smelting technique and of bellows in the region of Lazaina	Ellis (1838) refer to this text
Copalle, André		Voyage dans l'Intérieur de Madagascar et à la capitale du roi Radama pendant les années 1825-1826	BAM7	p.57	Reference to blacksmiths	
				p.90-91	Description of a smithing workshop implemented by the Ambaniandres people	
				p.94	Reference to iron smelting in Madagascar	
				p.95	Brief description of iron smelting	

Author	Date	Title	Book	Reference to iron		Comments
				Pages	Details	
Semerville		Souvenir de Sainte-Marie	Decary, Le voyage du lieutenant de vaisseau de Semerville à l'île sainte Marie en 1824, BAM 16 (1933)		Description of ore extraction	
Copland, Samuel	1822	History of the Island of Madagascar comprising a political account of the island, the religion, manners and customs of its inhabitants and its natural productions		p.102	Brief description of iron smelting	
Ellis, William	1838	History of Madagascar		p.5	Mention of iron ore	
				p.7	Mention of the existence of a smelting process in the Ankova, Imamo and Iron Mountain districts	
				p.89	Mention of iron as a natural ressource	
				p.124	Mention that people in Ankova make their own iron	
				p.197	Details on the workers subjected to the Merina state, especially for iron production	
				p.305-310	Detailed account of the iron metallurgy chaine opératoire	
				p.310-314	Detailed account on iron manufactured artefacts	
Jones, David	1839		In Jean Valette, BAM 45,1966	p.195	Jones denies the veracity of Ellis' descriptions of Malagasy iron metallurgy (1838)	
Lacombe	1840	Voyage à Madagascar et aux îles Comores (Tomes 2)		p.23	Reference to the Hova people who master both smelting and smithing techniques	
				p.103	Mention of iron ores north of Antananarivo	
				p.104	In the Bemarivo region, mention of iron smelting and smithing in the village of Sora-minti	
				p.109	Blacksmith on the Nossi-Vola island	
				p.143	Mention of blacksmith	
				p.146-147	Mention of blacksmith	
				p.161-162	Detailed account of iron smelting and smithing techniques in the Anta-mamamboundres country (near the Manamboundre river)	
				p.253	Engraving of a bloomery furnace	
Griffiths, David	1843	A History of Madagascar or a Historical Summary of the Island with a description of its Products, Trade, and Inhabitants; Its cruel customs and Repugnant Idol Worship Also A History of the Mission, Its Successes and Failures and the Persecution and Martyrdom of Christians from its inception in 1818 util 1843	In Campbell, 2012	p.272	Reference to iron ore	
				p.273	Listing of natural resources available in Madagascar	
Ellis, William	1858	Three visits to Madagascar during the years 1853-1854-1856		p.264-265	Detailled account of the iron metallurgy chaine opératoire	
Simonin	1862	Les richesses naturelles de l'île de Madagascar		p.4	Reference to the presence of iron ores	
Sibree, James	1873	Madagascar et ses habitants, journal d'un séjour de quatre ans sur l'île		p.65	Reference to the presence of iron ores	
				p.131	Listing of iron objects that can be bought on the Antananarivo market	
				p.216	Reference to blacksmiths	
				p.218-220	Description of iron smelting and smithing techniques	

Author	Date	Title	Book	Reference to iron		Comments
				Pages	Details	
Mullens	1875	Twelve Months in Madagascar		p.95	Reference to iron objects one can find in the Betsileo country	
				p.139	Mention of iron objects available on the Malagasy markets	
				p.179	Story about the Vazimbas who didn't have iron weapons which lead to their defeat against the Merinas	
				p.183	Reference to bloomery furnaces	Description taken from Ellis, not clear if he saw these furnace with his own eyes
				p.195	Iron ore close to the Itasy lake	
				p.219	Iron ore around Antananarivo (Amaronkay, Itasy lake and the iron moutain)	
Buet	1883	Madagascar la reine des îles africaines		p.81	Young Malagasy people trained as blacksmiths by the English	
				p.100	Iron working at Laborde's industrial complex	
				p.212	Reference to iron ore as a available ressource	
				p.235	Reference to smithing techniques to make farming tools	
				p.281	Engraving out of context of a bloomery furnace	Engraving from Ellis
				p.285	Brief description of the geology and mention of ferruginous sandstone	
D'Escamps, Henri	1884	Histoire et Géograohie de Madagascar		p.448-449	Description of iron smelting and smithing techniques	Description taken from Lacombe
Pauliat, Louis	1884	Madagascar		p.xi	Reference to iron as an available natural ressource	
Wills	1885	Native Products used in Malagasy Industries	AAMM IX	p.97	Description of available natural ressources and of iron smelting technique	Description taken from Ellis
Oliver, Samuel Pasfield	1886	Madagascar, An Historical and Descriptive Account of the Island annd its Former Dependencies		vol1 p.343	Location of iron ores	
				vol1 p.490-491	Reference to iron ore availability	
				vol1 p.492	Legend on the introduction of iron in Madagascar by the grandfather of Radama I and a quick biblioagraphical compilation on iron metallurgy	
				vol2 p.88	Detailed account of the iron metallurgy chaine opératoire	Taken from Ellis (1838)
				p.89	Description of a second iron smelting technique	He is mixing the description between smelting and smithing hearth found in Ellis
Le Chartier & Pellerin	1888	Madagascar depuis sa découverte jusqu'à nos jours		p.364	Brief description of a simple pit furnace	
Catat, Louis	1893	Voyage à Madagascar (1889-1890)		p.4	Reference to good iron ores in Madagascar	
				p.40-41	Decription of a smithing workshop in the region of Antananarivo	
				p.245	Mention that iron is imported to the west coast of Madagascar from Germany and England	
Breton	1898	La Fabrication du fer en Emyrne		p.686-693	Detailed description of iron metallurgy in Imerina (map of the district, drawing of a furnace)	
Vacher	1899	Les cultures et industries indigènes à Madagascar	NRE 5	p.543-549	Detailed account of ore extraction, iron smelting (two technologies mentioned), smithing and iron sales	
Lacroix	1922	Minéralogie du Madagascar		Vol2 p.72	General history of iron production in Madagascar	Taken from Grandidier
				Vol2 p.72-74	List of iron ore deposits	
				Vol2 p.170-171	Photograph of a smelting furnace (Plate 21)	
Decary, Raymond	1966	Coutumes guerrières et organisation militaire chez les anciens malgaches			Description of ore extraction	

Bibliography

Adelaar, A. (1995). Asian roots of the Malagasy; A linguistic perspective. *Bijdragen tot de Taal-, Land- en Volkenkunde*, 151(3):325-356.

Alipour, R., Rehren, T., and Martinón-Torres, M. (2021). Chromium crucible steel was first made in Persia. *Journal of Archaeological Science*, 127:105224.

Allan, J. W. (1979). *Persian Metal Technology 700-1300 AD*. Ithaca Press, for the Faculty of Oriental Studies and the Ashmolean Museum, University of Oxford Edition.

Allibert, C. (2008). Austronesian Migration and the Establishment of the Malagasy Civilization: Contrasted Readings in Linguistics, Archaeology, Genetics and Cultural Anthropology. *Diogenes*, 55(2):7–16.

Armistead, S. E., Collins, A. S., Merdith, A. S., Payne, J. L., Cox, G. M., Foden, J. D., Razakamanana, T. and De Waele, B. (2019). Evolving Marginal Terranes During Neoproterozoic Supercontinent Reorganization: Constraints from the Bemarivo Domain in Northern Madagascar. *Tectonics*, 38(6):2019–2035.

Barbosa, D. (1516). Description de Madagascar. In *Collection des Ouvrages Anciens Concernant Madagascar (COACM)*, Grandidier, A. and Grandidier, G., Dir., vol. 1:54. Comité de Madagascar, Paris.

Battistini, R. and Vérin, P. (1966). Irodo et la tradition vohémarienne. *Revue de Madagascar*, 36:XVII–XXXII.

Bauzyte, E. (2019). Making and trading iron in the Swahili World: An archaeometallurgical study of iron production technologies, their role, and exchange networks in 500-1500 CE coastal Tanzania. PhD, Aarhus University, Aarhus.

Bazargan, S. (2017). Leopold Bloom and William Ellis's Three Visits to Madagascar: Photography, Botany, and Race. *Joyce Studies Annual 2017*, 65–93.

Beaujard, P. (2003). Les arrivées austronésiennes à Madagascar: vagues ou continuum? *Etudes Océan Indien*, 35-36:59–147.

Beaujard, P. (2005). The Indian Ocean in Eurasian and African World-Systems Before the Sixteenth Century *Journal of World History*, 16(4):411–465.

Ben-Yosef, E. and Yagel, O. (2019). Calcium content in metallurgical slag as a proxy for fuel efficiency of ancient copper smelting technologies. *UISPP: the Journal of the International Union for Prehistoric and Protohistoric Sciences*, 2(1):66–76.

Bleekrode, S. A. (1857). *De ijzerslakken in Nederland en de ijzerbereiding in vroegeren tijd*, Amsterdam.

Boothby, R. (1646). Courte description ou découverte de la très fameuse île de Madagascar ou Saint-laurent, en Asie, à proximité de l'Inde Orientale (1644). In *Collection des Ouvrages Anciens Concernant Madagascar (COACM)*, Grandidier, A., Charles-Roux, Delhorbe, C., Froidevaux, H. and Grandidier, G., Dir., vol.3:74–96. Comité de Madagascar, Paris.

Breton, A. (1898). Fabrication du fer en Emyrne. *Notes, Reconnaissances et Exploration*, 3:675–699.

Brown, J. (1995). *Traditional Metalworking in Kenya*. Cambridge Monograph in African Archaeology 38, Oxbow books, Oxford.

Buchanan, F. (1807). *A Journey from Madras through the countries of Mysore, Canara, and Malabar*. 3 vol., London.

Campbell, G. (2005). *An Economic History of Imperial Madagascar 1750-7895 - the Rise and Fall of an Island Empire*. African Studies 106, Cambridge University Press, London.

Campbell, G. (2012). *David Griffiths and the Missionary "History of Madagascar"*. In Studies in Christian Mission 41, Academic Publisher.

Campbell, G. (2019). *Africa and the Indian Ocean World from Early Times to Circa 1900 (New Approaches to African History)*. Cambridge University Press, Cambridge.

Célis, G. R. (1989). La métallurgie traditionnelle au Burundi, au Rwanda et au Buha : Essai de synthèse. *Anthropos*, 83(1.3):25–46.

Célis, G. R. (1991). *Eisenhütten in Afrika: Beschreibung eines traditionellen Handwerks/Les fonderies africaines du fer : un grand métier disparu*. Sammlung 7: Afrika, Museum für Völkerkunde, Frankfurt am Main.

Charlton, M. F. (2015). The last frontier in 'sourcing': The hopes, constraints and future for iron provenance research. *Journal of Archaeological Science*, 56:210–220.

Charlton, M. F., Crew, P., Rehren, T. and Shennan, S. J. (2010). Explaining the evolution of ironmaking recipes - an example from northwest Wales. *Journal of Anthropological Archaeology*, 29(3):352–367.

Chauvicourt, J. and Chauvicourt, S. (1968). *Numismatique malgache, Fascicule 3, Les premières monnaies introduites à Madagascar*, Tananarive.

Chen, K. (2000). Ancient Iron Technology of Taiwan. PhD, Harvard University.

Childs, T. (1991). Style, technology, and iron smelting furnaces in Bantu-speaking Africa. *Journal of Anthropological Archaeology*, 10(4):332–359.

Chirikure, S. (2014). Geochemistry of Ancient Metallurgy: Examples from Africa and Elsewhere. In *Treatise on Geochemistry (2nd edition)*, Holland, H. D., Turekian, K. K., Eds., Elsevier, New York:169–189.

Chirikure, S. and Rehren, T. (2004). Ores, Furnaces, Slags, and Prehistoric Societies: Aspects of Iron Working in the Nyanga Agricultural Complex, AD 1300–1900. *African Archaeological Review*, 21(3):135–152.

Chittick, N. (1974). *Kilwa: an Islamic Trading City on the East African Coast*, vol. ii - the Finds. Bristish Institute in Eastern Africa, memoir n°5, Nairobi.

Chittick, N. (1984). *Manda: Excavations at an Island Port on the Kenya Coast*. British Insitute in Eastern Africa, memoir n°9, Nairobi.

Cline, W. (1937). *Mining and Metallurgy in Negro Africa*. Menasha, George Banta Publishing, General Series in Anthropology 5.

Clist, B. (1995). New field data on the ancient iron metallurgy of Madagascar. *Nyame Akuma*, 43:23–27.

Cornell, R. M. and Schwertmann, U. (2006). The Iron Oxides: Structure, Properties, Reactions, Occurrences and Uses. 2nd Edition, Wiley.

Coulaud, D. (1973). Les forgerons à la lisière forestière du Nord-Betsiléo (Madagascar). *Journal de la Société des Africanistes*, 43(2):235–242.

Coustures, M.-P., Béziat, D., Tollon, F., Domergue, C., Long, L. and Rebiscoul, A. (2003). The use of trace element analysis of entrapped slag inclusions to establish ore-bar iron links: Examples from two gallo-roman iron-making sites in France (les Martys, Montagne Noire, and les Ferrys, Loiret). *Archaeometry*, 45(4):599–613.

Craddock, P. T. (1985). History of the distillation of metals. *Bull. Met. Mus. Jpn. Inst. Met.*, 10:3–25.

Craddock, P. T. (1995). *Early Metal Mining and Production.* Edinburgh University Press, Edinburgh.

Crew, P. (2000). The influence of clay and charcoal ash on bloomery slags. In *Il Ferro Nelle Alpi - Iron in the Alps, Giacimenti Miniere e Metallurgia dall'Antichità al XVI Secolo, Atti del Convegno*, Tizzoni, C., tizzoni, M., Eds. Proceedings of The Conference, Bienno, Italy, 2–4 Ottobre 1998; Comune di Bienno: Bienno, Italy, 38–48.

Crossland, Z. (2001). Time and the anscestors: landscape survey in the Andrantsay region of Madagascar. *Antiquity*, 75(290):825–36.

David, N., Heimann, R., Killick, D. and Wayman, M. (1989). Between bloomery and blast furnace: Mafa iron-smelting technology in north Cameroon. *African Archaeological Review*, 7(1):183–208.

de Bry, J. T. (1598). *India Orientalis*, 13 vol., Frankfort am Main.

de Flacourt, E. (1661). *Histoire de la grande isle Madagascar, composée par le sieur de Flacourt, avec une relation de ce qui s'est passé les années 1655, 1656 et 1657*, Paris.

de Wit, M. J. (2003). Madagascar: Heads It's a Continent, Tails It's an Island. *Annual Review of Earth and Planetary Sciences*, 31(1):213–248.

Deer, W. A., Howie, R. A. and Zussman, J. (1992). *An Introduction to the Rock-Forming Minerals.* Pearson, 2nd edition.

Desaulty, A.-M., Mariet, C., Dillmann, P., Joron, J.-L. and Fluzin, P. (2008). A provenance study of iron archaeological artefacts by Inductively Coupled Plasma-Mass Spectrometry multi-elemental analysis. *Spectrochimica Acta Part B: Atomic Spectroscopy*, 63(11):1253–1262.

D'Escamps, H. (1884). *Histoire et Geographie de Madagascar.* Firmin-didot et cie, Paris.

Dewar, R. E., Radimilahy, C., Wright, H. T., Jacobs, Z., Kelly, G. O. and Berna, F. (2013). Stone tools and foraging in northern Madagascar challenge Holocene extinction models. *Proceedings of the National Academy of Sciences*, 110(31):12583–12588.

Dewar, R. E. and Wright, H. T. (1993). The culture history of Madagascar. *Journal of World Prehistory*, 7(4):417–466.

Dez, J. (1965). De l'influence arabe à Madagascar à l'aide de faits de linguistique. *Revue de Madagascar*, 34 (2eme trimestre):19–38.

Dillmann, P. and L'Héritier, M. (2007). Slag inclusion analyses for studying ferrous alloys employed in French medieval buildings: supply of materials and diffusion of smelting processes. *Journal of Archaeological Science*, 34(11):1810–1823.

Douglass, K., Hixon, S., Wright, H. T., Godfrey, L. R., Crowley, B. E., Manjakahery, B., Rasolondrainy, T., Crossland, Z. and Radimilahy, C. (2019). A critical review of radiocarbon dates clarifies the human settlement of Madagascar. *Quaternary Science Reviews*, 221:105878.

Drury, R. (1729). Madagascar or Robert Drury's journal during fifteen years captivity on that island. In *Collection des Ouvrages Anciens Concernant Madagascar (COACM)*, Grandidier, A., Charles-Roux, Delhorbe, C-, Froidevaux, H., and Grandidier, G., Dir., vol. 4:20–359. Comité de Madagascar, Paris.

Dupaigne, B. (1987). Les Maîtres du fer et du feu: Etude de la metallurgie du fer chez les Kouy du nord du Cambodge, dans le contexte historique et ethnographique de l'ensemble Khmer. PhD, EHESS, Paris.

Ellingham, H. J. T. (1944). Reducibility of oxides and sulphides in metallurgical processes. *Journal of the Society of Chemical Industry*, 63(5):125–160.

Ellis, W. (1838). *History of Madagascar*, 2 vol. Fisher, son and Co, London.

Ellis, W. (1858). *Three visits to Madagascar during the years 1853-1854-1856, including a journey to the capital with notices of the Natural History of the country and of present civilisation of the people*, 2 vol. Fisher Son and Co, London.

Eschenlohr, L. and Serneels, V. (1991). *Les bas fourneaux mérovingiens de Boécourt, Les Boulies (JU, Suisse)*. Office du patrimoine historique et Société jurassienne d'émulation, Cahier d'archéologie jurassienne 2.

Everard, R. (1686). Relation de trois années de souffrances dans l'île d'Assasa (Nosy Bé), près de Madagascar, au cours d'un voyage aux Indes en 1686. In *Collection des Ouvrages Anciens Concernant Madagascar (COACM)*, Grandidier, A., Charles-Roux, Delhorbe, C., Froidevaux, H. and Grandidier, G., Dir., vol.3:415. Comité de Madagascar, Paris

Gabler, S. C. (2005). Iron furnaces and future kings: Craft specialization and the emergence of political power in central Madagscar. PhD, University of Michigan.

Gallay, A., Huysecom, E., Mayor, A. and Gelbert, A. (2012). *Potières du Sahel à la decouverte des traditions céramiques de la Boucle du Niger (Mali)*. Golion: Infolio.

Gaudebout, P. and Vernier, E. (1941). Notes sur une campagne de fouilles à Vohémar - mission Rasikajy 1941. *Bulletin de l'Académie malgache*, 24:100–114.

Girbal, B. M. (2017). The Technological Context of Crucible Steel Production in Northern Telangana, India. PhD, University of Exeter.

Goldschmidt, V. M. (1937). The principles of distribution of chemical elements in minerals and rocks. The Seventh Hugo Müller lecture, delivered before the Chemical Society on March 17th, 1937. *Journal of the Chemical Society (resumed)*, 655–673.

Gommery, D., Ramanivosoa, B., Faure, M., Guérin, C., Kerloc'h, P., Sénégas, F. and Randrianantenaina, H. (2011). Les plus anciennes traces d'activités anthropiques de Madagascar sur des ossements d'hippopotames subfossiles d'anjohibe (province de Mahajanga). *Comptes Rendus - Palevol*, 10(4):271–278.

Gosselain, O. P. (2000). Materializing Identities: An African Perspective. *Journal of Archaeological Method and Theory*, 7(3):187–217.

Grandidier, A. and Grandidier, G. (1908). Ethnographie de Madagascar, 4 vol., In *Histoire physique, naturelle et politique de Madagascar*, 6 vol., Imprimerie nationale, Paris.

Griffin, W. D. (2009). The Matitanana Archaeological Project: Culture History and Social Complexity in the Seven Rivers Region of Southeastern Madagascar. PhD, University of Michigan.

Hamond, W. (1640). Paradoxe montrant que les habitants de Madagascar sont, au point de vue des biens temporels, le peuple le plus heureux de la terre. In *Collection des Ouvrages Anciens Concernant Madagascar (COACM)*, Grandidier, A., Charles-Roux, Delhorbe, C., Froidevaux, H. and Grandidier, G., Dir., vol.3:10–57. Comité de Madagascar, Paris.

Han, R. (2000). Tianma Qucun yizhi chutu tieqi de jianding, [The study of iron objects excavated from the Tianma-Qucun site]. In *Tianma Qucun 1980–1989, [The excavation report of Tianma-Qucun site (1980–1989)]*, Beijing University and Provincial Archaeology Inistitute Shanxi, Eds., 1178–1180.

Hansford, J. P., Lister, A. M., Weston, E. M. and Turvey, S. T. (2021). Simultaneous extinction of Madagascar's megaherbivores correlates with late Holocene human-caused landscape transformation. *Quaternary Science Reviews*, 263(1):106996.

Harrisson, T. and J. O'Connor, S. (1969). *Excavations of the prehistoric iron industry in West Borneo*. Data paper 72, Cornell University.

Hastie, J. (1817). *Journal of Mr. Hastie, govt. agent in Madagascar, from Sep. 8th, 1823 to 1 January, 1824*, volume 2.

Hegde, K. T. M. (1973). A model for understanding ancient Indian iron metallurgy. *Man*, 8(3):416–421.

Heurtebize, G. and Vérin, P. (1974). Premières découvertes sur l'ancienne culture de l'intérieur de l'Androy (Madagascar). Archéologie de la Vallée du Lambomaty sur la haute Manambovo. *Journal de la Société des Africanistes*, 44(2):113–121.

Horton, M. (1984). Early settlement of the Northern Swahili coast. PhD, University of Cambridge.

Horton, M., Boivin, N., Crowther, A., Gaskell, B., Radimilahy, C. and Wright, H. (2017). East Africa as a source for Fatimid rock crystal: workshops from Kenya to Madagascar. In *Gemstones in the First Millenium AD: Mines, Trade, Workshops and Symbolism*, Hilgner, A., Greiff, S., Quast, D., Eds. Proceeding of the conference, Mainz, Germany, 20–22 Ottobre 2015; Verlag des Römisch-Germanischen Zentralmuseum: Mainz, Germany, 103–118.

Humphris, J., Martinón-Torres, M., Rehren, T. and Reid, A. (2009). Variability in single smelting episodes – a pilot study using iron slag from Uganda. *Journal of Archaeological Science*, 36(2):359–369.

Hurles, M. E., Sykes, B. C., Jobling, M. A. and Forster, P. (2005). The Dual Origin of the Malagasy in Island Southeast Asia and East Africa: Evidence from Maternal and Paternal Lineages. *American Journal of Human Genetics*, 76(5):894–901.

Huysecom, E. (1996). Inagina, l'ultime maison du fer. Video Film, Format Beta SP, 52 minutes.

Ige, A. and Rehren, T. (2003). Black sand and iron stone: iron smelting in Modakeke, Ife, South Western Nigeria. *Institute for Archaeo-Metallurgical Studies*, 23:15–20.

Iles, L. and Martinón-Torres, M. (2009). Pastoralist iron production on the Laikipia Plateau, Kenya: wider implications for archaeometallurgical studies. *Journal of Archaeological Science*, 36(10):2314–2326.

Jackson, C. M., Booth, C. A. and Smedley, J. W. (2005). Glass by Design? Raw Materials, Recipes and Compositional Data. *Archaeometry*, 47(4):781–795.

Juleff, G. (2009). Technology and evolution: A root and branch view of Asian iron from first-millennium BC Sri Lanka to Japanese steel. *World Archaeology*, 41(4):557–577.

Killick, D. (2004). Social constructionist approaches to the study of technology. *World Archaeology*, 36(4):571–578.

Killick, D. (2009). Agency, dependency, and long-distance trade: East Africa and the Islamic World, ca. 700–1500 CE. In *Polities and Power: Archaeological Perspectives on the Landscapes of Early States*, Falconer S. E., Redman, C. L. Eds., University of Arizona Press, 179–207.

Killick, D. and Fenn, T. (2012). Archaeometallurgy: The Study of Preindustrial Mining and Metallurgy. *Annual Review of Anthropology*, 41(1):559–575.

Killick, D. and Gordon, R. (1988). The mechanism of iron production in the furnace. In *Proceedings of the 26th International Archaeometry Symposium*, Farquhar, R. M. , Hancock, R. G. V., Pavlish, L. A., Eds., Toronto, Canada, 16-20 May 1988. Toronto: University of Toronto, 120–123.

Killick, D. and Miller, D. (2014). Smelting of magnetite and magnetite-ilmenite iron ores in the northern Lowveld, South Africa, ca. 1000 CE to ca. 1880 CE. *Journal of Archaeological Science*, 43:239–255.

Kowalski, M., Spencer, P. and Neuschütz, D. (2000). *Slag Atlas*. Verein Deutscher Eisenhüttenleute (VDEh), Ed., Verlag Stahleisen GmbH, Düsseldorf, Germany.

Kronz, A. (2003). Ancient iron production compared to medieval techniques in Germany: Fayalitic slag and elemental mass balance. *Proceedings of the International Conference Archaeometallurgy in Europe*, Milano, Italy, 2003, Associazione Italiana di Metallurgia, 555–564.

Lacroix, A. (1922). *Minéralogie de Madagascar*, 2 vol. Challamel, A., Ed., Paris.

Lave, J. and Wenger, E. (1991). *Situated Learning: Legitimate Peripheral Participation*. Cambridge University Press.

Leguével de Lacombe, B.-F. (1840). *Voyage à Madagascar et aux îles Comores (1823 à 1830)*, Tome 2. Desessart, L., Ed., Paris.

Lemonnier, P. (1976). La description des chaînes opératoires: Contribution à l'analyse des systèmes techniques. *Techniques et Culture*, 1:100–151.

Leroi-Gourhan, A. (1964). *Le Geste et la Parole*. A. Michel, Paris.

Leroy, M., Merluzzo, P. and Le Carlier, C. (2015). *Archéologie du Fer en Lorraine. Minette et Production du Fer en Bas Fourneaux dans l'Antiquité et au Moyen-Age*. Fensch vallée edition.

Leroy, S. (2010). Circulation au moyen âge des matériaux ferreux issus des Pyrénées ariégeoises et de la Lombardie: Apport du couplage des analyses en éléments traces et multivariées. PhD, Belfort-Montbéliard.

Leroy, S., Bauvais, S., Delqué-Količ, E., Hendrickson, M., Josso, N., Dumoulin, J. P. and Soutif, D. (2020). First experimental reconstruction of an Angkorian iron furnace (13th–14th centuries CE): Archaeological and archaeometric implications. *Journal of Archaeological Science: Reports*, 34(Part A):102592.

Li, Y., Ma, C., Murakami, Y., Zhou, Z., Yang, Y. and Li, Y. (2019). Cast Iron Smelting and Fining: An Iron Smelting Site of the Eastern Han Dynasty in Xuxiebian, Sichuan Province, China. *Sungkyun Journal of East Asian Studies*, 19(1):91–111.

Lopez, E. (1585). Description de Madagascar. In *Collection des Ouvrages Anciens Concernant Madagascar (COACM)*, Grandidier, A. and Grandidier, G., Dir., vol.1:147. Comité de Madagascar, Paris.

MacPhee, R. D. E. and Burney, D. A. (1991). Dating of modified femora of extinct dwarf *Hippopotamus* from Southern Madagascar: Implications for constraining human colonization and vertebrate extinction events. *Journal of Archaeological Science*, 18(6):695–706.

Mangin, M. (Dir), Dabosi, F., Domergue, C., and Fluzin, P., Leroy, M., Merluzzo, P., Ploquin, A. and Serneels, V. (2004). *Le fer*. Editions errance, Paris.

Marschall, W. (1968). Metallurgie und frühe Besiedlungsgeschichte Indonesiens. *Ethnologica*, bd 4.

Martinelli, B. (1993). Fonderie ouest-africaines. Classement comparatif et tendances. *Techniques et Cultures*, 21:195–221.

Mayeur, M. (1785). Voyage au pays d'Ancove (1785). Transcripted by Dumaine, N., In *Bulletin de l'Académie malgache (1930)*, 12(2):28.

Melluso, L., Morra, V., Brotzu, P., Razafiniparany, A., Ratrimo, V. and Razafimahatratra, D. (1997). Geochemistry and Sr-isotopic composition of the late cretaceous flood basalt sequence of northern Madagascar: petrogenetic and geodynamic implications. *Journal of African Earth Sciences*, 24(3):371–390.

Melluso, L., Morra, V., Brotzu, P., Tommasini, S., Renna, M. R., Duncan, R. A., Franciosi, L. and D'amelio, F. (2005). Geochronology and Petrogenesis of the Cretaceous Antampombato–Ambatovy Complex and Associated Dyke Swarm, Madagascar. *Journal of Petrology*, 46(10):1963–1996.

Michler, M., collab. (2016). Grand est, Bas-Rhin (67), Weyershein, rue de la Gare - lotissement "les Hauts de la Zorn" : Dépot de crémation du bronze final et atelier de forge de la fin du premier age du fer en bordure d'une zone d'ensilage. Rapport de fouilles, INRAP Grand Est, Dijon.

Miller, D., Desai, N. and Lee-Thorp, J. (2000). Indigenous Gold Mining in Southern Africa: A Review. *South African Archaeological Society Goodwin Series*, 8:91–99.

Miller, D., Killick, D. and van der Merwe, N. J. (2001). Metal working in the northern Lowveld, South Africa A.D. 1000–1890. *Journal of Field Archaeology*, 28(3-4):401–417.

Misra, M. K., Ragland, K. W. and Baker, A. J. (1993). Wood ash composition as a function of furnace temperature. *Biomass and Bioenergy*, 4(2):103–116.

Monnier (1910). Notes relative à divers objets en pierre trouvés dans la province de Vohémar. *Bulletin de l'Académie malgache*, 8:141–143.

Morel, M. and Serneels, V. (2021). Interpreting the Chemical Variability of Iron Smelting Slag: A Case Study from Northeastern Madagascar. *Minerals*, 11(8):900.

Mouren and Rouaix (1913). Industrie ancienne des objets en pierre de Vohémar. *Bulletin de l'Académie malgache*, 12:3–13.

Nitsche, C., Schreurs, G. and Serneels, V. (2022). The Enigmatic Softstone Vessels of Northern Madagascar: Petrological Investigations of a Medieval Quarry. *Journal of Field Archaeology*, 1–18.

O'Connor, S. J. (1985). Metallurgy and Immortality at Candi Sukuh, Central Java. *Indonesia*, 39:53–70.

Oliver, S. P. (1886). *Madagascar, An Historical and Descriptive Account of the Island and its Former Dependencies*, 2 vol. Macmillan and co, New York.

Park, J. S. and Rehren, T. (2011). Large-scale 2nd to 3rd century ad bloomery iron smelting in Korea. *Journal of Archaeological Science*, 38(6):1180–1190.

Park, J. S. and Shinde, V. (2013). Iron technology of the ancient megalithic communities in the Vidarbha region of India. *Journal of Archaeological Science*, 40(11):3822–3833.

Parker Pearson, M. (2010). *Pastoralists, Warriors and Colonists: The Archaeology of Southern Madagascar*. BAR International Serie 2139, Archaeopress, Oxford.

Paynter, S., Crew, P., Blakelock, E. and Hatton, G. (2015). Spinel-rich slag and slag inclusions from a bloomery smelting and smithing experiment with a sideritic ore. *Historical Metallurgy*, 49(2):126–143.

Pearson, M. (2003). *The Indian Ocean*, Routledge, London.

Perlès, C. (1987). *Les industries lithiques taillées de Franchthi (Argolide, Grèce) : Présentation Générale et Industrie Paléolithique*. Indiana University Press, Terre Haute.

Pétrequin, P. and Pétrequin, A.-M. (2000). *Ecologie d'un outil : la hache de pierre en Irian Jaya (Indonésie)*. Monographie du CRA, CNRS edition.

Pfaffenberger, B. (1992). Social Anthropology of Technology. *Annual Review of Anthropology*, 21:491–516.

Pierron, D., Heiske, M., Razafindrazaka, H., Rakoto, I., Rabetokotany, N., Ravololomanga, B., Rakotozafy, L. M.-A., Mialy Rakotomalala, M., Razafiarivony, M., Rasoarifetra, B., Andriamampianina Raharijesy, M., Razafindralambo, L., Ramilisonina, Fanony, F., Lejamble, S., Thomas, O., Abdallah, A. M., Rocher, C., Arachiche, A., Tonaso, L., Pereda-loth, V., Schiavinato, S., Brucato, N., Ricaut, F.-X., Kusuma, P., Sudoyo, H., Ni, S., Boland, A., Deleuze, J.-F-, Beaujard, P., Grange, P., Adelaar, S., Stoneking, M., Rakotoarisoa, J.-A., Radimilahy, C., Letellier, T. (2017). Genomic landscape of human diversityacross Madagascar. In *Proceedings of the National Academy of Sciences*, 114(32), E6498–E6506.

Pleiner, R. (2000). *Iron in Archaeology: The European Bloomery Smelters*, Archeologický Ústav AV ČR: Praha, Czech.

Pradines, S. (2019). Islamic Archaeology in the Comoros: The Swahili and the Rock Crystal Trade with the Abbasid and Fatimid Caliphates. *Journal of Islamic Archaeology*, 6(1):109–135.

Pradines, S., Renel, H., Veyssier, D. and Zhao, B. (2016). Ironi Be (Dembeni, Mayotte). Rapport de mission 2015. *Nyame Akuma*, 85:44–56.

Radimilahy, C. (1988). L'Ancienne Métallurgie du Fer à Madagascar. *Cambridge Monographs in African Archaeology 28*, BAR International Serie 422.

Radimilahy, C. (1998). Mahilaka: An archaeological investigation of an early town in northwestern Madagascar. *Studies in African Archaeology 15*. PhD, Department of Archaeology and Ancient History, University of Uppsala.

Radimilahy, C. (2011). Réflexions sur la production pré-européenne du textile dans le nord de Madagascar. *Études Océan Indien*, 46-47:162–176.

Radimilahy, C. and Crossland, Z. (2015). Situating Madagascar: Indian Ocean dynamics and archaeological histories. *Azania: Archaeological Research in Africa*, 50(4):495–518.

Rakotoarisoa, J.-A. and Allibert, C. (2011). *Vohémar, cité-État malgache.* Inalco, Etudes Océan Indien, volume 46-47.

Rasamuel, D. (1984). L'ancien Fanongoavana. PhD, Université de Paris I.

Rasoarifetra, B. (2011). Les perles de Vohémar, origine et marqueurs culturels. *Études Océan Indien*, 46-47:178–193.

Rehren, T., Charlton, M., Chirikure, S., Humphris, J. and Veldhuijzen, H. A. (2007). *Decisions set in slag: the human factor in African iron smelting.* In *Metals and Mines - Studies in Archaeometallurgy*, LaNiece, S., Hook, D. R., Craddock, P. T., Eds. Archetype/British Museum: London, 211–218.

Robion-Brunner, C. (2008). Vers une histoire de la production du fer sur le plateau de Bandiagara (Pays Dogon, Mali) durant les empires précoloniaux : peuplement des forgerons et traditions sidérurgiques. PhD, Université de Genève.

Robion-Brunner, C. (2018). Pourquoi ton four n'est pas comme le mien ? Diversité technique dans la sidérurgie ancienne : le cas du Dendi (Bénin). *Journal des africanistes*, 88(2):16–39.

Robion-Brunner, C. (2020). What is the Meaning of the Extreme Variability of Ancient Ironworking in West Africa?: A Comparison between Four Case Studies. In *Mobile Technologies in the Ancient Sahara and Beyond*, Duckworth, C., Cuénod, A., Mattingly, D., Eds. Trans-Saharan Archaeology, Cambridge University Press, 290–314.

Rostoker, W. and Bronson, B. (1990). *Pre-Industrial Iron - Its Technology and Ethnology*, Archaeomaterials Monograph, N°1; Univ. Museum Pubns, Philadelphia, PA, USA.

Roux, V. (2009). Technological Innovations and Developmental Trajectories: Social Factors as Evolutionary Forces. In *Innovation in Cultural Systems: Contributions from Evolutionary Anthropology*, O'Brien, M. J., Shennan, S. J., Eds. The MIT Press, 217–234.

Sandelowsky, B. H. and Pendleton, W. C. (1969). Stone Tuyères from South West Africa. *The South African Archaeological Bulletin*, 24(93):14–20.

Schmidt, P. R. and Avery, D. H. (1978). Complex Iron Smelting and Prehistoric Culture in Tanzania. *Science*, 201(4361), 1085–1089

Schmidt, P. R. and Childs, T. S. (1985). Innovation and Industry during the Early Iron Age in East Africa: The KM2 and KM3 Sites of Northwest Tanzania. *The African Archaeological Review*, 3:53–94.

Schreurs, G. and Rakotoarisoa, J.-A. (2011). The archaeological site at Vohemar in a regional geographical and geological context. *Études Océan Indien*, 46-47.

Schucany, C. (2006). *Die römische Villa von Biberist-Spitalhof/SO (Grabungen 1982, 1983, 1986-1989). Untersuchungen im Wirtschaftsteil und Überlegungen zum Umland*, 3 vol. Remschalden, BAG Verlag.

Schwaner, M. (1853). *Beschruving van het Stroomgebied van den Barito.* Van Kampen, P. N., Ed., Amsterdam.

Seland, E. H. (2014). Archaeology of Trade in the Western Indian Ocean, 300 BC–AD 700. *Journal of Archaeological Research*, 22:367–402.

Serneels, V. (1993). Archéométrie des scories de fer: Recherche sur la sidérurgie ancienne en Suisse Romande. *Cahier d'Archéologie romande 61, 240.*

Serneels, V. (2002). Analyses chimiques des matières premières et des produits de l'opération de réduction dans le four basque de Agorregi. In *La ferrería y los molinos de Agorregi*, Crew, P., Crew, S, Dillmann, P., Fluzin, P., Herbach, R., Serneels, V., Simon, J. and Urteaga, M.M, Eds., Arkeologia 3, Gipuzkoako Foru Aldundia: Donostia-San Sebastián, Spain, 93–122.

Serneels, V. (2005). A propos de la qualité des fers produits par la méthode directe de réduction. In *L'acier en Europe avant Bessemer*, Dillmann, P., Pérez, L. and Verna, C., Eds., Proceedings of the International conference, Conservatoire national des Arts et Métiers, 8-10 december 2005, Paris. CNRS/Université de Toulouse-Le Mirail, Collection Méridienne, Série Histoire et Technique, 73-93.

Serneels, V., Donadini, F., Kiénon-Kaboré, H. T., Koté, L., Kouakou Kouassi, S., Ramseyer, D. and Simporé, L. (2014). Origine et développement de la métallurgie du fer au Burkina Faso et en Côte d'Ivoire. Avancement des recherches en 2013 et quantification des vestiges de Korsimoro (Burkina Faso). *SLSA Jahresbericht - Rapport annuel – Annual report 2013*, 65–112.

Serneels, V., Fialin, M. and Katona Serneels, I. (2019a). Minor elements and iron smelting slag. In *Proceedings of the 5th International Conference Archaeometallurgy in Europe*, Török, B., Giumlia-Mair, A., Eds. June 2019, Milskolc, Hungary, 86.

Serneels, V., Jaony, W. C., Morel, M., Nitsche, C., Radimilahy, C., Rakotoarsisoa, J.-A., Rasoarifetra, B., Schreurs, G. and Velomora, S. (2021). Pierre et fer à Madagascar (4) - Nouvelles données sur l'exploitation du territoire. *SLSA Jahresbericht – Rapport annuel – Annual report 2020*, 219–252.

Serneels, V., Jobin, P., Kiénon-Kaboré, H. T., Koté, L., Kouakou Kouassi, S., Ramseyer, D., Thiombiano-Ilboudo, E. and Simporé, L. (2015). Origine et développement de la métallurgie du fer au Burkina Faso et en Côte d'Ivoire. Seconde campagne dans la région de Kaniasso (Folon, Côte d'Ivoire) et autres recherches. *SLSA Jahresbericht - Rapport annuel – Annual report 2014*, 23–60.

Serneels, V., Morel, M., Nitsche, C., Radimilahy, C., Rakotoarsisoa, J.-A., Rasoarifetra, B. and Schreurs, G. (2018). Pierre et fer à Madagascar (1) – Vestiges sidérurgiques de Benavony et de la rivière Matavy. *SLSA Jahresbericht – Rapport annuel – Annual report 2017*, 109–156.

Serneels, V., Morel, M., Nitsche, C., Radimilahy, C., Rakotoarsisoa, J.-A., Rasoarifetra, B., Schreurs, G. and Velomora, S. (2019b). Pierre et fer à Madagascar (2) – Les scories d'Amboronala et les carrières de Milanoa. *SLSA Jahresbericht – Rapport annuel – Annual report 2018*, 313–366.

Serneels, V., Morel, M., Nitsche, C., Radimilahy, C., Rakotoarisoa, J.-A., Schreurs, G. and Velomora, S. (2020). Pierre et fer à Madagascar (3) - La carrière de Bobalila. *SLSA Jahresbericht – Rapport annuel – Annual report 2019*, 291–352.

Serneels, V. and Perret, S. (2003). Quantification of smithing activities based on the investigation of slag and other material remains. In *Proceedings of the 5th International Conference Archaeometallurgy in Europe*, Associazione Italiana di Metallurgia, 24-26 september, Milano, Italy, 469–478.

Sibree, J. (1873). *Madagascar et ses habitants, journal d'un séjour de quatre ans dans l'île*, Toulouse.

Sinclair, P. J. J., Törnblom, M., Bohm, C., Sigvallius, B. and Hulthen, B. (1988). *Analyses of slag, iron, ceramics and animal bones from excavations in Mozambique.* Central board of National Antiquities, Studies in African archaeology, 2nd edition.

Soulignac, R. (2017). Les Scories de Forge du Pays Dogon, Mali: Entre Ethnoarcheologie, Archeologie Experimentale et Archeometrie. PhD, Université de Fribourg.

Storey, M., Mahoney, J. J. and Saunders, A. D. (1997). Cretaceous Basalts in Madagascar and the Transition between Plume and Continental Lithosphere Mantle Sources. In *Large Igneous Provinces: Continental, Oceanic, and Planetary Flood Volcanism*, Mahoney, J. J. and Coffin, M. F., Eds., Geophysical Monograph Series, 100:95–122.

Swan, L. (1994). Early Gold Mining on the Zimbabwean Plateau. *Studies in African Archaeology 9*, Societas Archaeologica Uppsliensis, Uppsala.

Tardy, Y. (1992). Diversity and terminology of lateritic profiles. In *Wathering, Soils and Paleosols*, Martini, I.P. and Chesworth, W., Eds., Chap.15. Developments in Earth Surface Processes, 2:379–405.

Tofanelli, S., Bertoncini, S., Castrı, L., Luiselli, D., Calafell, F., Donati, G., Paoli, G. and Castrì, L. (2009). On the Origins and Admixture of Malagasy: New Evidence from High-Resolution Analyses of Paternal and Maternal Lineages. *Molecular Biology and Evolution*, 26(9):2109–2124.

Tripathi, V. (2014). Recently Discovered Iron Working Site in Vindhya-Kaimur Region, India. *ISIJ International*, 54(5):1010–1016.

Truffaut, E. (2014). Steelmaking in a bloomery furnace: Behavior of manganese. Research on the ferrum noricum process. In *Early Iron in Europe*, Cech, B. and Rehren, T., Eds., Monique mergoil, Monographie instrumentum edition, 185–198.

Tucker, R. D., Roig, J. Y., Moine, B., Delor, C. and Peters, S. G. (2014). A geological synthesis of the Precambrian shield in Madagascar. *Journal of African Earth Sciences*, 94:9–30.

Tylecote, R., Ghaznavi, H. and Boydell, P. (1977). Partitioning of trace elements between the ores, fluxes, slags and metal during the smelting of copper. *Journal of Archaeological Science*, 4(4):305–333.

Vacher, Lt., (1899). Les cultures et industries indigènes à Madagascar. In *Notes, Reconnaissances et Exploration*, 5:543–549.

Valette, J. (1966). L'histoire de Madagascar de William Ellis jugée par David Jones. In *Bulletin de l'Académie malgache*, 44(2):193–197.

Van Gosen, B. S., Fey, David, L., Shah, A. K., Verplanck, Philip, L. and Hoefen, T. M. (2014). *Deposit Model for Heavy-Mineral Sands in Coastal Environments Mineral Deposit Models for Resource Assessment, Scientific Investigations Report 2010–5070–L*. Central Mineral and Environmental Resources Science Center, U.S. Geological Survey edition, Reston, VA, USA.

Vega, E., Dillmann, P. and Fluzin, P. (2002). Contribution à l'étude de fers phosphoreux en sidérurgie ancienne. *Revue d'Archéométrie*, 26(1):197–208.

Veldhuijzen, H. A. and Rehren, T. (2007). Slag and the city: Early iron production at Tell Hammeh, Jordan, and Tel Beth-Shemesh, Israel. In *Metals and Mines-Studies in Archaeometallurgy*, La Niece, S., Hook, D. and Craddock, P., Eds. Archetype/British Museum, 189–201.

Vérin, P. (1986). *The History of Civilisation in North Madagascar*. A.A. Balkema: Rotterdam, The Netherlands.

Vernier, E. and Millot, J. (1971). Archéologie malgache, comptoirs musulmans. *Catalogues du Musée de l'Homme*, Paris.

Wagner, D. B. (2008). Ferrous Metallurgy. In *Science and Civilisation in China, Vol. 5, Chemistry and Chemical Technology, Part 11.* Cambridge University Press.

Wills, J. (1885). Native Products Used in Malagasy Industries. In *Antananarivo Annual and Madagascar Magazine (AAMM)*, Sibree, J. and Baron, R., Eds. Comité de Madagascar, 9:97–98.

Wright, H. T. (1992). Early Islam, Oceanic Trade and Town Development on Nzwani: The Comorian Archipelago in the XIth-XVth Centuries AD. *Azania: Archaeological Research in Africa*, 27(1):81–128.

Wright, H. T. and Fanony, F. (1992). The Evolution of Settlement Systems in the Mananara River Valley: Archaeological Reconnaissances of 1983-1990. *Taloha*, 11:16–64.

Wright, H. T., Rakotoarisoa, J.-A., Heurtebize, G. and Vérin, P. (1993). The Evolution of Settlement Systems in the Efaho River Valley, Anosy: A Preliminary Report on the Archaeological Reconnaissances of 1983-1986. *Indo-Pacific Prehistory Association Bulletin*, 12:2–20.

Wyne-Jones S., LaViolette A. (Eds). (2018). *The Swahili World.* Routledge.

Zhao, B. (2011). Vers une expertise plus fine et une approche plus historique de la céramique chinoise de la nécropole de Vohémar. *Etudes Océan Indien*, 46-47:92–103.

Zitzmann, A. (1977). *The Iron Ore Deposits of Europe and Adjacent Areas.* E. Schweizerbart'sche Verlagsbuchhandlung, Bundesanstalt für Geowissenschaften und Rohstoffe, Hannover.